Springer Proceedings in Physics

Volume 202

The series Springer Proceedings in Physics, founded in 1984, is devoted to timely reports of state-of-the-art developments in physics and related sciences. Typically based on material presented at conferences, workshops and similar scientific meetings, volumes published in this series will constitute a comprehensive up-to-date source of reference on a field or subfield of relevance in contemporary physics. Proposals must include the following:

- name, place and date of the scientific meeting
- a link to the committees (local organization, international advisors etc.)
- scientific description of the meeting
- list of invited/plenary speakers
- an estimate of the planned proceedings book parameters (number of pages/ articles, requested number of bulk copies, submission deadline).

More information about this series at http://www.springer.com/series/361

Tetsuya Kawachi · Sergei V. Bulanov
Hiroyuki Daido · Yoshiaki Kato
Editors

X-Ray Lasers 2016

Proceedings of the 15th International Conference on X-Ray Lasers

Springer

Editors
Tetsuya Kawachi
Kansai Photon Science Institute
National Institutes for Quantum and Radiological Science and Technology
Kyoto
Japan

Hiroyuki Daido
Naraha Remote Technology Development Center
Japan Atomic Energy Agency
Fukushima
Japan

Sergei V. Bulanov
Kansai Photon Science Institute
National Institutes for Quantum and Radiological Science and Technology
Kyoto
Japan

Yoshiaki Kato
Graduate School for Creation of New Photonics Industries
Shizuoka
Japan

ISSN 0930-8989 ISSN 1867-4941 (electronic)
Springer Proceedings in Physics
ISBN 978-3-319-73024-0 ISBN 978-3-319-73025-7 (eBook)
https://doi.org/10.1007/978-3-319-73025-7

Library of Congress Control Number: 2017962107

Printed on acid-free paper

This Springer imprint is published by Springer Nature
The registered company is Springer International Publishing AG
The registered company address is: Gewerbestrasse 11, 6330 Cham, Switzerland

Preface

This volume comprises the papers presented at the 15th International Conference on X-Ray Lasers (ICXRL 2016) held in Nara during May 22—27, 2016, under organization by the Kansai Photon Science Institute (KPSI) of the National Institutes for Quantum and Radiological Science and Technology (QST). Nara was the capital of Japan in the eighth century, a small but beautiful city filled with tranquility where people are living with old temples and shrines. The conference venue was "Nara Kasugano International Forum", where oral presentations were made on the traditional *Noh* stage.

The International Conference on X-Ray Lasers is a series of biennial meetings, which have been held since 1986 in many cities over the world as summarized in Table 1. The first meeting was held in Aussois, France, in 1986 as The International

Table 1 The International Conference on X-Ray Lasers from 1986 to 2017

No.	Year	City	Country	Chair(s)
1	1986	Aussois	France	P. Jaegle, A. Sureau
2	1990	York	UK	G. J. Tallents
3	1992	Schliersee	Germany	E. E. Fill
4	1994	Williamsburg	USA	D. C. Eder, D. L. Matthews
5	1996	Lund	Sweden	S. Swanberg, C.-G. Wahlstrom
6	1998	Kyoto	Japan	Y. Kato, H. Takuma, H. Daido
7	2000	Saint-Malo	France	G. Jamelot, C. Moeller, A. Klisnik
8	2002	Aspen	USA	J. J. Rocca, J. Dunn, S. Suckewer
9	2004	Beijing	China	J. Zhang
10	2006	Berlin	Germany	P. V. Nickles, K. A. Januelwicz
11	2008	Belfast	UK	C. Lewis, D. Riley
12	2010	Gwangju	Korea	J.-M. Lee, C.-H. Nam, K. A. Januelwicz
13	2012	Paris	France	S. Sebban, J. Gauthier, D. Ross, P. Zeitoun
14	2014	Fort Collins	USA	J. Rocca, C. Menoni, M. C. Marconi
15	2016	Nara	Japan	T. Kawachi, S. Bulanov, H. Daido, Y. Kato

Colloquium on X-Ray Lasers, and it was renamed to The International Conference on X-Ray Lasers from the third meeting. Throughout its 30 years of history, this conference has been a primary international forum for presenting the latest advances in the development of X-ray lasers and their applications. This conference has also covered alternative methods to generate coherent X-ray radiation. The dynamic advances in X-ray lasers and related fields can be overviewed by visiting the topics presented in previous Conferences which are recorded in the Conference Proceedings.

The ICXRL 2016 was attended by more than 120 participants from 17 countries. In total 117 papers were presented, comprising 44 invited, 21 contributed oral and 52 poster presentations. The papers compiled in this volume highlights the rapid progress in recent years, including high average power table-top soft X-ray lasers driven by LD-pumped solid state lasers, X-ray free-electron lasers, new schemes for coherent X-ray generation by laser-driven particle accelerators, and laser-Compton γ-ray sources for nuclear photonics. It also covers applications of compact soft X-ray sources to nano-scale imaging and EUV lithography, investigation of ultra-fast dynamics with harmonic sources, and high resolution X-ray coherent phase- and diffraction-imaging.

In addition to the scientific exchanges, the conference participants enjoyed an excursion to the north-west of Kyoto with the Hozu-gawa River Boat Ride along a valley surrounded by new green leaves, and the Conference Banquet including the tasting of various kinds of *Sake* from a local brewery. At this Banquet, it was announced that the next Conference will be held in Prague, Czech Republic in 2018.

The 15th International Conference on X-Ray Lasers was supported by The Ministry of Education, Culture, Sports, Science and Technology of Japan, the National Institutes for Quantum and Radiological Science and Technology, the Japan Atomic Energy Agency, The Laser Society of Japan, The Japan Society of Plasma Science and Nuclear Fusion Research, The Physical Society of Japan, and The Graduate School for the Creation of New Photonics Industries. We also appreciate most valuable support from a number of companies.

We would like to thank the International Advisory Board for helping to organize the very up to date scientific program. We would also like to thank the Local Organizing Committee for making this conference valuable for exchanges among the international communities. Special thanks to the local staff, Masaharu Nishikino, Akira Sasaki, Masataka Kado, Satoshi Orimo, Yasushi Tamura, Koichi Umetani, and Kahoko Watanabe of KPSI, for their excellent work in organizing and helping to run the meeting smoothly. We also thank Wataru Utsumi, Director of KPSI, and Kiminori Kondo, Head of the Department of Advanced Photon Research, for their help and useful suggestions.

Nara, Japan

Tetsuya Kawachi
Sergei V. Bulanov
Hiroyuki Daido
Yoshiaki Kato

Contents

Contributors

J. Abdallah Theoretical Division, Los Alamos National Laboratory, Los Alamos, USA

T. Abe Gigaphoton Inc, Shinomiya Hiratsuka Kanagawa, Japan

T. Akagi KEK, Oho, Tsukuba, Ibaraki, Japan

T. R. Akhmedzhanov Texas A&M University, College Station, TX, USA

B. Albertazzi Graduate School of Engineering, Osaka University, Suita, Osaka, Japan; LULI, École Polytechnique, CNRS, CEA, UPMC, Palaiseau, France

M. Albrecht IoP ASCR, ELI Beamlines Project, Prague 8, Czech Republic; FNSPE of the CTU in Prague, Prague, Czech Republic

M. Z. Alkhimova Joint Institute for High Temperatures, Russian Academy of Sciences, Moscow, Russia; National Research Nuclear University MEPhI, Moscow, Russia

R. Amano Department of Innovation System Engineering & CORE, Utsunomiya University, Utsunomiya, Tochigi, Japan

S. Amano Laboratory of Advanced Science and Technology for Industry, University of Hyogo, Ako, Hyogo, Japan

A. Andreev Max Born Institute, Berlin, Germany; ELI-ALPS, Szeged, Hungary

V. A. Antonov Institute of Applied Physics RAS, Nizhny Novgorod, Russia

G. Arai Department of Innovation System Engineering & CORE, Utsunomiya University, Utsunomiya, Tochigi, Japan

S. Araki KEK, Oho, Tsukuba, Ibaraki, Japan

O. Muray Ricardo Arturo Graduate School of Engineering, Osaka University, Suita, Osaka, Japan; Photon Pioneers Center, Osaka University, Suita, Osaka, Japan

I. A. Artyukov L.P. Lebedev Physical Institute, Moscow, Russia

V. Aslanyan York Plasma Institute Department of Physics, University of York, York, UK

M. Ayele Institute of Optoelectronics, Military University of Technology, Warsaw, Poland

M. Baba Saitama Medical University, Saitama, Japan

Gennadiy Bagdasarov Keldysh Institute of Applied Mathematics RAS, Moscow, Russia; National Research Nuclear University MEPhI (Moscow Engineering Physics Institute), Moscow, Russia

R. Banici CETAL, INFLPR, Măgurele, Romania

F. Barbato Federal Laboratories for Materials Science & Technology (Empa), Dübendorf, Switzerland

A. Bartnik Institute of Optoelectronics, Military University of Technology, Warsaw, Poland

C. Baumgarten Colorado State University, Fort Collins, CO, USA

E. R. Bernstein Department of Chemistry, Colorado State University, Fort Collins, CO, USA

M. Berrill Oak Ridge National Laboratory, Oak Ridge, TN, USA

A. Blechschmidt Berlin Laboratory for innovative X-ray Technologies (BLiX), Berlin, Germany; Institut für Optik und Atomare Physik (IOAP), Technische Universität Berlin, Berlin, Germany

D. Bleiner Federal Laboratories for Materials Science & Technology (Empa), Dübendorf, Switzerland; EMPA Materials Science and Technology, St. Gallen, Switzerland

K. Bohaček IoP ASCR, Prague 8, Czech Republic; LOA, ENSTA, Chemin de La Hunière, Palaiseau, France

Alexey Boldarev Keldysh Institute of Applied Mathematics RAS, Moscow, Russia; National Research Nuclear University MEPhI (Moscow Engineering Physics Institute), Moscow, Russia

N. Booth CLF, STFC Rutherford Appleton Laboratory, Didcot, Oxfordshire, UK

Charles Bourassin-Bouchet Institut des Sciences Moléculaires d'Orsay, UMR 8214, CNRS, Université Paris-Sud, Orsay Cedex, France; Synchrotron SOLEIL, Université Paris-Saclay, Saint-Aubin, Gif-sur-Yvette, France

S. V. Bulanov Kansai Photon Science Institute, National Institutes for Quantum and Radiological Science and Technology (QST), Kizugawa, Kyoto, Japan; Moscow Institute of Physics and Technology (State University), Dolgoprudnyi, Moscow, Russia; A. M. Prokhorov Institute of General Physics of the Russian Academy of Sciences, Moscow, Russia

Stepan Sergeevich Bulanov Lawrence Berkeley National Laboratory, Berkeley, CA, USA

A. S. Busarov L.P. Lebedev Physical Institute, Moscow, Russia

N. M. H. Butler Department of Physics, SUPA, University of Strathclyde, Glasgow, UK

W. Chao Lawrence Berkeley Lab, Center for X-Ray Optics, Berkeley, CA, USA

U. Chaulagain IoP ASCR, Prague 8, Czech Republic

A. Choukourov Department of Macromolecular Physics, Faculty of Mathematics and Physics, Charles University in Prague, Prague, Czech Republic

G. Cojocaru CETAL, INFLPR, Măgurele, Romania

J. Colgan Theoretical Division, Los Alamos National Laboratory, Los Alamos, USA

M. Cotelo Instituto de Fusión Nuclear, E.T.S.I. Industriales, Universidad Politécnica de Madrid, Madrid, Spain

D. C. Crick Department of Microbiology, Immunology and Pathology, Colorado State University, Fort Collins, CO, USA

J. Czwartos Institute of Optoelectronics, Military University of Technology, Warsaw, Poland

R. Dabu National Institute for Nuclear Physics-HH, ELI-NP, Măgurele, Romania

Hugo Dacasa Laboratoire d'Optique Appliquée, UMR 7639 ENSTA-CNRS-Ecole Polytechnique, Chemin de la Hunière, Palaiseau, France

H. Daido Naraha Remote Technology Development Center, JAEA, Naraha Futaba, Fukushima, Japan

R. J. Dance Department of Physics, SUPA, University of Strathclyde, Glasgow, UK

R. Deguchi Institute of Laser Engineering, Osaka University, Suita, Osaka, Japan

A. Dehlinger Berlin Laboratory for innovative X-ray Technologies (BLiX), Berlin, Germany; Institut für Optik und Atomare Physik (IOAP), Technische Universität Berlin, Berlin, Germany

S. Demmler Institute of Applied Physics, Jena University, Jena, Germany

T.-H. Dinh Kansai Photon Science Institute, National Institutes for Quantum and Radiological Science and Technology, Kyoto, Kizugawa, Japan; Faculty of Engineering and Center for Optical Research and Education (CORE), Department of Electrical and Electronic Engineering, Department of Innovation System Engineering & CORE, Utsunomiya University, Utsunomiya, Tochigi, Japan

N. Dover Kansai Photon Science Institute, National Institutes for Quantum and Radiological Science and Technology (QST), Kizugawa, Kyoto, Japan

T. Ejima Institute of Multidisciplinary Research for Advanced Materials, Tohoku University, Sendai, Japan

K. Endo Toyama Co., Ltd, Yamakita, Kanagawa, Japan

E. Esarey Lawrence Berkeley National Laboratory, Berkeley, CA, USA

J. C. Escudero Instituto de Fusión Nuclear, E.T.S.I. Industriales, Universidad Politécnica de Madrid, Madrid, Spain

T. Zh. Esirkepov Kansai Photon Science Institute, National Institutes for Quantum and Radiological Science and Technology (QST), Kizugawa, Kyoto, Japan

A. Ya. Faenov Joint Institute of High temperature, Russian Academy of Science, Moscow, Russia; Open and Transdisciplinary Research Initiatives, Osaka University, Suita, Osaka, Japan; Institutes for Academic Initiatives, Osaka University, Suita, Osaka, Japan

M. Fang State Key Laboratory of High Field Laser Physics, Shanghai Institute of Optics and Fine Mechanics, CAS, Shanghai, China

H. Fiedorowicz Institute of Optoelectronics, Military University of Technology, Warsaw, Poland

O. Frolov Pulse Plasma Systems Department, Institute of Plasma Physics of the Czech Academy of Sciences, Prague, Czech Republic

Y. Fu RIKEN Center for Advanced Photonics, Wako, Saitama, Japan

Y. Fujii Department of Innovation System Engineering & CORE, Utsunomiya University, Utsunomiya, Tochigi, Japan

Y. Fukuda Kansai Photon Science Institute, National Institutes for Quantum and Radiological Science and Technology (QST), Kizugawa, Kyoto, Japan

Vladimir Gasilov Keldysh Institute of Applied Mathematics RAS, Moscow, Russia; National Research Nuclear University MEPhI (Moscow Engineering Physics Institute), Moscow, Russia

Julien Gautier IoP ASCR, ELI Beamlines Project, Prague 8, Czech Republic; Laboratoire d'Optique Appliquée, UMR 7639 ENSTA-CNRS-Ecole Polytechnique, Chemin de la Hunière, Palaiseau, France; FNSPE of the CTU in Prague, Prague, Czech Republic

C. G. R. Geddes Lawrence Berkeley National Laboratory, Berkeley, CA, USA

A. Goltsov Texas A&M University, College Station, TX, USA; Princeton University, Princeton, NJ, USA

R. Gonin Laserix, Université Paris-Sud, Orsay Cedex, France

Anthony Gonsalves Lawrence Berkeley National Laboratory, Berkeley, CA, USA

A. González Instituto de Fusión Nuclear, E.T.S.I. Industriales, Universidad Politécnica de Madrid, Madrid, Spain

J. Green CLF, STFC Rutherford Appleton Laboratory, Didcot, Oxfordshire, UK

T. Green Department of Electrical & Computer Engineering, Colorado State University, Fort Collins, CO, USA

C. Gregory CLF, STFC Rutherford Appleton Laboratory, Didcot, Oxfordshire, UK

W. Grizolli MAX IV Laboratory, Lund University, Lund, Sweden

A. Grum-Grzhimailo Skobeltsyn Institute of Nuclear Physics, Moscow State University, Moscow, Russia

A. Guggenmos Ludwig-Maximilians-Universität München, Garching, Germany; Max-Planck-Institut für Quantenoptik, Garching, Germany

H. Habara Graduate School of Engineering, Osaka University, Suita, Osaka, Japan

S. Hädrich Helmholtz Institute Jena, Jena, Germany; Institute of Applied Physics, Jena University, Jena, Germany

R. Hajima National Institutes for Quantum and Radiological Science and Technology (QST), Tokai, Naka, Ibaraki, Japan

M. Harmand IMPMC, Université Pierre et Marie Curie, CNRS, Paris, France

J. R. Harries National Institutes for Quantum and Radiological Science and Technology, SPring-8, Sayo, Hyogo, Japan

N. Hartely Graduate School of Engineering, Osaka University, Suita, Osaka, Japan

N. Hasegawa Kansai Photon Science Institute, National Institutes for Quantum and Radiological Science and Technology (QST), Kizugawa, Kyoto, Japan

T. Hatano Institute of Multidisciplinary Research for Advanced Materials, Tohoku University, Sendai, Japan

M. Hatayama NTT Advanced Technology Corporation, Atsugi, Kanagawa, Japan

T. Hayashi Department of Mechanical Engineering, Kyushu University, Fukuoka, Japan

Y. Hayashi Kansai Photon Science Institute, National Institutes for Quantum and Radiological Science and Technology (QST), Kizugawa, Kyoto, Japan

F. Hennies MAX IV Laboratory, Lund University, Lund, Sweden

T. Higashiguchi Department of Innovation System Engineering & CORE, Utsunomiya University, Utsunomiya, Tochigi, Japan; Faculty of Engineering and Center for Optical Research and Education (CORE), Department of Electrical and Electronic Engineering, Utsunomiya University, Utsunomiya, Tochigi, Japan

P. Homer ELI-Beamlines, Institute of Physics v.v.i., Academy of Sciences of the Czech Republic, Prague 8, Czech Republic

Y. Honda KEK, Oho, Tsukuba, Ibaraki, Japan; Science of the Graduate University for Advanced Studies (SOKENDAI), Miura District, Kanagawa, Japan

T. Hori Gigaphoton Inc, Shinomiya Hiratsuka Kanagawa, Japan

V. Horný IoP ASCR, Prague 8, Czech Republic; IPP ASCR, Prague 8, Czech Republic; LOA, ENSTA, Chemin de La Hunière, Palaiseau, France

Erik R. Hosler GLOBALFOUNDRIES, Malta, NY, USA

J. Hrebicek ELI-Beamlines, Institute of Physics v.v.i., Academy of Sciences of the Czech Republic, Prague 8, Czech Republic

Ke Hu Department of Physics, Institute for Fusion Theory and Simulation (IFTS), Zhejiang University, Hangzhou, China

K. Huang Kansai Photon Science Institute, National Institutes for Quantum and Radiological Science and Technology (QST), Kizugawa, Kyoto, Japan

S. Ichimaru NTT Advanced Technology Corporation, Atsugi, Kanagawa, Japan

C. Ikeda Department of Mechanical Science Engineering, Hiroshima University, Hiroshima, Japan

T. Imazono National Institutes for Quantum and Radiological Science and Technology, Kizugawa, Kyoto, Japan

N. Inogamov Landau Institute for Theoretical Physics of Russian Academy of Sciences, Chernogolovka, Moscow Region, Russia

K. Inokuma NTT Advanced Technology Corporation, Atsugi, Kanagawa, Japan

Y. Inubushi JASRI/SPring-8, Sayo, Hyogo, Japan

T. Ishikawa JASRI/SPring-8, Sayo, Hyogo, Japan; RIKEN Harima Institute, Sayo, Hyogo, Japan

M. Ishino National Institutes for Quantum and Radiological Science and Technology, Kizugawa, Kyoto, Japan

A. M. Ito Department of Helical Plasma Research, National Institute for Fusion Science (NIFS), Toki City, Gifu, Japan

H. Iwayama UVSOR, IMS, Myodaiji, Okazaki, Aichi, Japan

R. Jarocki Institute of Optoelectronics, Military University of Technology, Warsaw, Poland

T. Jitsuno Insititute of Laser Engineering, Osaka University, Osaka, Japan

R. Jung Max Born Institute, Berlin, Germany

N. Kamachi Toyama Co., Ltd, Yamakita, Kanagawa, Japan

M. Kanasaki Kansai Photon Science Institute, National Institutes for Quantum and Radiological Science and Technology, Kyoto, Japan

M. Kando Kansai Photon Science Institute, National Institutes for Quantum and Radiological Science and Technology (QST), Kizugawa, Kyoto, Japan

F. Kannari Keio University, Yokohama, Japan

Y. Kato The Graduate School for the Creation of New Photonics Industries, Nishiku, Hamamatsu, Shizuoka, Japan

T. Kawachi Kansai Photon Science Institute, National Institutes for Quantum and Radiological Science and Technology (QST), Kizugawa, Kyoto, Japan

Y. Kawasuji Gigaphoton Inc, Shinomiya Hiratsuka Kanagawa, Japan

H. Kawazome Department of Information Engineering, National Institute of Technology, Kagawa College, Mitoyo, Kagawa, Japan

V. Yu. Kharin Helmholtz Institute Jena, Jena, Germany

U. Kleineberg Ludwig-Maximilians-Universität München, Garching, Germany; Max-Planck-Institut für Quantenoptik, Garching, Germany

Kyung Taec Kim Center for Relativistic Laser Science, Institute for Basic Science, Gwangju, Korea; Department of Physics and Photon Science, Gwangju Institute of Science and Technology, Gwangju, Korea

Hyung Taek Kim Center for Relativistic Laser Science, Institute for Basic Science, Gwangju, Korea; Advanced Photonics Research Institute, Gwangju Institute of Science and Technology, Gwangju, Korea

H. Kimura JASRI/SPring-8, Sayo, Hyogo, Japan

H. Kiriyama Kansai Photon Science Institute, National Institutes for Quantum and Radiological Science and Technology (QST), Kizugawa, Kyoto, Japan

K. Kitano Vacuum and Optical Instruments, Tokyo, Japan

O. Kocharovskaya Texas A&M University, College Station, TX, USA

R. Kodama Graduate School of Engineering, Osaka University, Suita, Osaka, Japan; Open and Transdisciplinary Research Initiatives, Osaka University, Suita, Osaka, Japan; Photon Pioneers Center, Osaka University, Suita, Osaka, Japan

T. Kodama Gigaphoton Inc, Shinomiya Hiratsuka Kanagawa, Japan

A. O. Kolesnikov P. N. Lebedev Physical Institute, Russian Academy of Sciences, Moscow, Russia; Moscow Institute of Physics and Technology (State University), Dolgoprudnyi, Moscow, Russia

M. Koenig Graduate School of Engineering, Osaka University, Suita, Osaka, Japan; LULI, École Polytechnique, CNRS, CEA, UPMC, Palaiseau, France

J. K. Koga Kansai Photon Science Institute, National Institutes for Quantum and Radiological Science and Technology (QST), Kizugawa, Kyoto, Japan

K. Kolacek Pulse Plasma Systems Department, Institute of Plasma Physics of the Czech Academy of Sciences, Prague, Czech Republic

A. Kon Kansai Photon Science Institute, National Institutes for Quantum and Radiological Science and Technology (QST), Kizugawa, Kyoto, Japan; Japan Synchrotron Radiation Research Institute (JASRI), Sayo, Hyogo, Japan

K. Kondo Kansai Photon Science Institute, National Institutes for Quantum and Radiological Science and Technology (QST), Kizugawa, Kyoto, Japan

S. Kondo Kansai Photon Science Institute, National Institutes for Quantum and Radiological Science and Technology, Kyoto, Japan

Y. Kondo Faculty of Engineering and Center for Optical Research and Education (CORE), Department of Electrical and Electronic Engineering, Utsunomiya University, Utsunomiya, Tochigi, Japan

Georg Korn Institute of Physics ASCR, v.v.i. (FZU), ELI-Beamlines Project, Prague, Czech Republic

A. A. Kornilova Moscow State University, Moscow, Russia

J. Kostecki Institute of Optoelectronics, Military University of Technology, Warsaw, Poland

A. Kosuge KEK, Oho, Tsukuba, Ibaraki, Japan

H. Kotaki Kansai Photon Science Institute, National Institutes for Quantum and Radiological Science and Technology (QST), Kizugawa, Kyoto, Japan

M. Kozlová IoP ASCR, ELI Beamlines Project, Prague 8, Czech Republic; IPP ASCR, Prague 8, Czech Republic

M. Krůs IoP ASCR, ELI Beamlines Project, Prague 8, Czech Republic; IPP ASCR, Prague 8, Czech Republic

S. Kuma Atomic, Molecular, and Optical Physics Laboratory, RIKEN, Wako, Saitama, Japan

T. Kumada Quantum Beam Science Directorate, Kansai Photon Science Institute, Japan Atomic Energy Agency, Umemidai, Kizugawa, Kyoto, Japan; Materials Sciences Research Center, Japan Atomic Energy Agency, Ibaraki, Japan

S. Kurokawa Department of Mechanical Engineering, Kyushu University, Fukuoka, Japan

N. Kuwabara Toyama Co., Ltd, Yamakita, Kanagawa, Japan

A. A. Kuzin Moscow Institute of Physics and Technology (State University), Dolgoprudnyi, Moscow, Russia; Institute for Spectroscopy Russian Academy of Sciences, Troitsk, Moscow, Russia

I. Kuznetsov Department of Electrical & Computer Engineering, Colorado State University, Fort Collins, CO, USA

C. Kyaw Engineering Research Center for EUV Science and Technology and Electrical and Computer Engineering Department, Colorado State University, Fort Collins, CO, USA

J. Laksman MAX IV Laboratory, Lund University, Lund, Sweden

W. P. Leemans Lawrence Berkeley National Laboratory, Berkeley, CA, USA

B. Lei Helmholtz Institute Jena, Jena, Germany

Y. X. Leng State Key Laboratory of High Field Laser Physics, Shanghai Institute of Optics and Fine Mechanics, CAS, Shanghai, China

Tadzio Levato Institute of Physics ASCR, v.v.i. (FZU), ELI-Beamlines Project, Prague, Czech Republic

W. Li Engineering Research Center for EUV Science and Technology and Electrical and Computer Engineering Department, Colorado State University, Fort Collins, CO, USA

R. X. Li State Key Laboratory of High Field Laser Physics, Shanghai Institute of Optics and Fine Mechanics, CAS, Shanghai, China; IFSA Collaborative Innovation Centre, Shanghai Jiao Tong University, Shanghai, China

W. T. Li State Key Laboratory of High Field Laser Physics, Shanghai Institute of Optics and Fine Mechanics, CAS, Shanghai, China

J. Limpert Helmholtz Institute Jena, Jena, Germany; Institute of Applied Physics, Jena University, Jena, Germany; Fraunhofer Institute for Applied Optics and Precision Engineering, Jena, Germany

J. Q. Liu State Key Laboratory of High Field Laser Physics, Shanghai Institute of Optics and Fine Mechanics, CAS, Shanghai, China

J. S. Liu State Key Laboratory of High Field Laser Physics, Shanghai Institute of Optics and Fine Mechanics, CAS, Shanghai, China; IFSA Collaborative Innovation Centre, Shanghai Jiao Tong University, Shanghai, China

S.-Y. Liu Toyama Co., Ltd, Yamakita, Kanagawa, Japan

Benoît Mahieu Laboratoire d'Optique Appliquée, UMR 7639 ENSTA-CNRS-Ecole Polytechnique, Chemin de la Hunière, Palaiseau, France

T. Makimura Institute of Applied Physics, University of Tsukuba, Tsukuba, Japan

E. Malm Synchrotron Radiation Research, Lund University, Lund, Sweden

M. C. Marconi Engineering Research Center for EUV Science and Technology and Electrical and Computer Engineering Department, Colorado State University, Fort Collins, CO, USA

Daniele Margarone Institute of Physics ASCR, v.v.i. (FZU), ELI-Beamlines Project, Prague, Czech Republic

T. Maruyama NTT Advanced Technology Corporation, Atsugi, Kanagawa, Japan

Y. Matsukawa Department of Mechanical Engineering, Kyushu University, Fukuoka, Japan

K. Matsunaga Department of Mechanical Engineering, Kyushu University, Fukuoka, Japan

T. Matsuoka Open and Transdisciplinary Research Initiatives, Osaka University, Suita, Osaka, Japan

S. Matsuyama Graduate School of Engineering, Osaka University, Suita, Osaka, Japan

P. McKenna Department of Physics, SUPA, University of Strathclyde, Glasgow, UK

C. S. Menoni Engineering Research Center for EUV Science and Technology and Electrical and Computer Engineering Department, Colorado State University, Fort Collins, CO, USA

K. Midorikawa RIKEN Center for Advanced Photonics, Wako, Saitama, Japan

K. Mikami Kansai Photon Science Institute, National Institutes for Quantum and Radiological Science and Technology, Kyoto, Japan

Y. Minami Institute for Solid State Physics, The University of Tokyo, Chiba, Japan

S. Mironov Institute of Applied Physics RAS, Nizhny Novgorod, Russia

A. Mitrofanov Joint Institute for High Temperatures, RAS, Moscow, Russia

T. Miura HiLASE Centre, Dolní Břežany, Czech Republic

H. Mizoguchi Gigaphoton Inc, Oyama Tochigi, Japan

M. Mori National Institutes for Quantum and Radiological Science and Technology (QST), Tokai, Naka, Ibaraki, Japan

S. Mori Center for Advanced Microscopy and Spectroscopy, IMRAM, Tohoku University, Sendai, Japan

Kengo Moribayashi National Institutes for Quantum and Radiological Science and Technology, Kizugawa-City, Japan

A. Morozov Princeton University, Princeton, NJ, USA

R. Motokawa Materials Sciences Research Center, Japan Atomic Energy Agency, Ibaraki, Japan

S. Motokoshi Institute for Laser Technology, Osaka, Japan

G. Mourou DER-IZEST, École Polytechnique, Palaiseau Cedex, France

R. Nagai National Institutes for Quantum and Radiological Science and Technology (QST), Tokai, Naka, Ibaraki, Japan

D. Nakamura Graduate School of Information Sciences and Electrical Engineering, Kyushu University, Fukuoka, Japan

N. Nakanii Kansai Photon Science Institute, National Institutes for Quantum and Radiological Science and Technology (QST), Kizugawa, Kyoto, Japan

H. Nakarai Gigaphoton Inc, Shinomiya Hiratsuka Kanagawa, Japan

Chang Hee Nam Center for Relativistic Laser Science, Institute for Basic Science, Gwangju 61005, Korea; Department of Physics and Photon Science, Gwangju Institute of Science and Technology, Gwangju 61005, Korea

S. Namba Graduate School of Engineering, Hiroshima University, Higashihiroshima, Hiroshima, Japan; Department of Mechanical Science Engineering, Hiroshima University, Hiroshima, Japan

D. Neely Central Laser Facility, Rutherford Appleton Laboratory, STFC, Chilton, Didcot, Oxon, UK; Department of Physics, SUPA, University of Strathclyde, Glasgow, UK

V. Nefedova IoP ASCR, ELI Beamlines Project, Prague 8, Czech Republic; FNSPE of the CTU in Prague, Prague, Czech Republic

D. V. Negrov Moscow Institute of Physics and Technology (State University), Dolgoprudnyi, Moscow, Russia

J. Nejdl ELI-Beamlines, Institute of Physics v.v.i., Academy of Sciences of the Czech Republic, Prague 8, Czech Republic

Joseph Nilsen Lawrence Livermore National Laboratory, Livermore, CA, USA

K. Nishihara Institute for Laser Engineering, Osaka University, Osaka, Japan

T. Nishikawa Graduate School of Natural Science and Technology, Okayama, Japan

M. Nishikino Kansai Photon Science Institute, National Institutes for Quantum and Radiological Science and Technology (QST), Kizugawa, Kyoto, Japan; Quantum Beam Science Directorate, Kansai Photon Science Institute, Japan Atomic Energy Agency, Umemidai, Kizugawa, Kyoto, Japan

H. Nishimura Institute of Laser Engineering, Osaka University, Suita, Osaka, Japan

K. Nishitani Kansai Photon Science Institute, National Institutes for Quantum and Radiological Science and Technology (QST), Kizugawa, Kyoto, Japan; Interdisciplinary Graduate School of Engineering, Kyushu University, Kasuga, Fukuoka, Japan

M. Nishiuchi Kansai Photon Science Institute, National Institutes for Quantum and Radiological Science and Technology (QST), Kizugawa, Kyoto, Japan

K. M. Nowak Gigaphoton Inc, Shinomiya Hiratsuka Kanagawa, Japan

M. Ogawa Institute of Applied Physics, University of Tsukuba, Tsukuba, Japan

K. Ogura Kansai Photon Science Institute, National Institutes for Quantum and Radiological Science and Technology (QST), Kizugawa, Kyoto, Japan

T. Ohchi NTT Advanced Technology Corporation, Atsugi, Kanagawa, Japan

N. Ohnishi Department of Aerospace Engineering, Tohoku University, Sendai, Japan

S. Ohta Vacuum and Optical Instruments, Tokyo, Japan

H. Okada Kansai Photon Science Institute, National Institutes for Quantum and Radiological Science and Technology, Kyoto, Japan

T. Okada Graduate School of Information Sciences and Electrical Engineering, Kyushu University, Fukuoka, Japan

S. Okazaki Gigaphoton Inc, Oyama Tochigi, Japan

S. Oku NTT Advanced Technology Corporation, Atsugi, Kanagawa, Japan

E. Oliva Instituto de Fusión Nuclear, E.T.S.I. Industriales, Universidad Politécnica de Madrid, Madrid, Spain

Olga Olkhovskaya Keldysh Institute of Applied Mathematics RAS, Moscow, Russia

Y. Ono Department of Electrical and Electronic Engineering, Utsunomiya University, Utsunomiya, Japan

T. Otobe Quantum Beam Science Directorate, Kansai Photon Science Institute, Japan Atomic Energy Agency, Umemidai, Kizugawa, Kyoto, Japan; Kansai Photon Science Institute, National Institutes for Quantum and Radiological Science and Technology, Kyoto, Japan

N. Ozaki Graduate School of Engineering, Osaka University, Suita, Osaka, Japan; Photon Pioneers Center, Osaka University, Suita, Osaka, Japan

M. Pedicone Colorado State University, Fort Collins, CO, USA

B. Pfau Max-Born Institute, Berlin, Germany

K. Ta Phuoc IoP ASCR, Prague 8, Czech Republic; FNSPE of the CTU in Prague, Prague, Czech Republic

S. A. Pikuz Joint Institute for High Temperatures, RAS, Moscow, Russia; National Research Nuclear University MEPhI, Moscow, Russia

T. A. Pikuz Graduate School of Engineering, Osaka University, Suita, Osaka, Japan; Joint Institute for High Temperatures of the Russian Academy of Sciences, Moscow, Russia

P. Pira Department of Spectroscopy, J. Heyrovsky Institute of Physical Chemistry of the Czech Academy of Sciences, Prague, Czech Republic

A. S. Pirozhkov Kansai Photon Science Institute, National Institutes for Quantum and Radiological Science and Technology (QST), Kizugawa, Kyoto, Japan

N. L. Popov L.P. Lebedev Physical Institute, Moscow, Russia

C. Pratsch Helmholtz-Zentrum Berlin (HZB), Berlin, Germany

R. Qi State Key Laboratory of High Field Laser Physics, Shanghai Institute of Optics and Fine Mechanics, CAS, Shanghai, China

Z. Y. Qin State Key Laboratory of High Field Laser Physics, Shanghai Institute of Optics and Fine Mechanics, CAS, Shanghai, China

G. Račiukaitis Center for Physical Sciences and Technology, Vilnius, Lithuania

E. N. Ragozin P. N. Lebedev Physical Institute of the Russian Academy of Sciences, Moscow, Russia; Moscow Institute of Physics and Technology (State University), Dolgoprudnyi, Moscow Region, Russia

B. A. Reagan Colorado State University, Fort Collins, CO, USA

L. Rimgaila Center for Physical Sciences and Technology, Vilnius, Lithuania

J. J. Rocca Engineering Research Center for EUV Science and Technology and Electrical and Computer Engineering Department, Colorado State University, Fort Collins, CO, USA

W. Rockward Department of Physics, Morehouse College, Atlanta, GA, USA

A. Rockwood Colorado State University, Fort Collins, CO, USA

N. N. Rosanov Vavilov State Optical Institute, Saint-Petersburg, Russia

A. K. Rossall York Plasma Institute Department of Physics, University of York, York, UK

J. Rothhardt Helmholtz Institute Jena, Jena, Germany; Institute of Applied Physics, Jena University, Jena, Germany

B. Rus ELI-Beamlines, Institute of Physics v.v.i., Academy of Sciences of the Czech Republic, Prague 8, Czech Republic

S. G. Rykovanov Helmholtz Institute Jena, Jena, Germany

I. Saber Institute of Optoelectronics, Military University of Technology, Warsaw, Poland

A. Sagisaka Kansai Photon Science Institute, National Institutes for Quantum and Radiological Science and Technology (QST), Kizugawa, Kyoto, Japan

T. Saitou Gigaphoton Inc, Shinomiya Hiratsuka Kanagawa, Japan

H. Sakaki Kansai Photon Science Institute, National Institutes for Quantum and Radiological Science and Technology (QST), Kizugawa, Kyoto, Japan

R. Sankari MAX IV Laboratory, Lund University, Lund, Sweden

A. Sasaki Kansai Photon Science Institute, National Institutes for Quantum and Radiological Science and Technology, Kizugawa-shi Kyoto, Japan

A. Sasanuma Department of Innovation System Engineering & CORE, Utsunomiya University, Utsunomiya, Tochigi, Japan

P. V. Sasorov M. V. Keldysh Institute of Applied Mathematics, Russian Academy of Sciences, Moscow, Russia

Y. Sato Graduate School of Engineering, Osaka University, Suita, Osaka, Japan

J. Schmidt Pulse Plasma Systems Department, Institute of Plasma Physics of the Czech Academy of Sciences, Prague, Czech Republic

C. B. Schroeder Lawrence Berkeley National Laboratory, Berkeley, CA, USA

M. O. Scully Texas A&M University, College Station, TX, USA; Princeton University, Princeton, NJ, USA

S. Sebban IoP ASCR, ELI Beamlines Project, Prague 8, Czech Republic; Laboratoire d'Optique Appliquée, UMR 7639 ENSTA-CNRS-Ecole Polytechnique, Chemin de la Hunière, Palaiseau, France

C. Seim Max-Born Institute, Berlin, Germany; Berlin Laboratory for innovative X-ray Technologies (BLiX), Berlin, Germany

K. Sekiguchi Kansai Photon Science Institute, National Institutes for Quantum and Radiological Science and Technology (QST), Kizugawa, Kyoto, Japan

M. Serbanescu CETAL, INFLPR, Măgurele, Romania

A. N. Shatokhin P. N. Lebedev Physical Institute, Russian Academy of Sciences, Moscow, Russia; Moscow Institute of Physics and Technology (State University), Dolgoprudnyi, Moscow, Russia

E. Shigemasa UVSOR, IMS, Myodaiji, Okazaki, Aichi, Japan

Y. Shiraishi Gigaphoton Inc, Shinomiya Hiratsuka Kanagawa, Japan

T. Shizuma National Institutes for Quantum and Radiological Science and Technology (QST), Tokai, Naka, Ibaraki, Japan

V. N. Shlyaptsev Colorado State University, Fort Collins, CO, USA

I. Yu. Skobelev Joint Institute for High Temperatures, Russian Academy of Sciences, Moscow, Russia; National Research Nuclear University MEPhI, Moscow, Russia

W. Skrzeczanowski Institute of Optoelectronics, Military University of Technology, Warsaw, Poland

Z. H. Song State Key Laboratory of Intense Pulsed Radiation Simulation and Effect, Northwest Institute of Nuclear Technology, Xi'an, China

G. Soumagrne Gigaphoton Inc, Shinomiya Hiratsuka Kanagawa, Japan

C. Späth Ludwig-Maximilians-Universität München, Garching, Germany; Max-Planck-Institut für Quantenoptik, Garching, Germany

C. Spielmann Institute of Optics and Quantum Electronics, Jena University, Jena, Germany; Helmholtz Institute Jena, Jena, Germany

V. Stankevič Center for Physical Sciences and Technology, Vilnius, Lithuania

H. Stiel Max Born Institute, Berlin, Germany

J. Straus Pulse Plasma Systems Department, Institute of Plasma Physics of the Czech Academy of Sciences, Prague, Czech Republic

S. Suckewer Princeton University, Princeton, NJ, USA

A. Suda Tokyo University of Science, Chiba, Japan

T. Suemoto Toyota Physical and Chemical Research Institute, Aichi, Japan; Institute for Solid State Physics, The University of Tokyo, Chiba, Japan

A. Sunahara Institute for Laser Technology, Osaka, Japan

K. Sawada RIKEN Center for Advanced Photonics, Wako, Saitama, Japan; Keio University, Yokohama, Japan

M. Szczurek Institute of Optoelectronics, Military University of Technology, Warsaw, Poland

A. Takahashi Department of Health Sciences, School of Medicine, Kyushu University, Fukuoka, Japan

E. J. Takahashi RIKEN Center for Advanced Photonics, Wako, Saitama, Japan

Y. Takaya Department of Mechanical Engineering, Osaka University, Suita, Osaka, Japan

H. Takenaka Toyama Co., Ltd, Yamakita, Kanagawa, Japan

G. J. Tallents York Plasma Institute Department of Physics, University of York, York, UK

T. Tamura Faculty of Engineering and Center for Optical Research and Education (CORE), Department of Electrical and Electronic Engineering, Utsunomiya University, Utsunomiya, Tochigi, Japan

Y. Tamaru RIKEN Center for Advanced Photonics, Wako, Saitama, Japan; Tokyo University of Science, Chiba, Japan

H. Tanaka Gigaphoton Inc, Shinomiya Hiratsuka Kanagawa, Japan

K. A. Tanaka Graduate School of Engineering, Osaka University, Suita, Osaka, Japan

N. Tanaka Institute of Laser Engineering, Osaka University, Suita, Osaka, Japan

Y. Tange JASRI/SPring-8, Sayo, Hyogo, Japan

N. Terunuma KEK, Oho, Tsukuba, Ibaraki, Japan; Science of the Graduate University for Advanced Studies (SOKENDAI), Miura District, Kanagawa, Japan

T. Togashi JASRI/SPring-8, Sayo, Hyogo, Japan

V. Tomkus Center for Physical Sciences and Technology, Vilnius, Lithuania

K. Tono JASRI/SPring-8, Sayo, Hyogo, Japan

A. Torrisi Institute of Optoelectronics, Military University of Technology, Warsaw, Poland

M. Toyoda Center for Advanced Microscopy and Spectroscopy, IMRAM, Tohoku University, Sendai, Japan

E. Tubman Department of Physics, York Plasma Institute, University of York, York, UK

J. Tümmler Max Born Institute, Berlin, Germany

A. Tünnermann Helmholtz Institute Jena, Jena, Germany; Institute of Applied Physics, Jena University, Jena, Germany; Fraunhofer Institute for Applied Optics and Precision Engineering, Jena, Germany

R. Ungureanu CETAL, INFLPR, Măgurele, Romania

H. Urai Institute of Applied Physics, University of Tsukuba, Tsukuba, Japan

J. Urakawa KEK, Oho, Tsukuba, Ibaraki, Japan

P. Velarde Instituto de Fusión Nuclear, E.T.S.I. Industriales, Universidad Politécnica de Madrid, Madrid, Spain

S. Vicens Instituto de Fusión Nuclear, E.T.S.I. Industriales, Universidad Politécnica de Madrid, Madrid, Spain

A. V. Vinogradov L.P. Lebedev Physical Institute, Moscow, Russia

E. A. Vishnyakov P. N. Lebedev Physical Institute, Russian Academy of Sciences, Moscow, Russia

N. V. Vysotina Vavilov State Optical Institute, Saint-Petersburg, Russia

V. I. Vysotskii Kiev National Shevchenko University, Kiev, Ukraine

M. V. Vysotskyy Kiev National Shevchenko University, Kiev, Ukraine

P. Wachulak Institute of Optoelectronics, Military University of Technology, Warsaw, Poland

N. Wada Institute of Laser Engineering, Osaka University, Suita, Osaka, Japan

C. Wang State Key Laboratory of High Field Laser Physics, Shanghai Institute of Optics and Fine Mechanics, CAS, Shanghai, China

J. W. Wang Helmholtz Institute Jena, Jena, Germany

S. Wang Colorado State University, Fort Collins, CO, USA

W. T. Wang State Key Laboratory of High Field Laser Physics, Shanghai Institute of Optics and Fine Mechanics, CAS, Shanghai, China

Y. Wang Colorado State University, Fort Collins, CO, USA

Y. Watanabe Gigaphoton Inc, Shinomiya Hiratsuka Kanagawa, Japan

F. L. Wei State Key Laboratory of Intense Pulsed Radiation Simulation and Effect, Northwest Institute of Nuclear Technology, Xi'an, China

A. West York Plasma Institute Department of Physics, University of York, York, UK

J. Wheeler DER-IZEST, École Polytechnique, Palaiseau Cedex, France; National Institute for Nuclear Physics-HH, ELI-NP, Măgurele, Romania

S. Wilson York Plasma Institute Department of Physics, University of York, York, UK

Obert R. Wood II GLOBALFOUNDRIES, Malta, NY, USA

N. Woolsey Department of Physics, York Plasma Institute, University of York, York, UK

Hui-Chun Wu Department of Physics, Institute for Fusion Theory and Simulation (IFTS), Zhejiang University, Hangzhou, China; IFSA Collaborative Innovation Center, Shanghai Jiao Tong University, Shanghai, China

Ł. Węgrzyński Institute of Optoelectronics, Military University of Technology, Warsaw, Poland

Y. Xu State Key Laboratory of High Field Laser Physics, Shanghai Institute of Optics and Fine Mechanics, CAS, Shanghai, China

Z. Z. Xu State Key Laboratory of High Field Laser Physics, Shanghai Institute of Optics and Fine Mechanics, CAS, Shanghai, China; IFSA Collaborative Innovation Centre, Shanghai Jiao Tong University, Shanghai, China

M. Yabashi JASRI/SPring-8, Sayo, Hyogo, Japan; RIKEN Harima Institute, Sayo, Hyogo, Japan

T. Yabuuchi RIKEN Harima Institute, Sayo, Hyogo, Japan

T. Yamada Gigaphoton Inc, Shinomiya Hiratsuka Kanagawa, Japan

Y. Yamamoto Kansai Photon Science Institute, National Institutes for Quantum and Radiological Science and Technology, Kyoto, Japan

K. Yamauchi Graduate School of Engineering, Osaka University, Suita, Osaka, Japan

T. Yamazaki Gigaphoton Inc, Shinomiya Hiratsuka Kanagawa, Japan

T. Yanagida Gigaphoton Inc, Shinomiya Hiratsuka Kanagawa, Japan

M. Yanagihara Center for Advanced Microscopy and Spectroscopy, IMRAM, Tohoku University, Sendai, Japan

K. Yasuda Graduate School of Engineering, Osaka University, Suita, Osaka, Japan

L. Yin Colorado State University, Fort Collins, CO, USA

A. Yogo Institute of Laser Engineering, Osaka University, Suita, Osaka, Japan

H. Yokoo Department of Mechanical Engineering, Kyushu University, Fukuoka, Japan

C. H. Yu State Key Laboratory of High Field Laser Physics, Shanghai Institute of Optics and Fine Mechanics, CAS, Shanghai, China

H. Yumoto JASRI/SPring-8, Sayo, Hyogo, Japan

Hyeok Yun Center for Relativistic Laser Science, Institute for Basic Science, Gwangju 61005, Korea

Philippe Zeitoun Laboratoire d'Optique Appliquée, UMR 7639 ENSTA-CNRS-Ecole Polytechnique, Chemin de la Hunière, Palaiseau, France

M. Zenba NTT Advanced Technology Corporation, Atsugi, Kanagawa, Japan

Z. J. Zhang State Key Laboratory of High Field Laser Physics, Shanghai Institute of Optics and Fine Mechanics, CAS, Shanghai, China

A. Zhidkov Open and Transdisciplinary Research Initiatives, Osaka University, Suita, Osaka, Japan

M. Zürch Chemistry Department, University of California, Berkeley, USA; Institute of Optics and Quantum Electronics, Jena University, Jena, Germany; Helmholtz Institute Jena, Jena, Germany

Part I
Plasma Based X-Ray Lasers and Applications

Chapter 1
Overview of Laser-Driven Coherent Short-Wavelength Sources at PALS and ELI Beamlines

J. Nejdl, M. Kozlová, V. Nefedova, M. Albrecht, M. Krůs, J. Gautier and S. Sebban

Abstract Recent activities at PALS research centre dedicated to the development of coherent short-wavelength radiation sources are presented. The generation mechanisms discussed in this paper include high-order harmonic generation and transient plasma X-ray laser with solid target both using 20 TW Ti:Sapphire laser chain, and quasi-steady state plasma X-ray laser-driven by kJ sub-ns iodine laser system. Besides the ongoing research at PALS we also discuss the conceptual design of future high-order harmonic beamline employing 1 kHz 100 mJ 20 fs laser system at upcoming ELI Beamlines facility.

1.1 Introduction

Exploring nature at smaller spatial scales and shorter time scales using electromagnetic radiation requires shortening both the wavelength of the probe and the pulse duration. The large-scale facilities such as modern synchrotrons and free electron lasers can only partially cover the increasing demands for such light beams due to their limited access stemming from relatively high cost. Moreover, the precise synchronization of X-ray pulses produced this way with pump pulses in ultrafast pump-probe experiments is rather demanding due to the intrinsic temporal jitter of the accelerated electron bunch that generates the radiation. These are the

J. Nejdl (✉) · M. Kozlová · V. Nefedova · M. Albrecht · M. Krůs · J. Gautier · S. Sebban
IoP ASCR, ELI Beamlines Project, Na Slovance 2, 182 00 Prague 8, Czech Republic
e-mail: nejdl@fzu.cz

J. Nejdl · M. Kozlová · M. Krůs
IPP ASCR, Za Slovankou 3, 182 00 Prague 8, Czech Republic

J. Gautier · S. Sebban
LOA, ENSTA, Chemin de La Hunière, 91761 Palaiseau, France

V. Nefedova · M. Albrecht
FNSPE of the CTU in Prague, Prague, Czech Republic

T. Kawachi et al. (eds.), *X-Ray Lasers 2016*, Springer Proceedings in Physics 202, https://doi.org/10.1007/978-3-319-73025-7_1

reasons why looking for alternative sources is of much interest to the broad scientific community.

Some applications require radiation with high degree of coherence, because of its well predictable behaviour, its ability to be focused down to diffraction limited spot, and pulse durations often limited only by the spectral content of the radiation.

The advantage of laser-driven sources of X-ray radiation is, beside their compactness, the possibility to split the driving laser pulse and use it for sample excitation in a pump-probe experiment with temporal jitter limited only by the mechanical stability of the beam paths.

One way to create a source of coherent short-wavelength radiation is to generate an X-ray laser (XRL) by achieving population inversion in a long column of highly ionized species by linear focusing of a driving laser beam [1, 2]. Up to now, the electron collisional excitation turned out to be the most efficient process of creating the population inversion.

The narrow spectral line of radiation emitted on the axis of the column by amplified spontaneous emission (ASE), whose bandwidth is given by upper level lifetime and inhomogeneous Doppler broadening due to nonzero ion temperature, ensures relatively long coherence lengths [3]. The transverse coherence of the ASE-based XRLs on the other side is mostly given by its wavelength and by the geometry of the column. One can easily define the Fresnel number F of the gain region as

$$F = \frac{a^2}{\lambda L}, \tag{1.1}$$

with a being the transverse size of the column, L its length, and λ the wavelength of emitted radiation. Not considering the effects of refraction, one can expect the beam to be spatially coherent, if $F \leq 1$ However, as in some cases this condition is hard to be fulfilled, an alternative approach to improve the coherence of the XRL is to use them as amplifiers for high-quality XUV beams [4, 5] generated by high-order harmonic generation (HHG).

The process of HHG is based on the highly nonlinear interaction of intense laser pulse with matter. On a microscopic level it can be explained by a simple three-step model [6] where the valence electron is first ionized by strong laser field, it gets accelerated in this field as a free charged particle and subsequently recombines with its parent ion. Up to maximum energy (cut-off energy), there are two possible electron trajectories having the same kinetic energy during recombination: the short trajectory and the long trajectory referring to the length of the time interval between ionization and recombination. Macroscopically, one is interested in the investigation of the coherent build-up of the generated field in the medium called phase-matching (PM) [7, 8]. The nonlinear propagation of the driving laser pulse and linear propagation of the short-wavelength field in the generating medium can be assessed only with the use of complex numerical models. Nevertheless, applying many simplifications one can extract the contribution of different effects such as dispersion of neutral gas and plasma, geometrical phase effects, intensity dependent dipole phase or absorption in the media [9].

This paper describes the recent activities in the fields of plasma X-ray lasers and high-order harmonic generation at PALS research centre and introduces the conceptual design of a high-flux high-order harmonic beamline developed within the ELI beamlines project.

1.2 Laser-Driven Short-Wavelength Sources at PALS

1.2.1 Plasma X-ray Lasers

PALS research centre is currently equipped with two high-power laser drivers: ASTERIX IV laser system [10] operating at 1.3 μm wavelength and generating 350 ps long pulses with energy up to 1 kJ with one shot every 30 min, and a CPA-based 20 TW Ti:sapphire laser chain generating 45 fs pulses with energy up to 1 J at repetition rate of 10 Hz [11]. Both of these drivers are suitable to drive particular scheme of X-ray laser.

The collisionally pumped quasi-steady state (QSS) Ne-like zinc XRL emitting at 21.2 nm [12] has been routinely used as a user beamline at the PALS facility. It has proven to be a unique tool in hot dense plasma investigations [13, 14], in studies related to laboratory astrophysics [15], short-wavelength radiation ablation studies of materials [16], or radiation biophysics [17].

The detailed measurements of spectral linewidth of the Zn XRL revealed that there is a potential to use this pumping scheme to create an XRL amplifier capable of amplifying pulses shorter than 1 ps [18], which was considered as a limit for this kind of source until recently reported results with XRL driven by optical field ionization in dense gaseous targets [19].

The latest effort devoted to source development was oriented to the transient collisional excitation (TCE) scheme with the 1 J, 10 Hz Ti:sapphire laser system employing grazing incidence pumping (GRIP). The lasing action of Ni-like molybdenum at 18.9 nm close to saturation was already reported in [20]. Recently we have focused on analysis of the speckle statistics in order to better understand the coherence properties of the beam [21]. Typical speckle patterns of this laser for the two driving geometries (with grazing angles 25° and 20°) are depicted in Fig. 1.1.

As the two laser drivers are already temporally synchronizable, the ultimate goal of the XRL source development would be to combine the HHG seed with TCE preamplifier and QSS main amplifier in order to obtain GW-class soft X-ray pulses. Given the limited driving power it is, however, not clear which of the driving schemes is the most promising to work in both TCE and QSS pumping regimes.

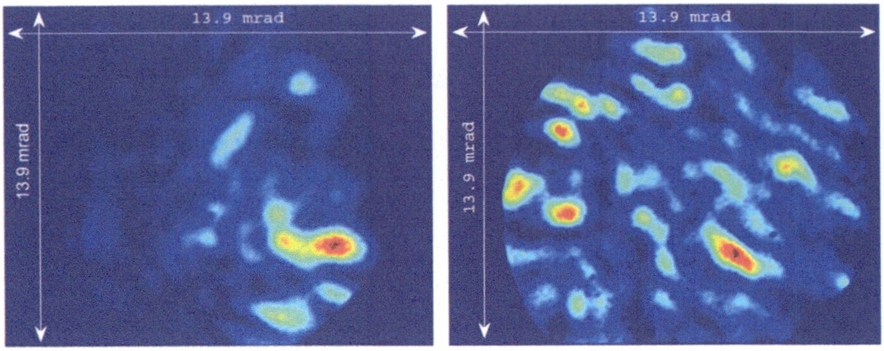

Fig. 1.1 Far-field intensity distribution of the Ne-like Mo XRL beam at pumping geometries with grazing angle 25° (left) and 20° (right); the length of the laser focus was 4.3 mm and 3.5 mm, respectively

1.2.2 Development of High-Energy High-Order Harmonic Source

We have performed an extensive study of scaling of the output parameters of HHG from noble gases in the optimum phase-matching conditions with respect to f-number of the driving laser beam using 50 fs Ti:Sapphire laser chain delivering either 0.5 mJ on target at 1 kHz repetition rate or up to 0.5 J at 10 Hz. Using this wide range of driving laser energies we have performed a series of experiments with different focusing configurations [22].

We observed generation cuttoff around 27 nm in the case of Ar, while the cuttoff for Ne and He was extending beyond our spectrometer range, i.e. it was probably below 9 nm. A typical spectrum obtained from generation in He with 5 m long focusing and 300 nm thick Zr filter serving to cut the laser light is shown in Fig. 1.2.

Fig. 1.2 Typical spectrum of harmonics generated in 44 mm long gas cell filled with He driven by $F_{\#} = 200$ beam with 50 fs pulses of energy 125 mJ. The spectrum was obtained using flat-field spectrometer with variable line spacing concave grating and a back-illuminated CCD camera

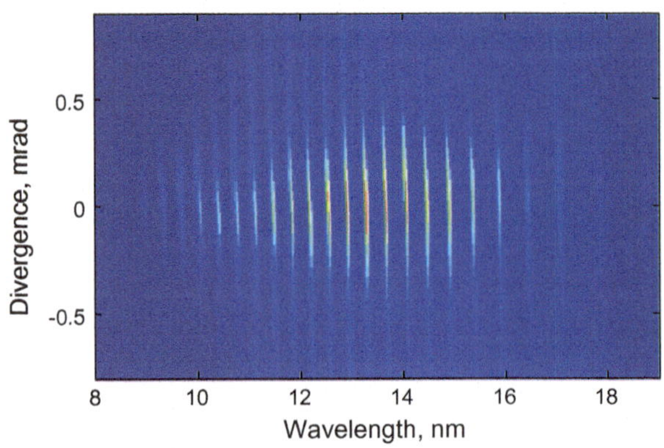

Table 1.1 Scaling of experimental parameters in different focusing configurations for optimum generation with Ar gas cell

Focal length (m)	Energy on target (mJ)	$F_{\#}$	Divergence (mrad)	Cell length (mm)	Gas pressure (mbar)	XUV energy (nJ)	Efficiency
0.5	0.5	50	2.5	4	40	0.7	1×10^{-6}
2	50	150	0.7	50	15	15	3×10^{-7}
5	50	300	0.3	80	6	40	1×10^{-6}

Although the exact conditions were slightly different for different experimental setups (e.g. pulse shape due to dispersion in glass of the focusing lens) general tendencies of the scaling of output parameters can be drawn.

The detailed experimental parameters for optimum generation in Ar are listed in Table 1.1. One can note that in the case of $f = 2$ m the generation was most likely beyond the saturation limit (the medium was over-ionized due to high driving intensity). That is why the efficiency of the generation is lower than in the other cases.

The divergence of the central part of the HHG beam (usually attributed to the short trajectory) scales as $1/F_{\#}$. The cell length, i.e. the maximum length of the media, where PM still applies, is roughly given by the Rayleigh range that is proportional to $(F_{\#})^2$. The optimum gas pressure scales as $(F_{\#})^{-2}$, and the output energy as $(F_{\#})^2$. In order to achieve optimum driving intensity, the input energy needs to scale also as $(F_{\#})^2$, which means that the conversion efficiency at optimum PM conditions is independent of the f-number as predicted theoretically [23].

One way to increase the conversion efficiency would be either to modify the generating medium and apply a technique called quasi-phase matching [24], or to shorten the wavelength of the driving laser on the expense of shifting the harmonic cut-off to longer wavelengths, as the microscopic efficiency scales roughly as $\lambda^{-(5-6)}$ [25].

1.3 Conceptual Design of High-Order Harmonic Beamline at ELI Beamlines

The ELI (Extreme Light Infrastructure) project is a part of the European plan to build the next generation of large research facilities. ELI Beamlines as a cutting-edge laser facility is currently being constructed near Prague, Czech Republic. The primary focus of this facility will be on applications of laser-driven ultrafast secondary sources of radiation and accelerated particles for user-based research. That is why the laser technology that will be available at ELI beamlines is aiming to develop high-repetition rate high peak-power femtosecond laser pulses.

One of the user beamlines is designed to provide tunable ultrashort coherent pulses of VUV to soft X-ray radiation through HHG in gases. It is designed to be driven by the 'L1' laser that is an OPCPA system with 1 kHz repetition rate delivering <20 fs pulses with energy up to 100 mJ [26].

This HHG beamline is based on the interaction of intense laser pulse with gaseous targets in a loose focusing geometry (with maximum f-number reaching 1000), in order to achieve maximum conversion efficiency of the available laser power into the XUV beam. Depending on the requested wavelength the proper noble gas as generating medium will be used together with optimum focusing geometry to maximize the XUV flux. The following XUV diagnostics will be implemented to characterize the output beam:

- *Spectrometer* consisting of a toroidal mirror, a slit, a set of spherical reflective gratings with variable line spacing, and a back-thinned CCD camera. The optical system is designed to compensate for astigmatism of the grating and the slit enables increasing the spectral resolution on expense of sensitivity of the device.
- *Pulse energy measurement* with absolute offline mode using calibrated silicon photodiode placed directly to the XUV beam and relative online mode where we will measure the charge of the photocurrent pulses from the thin metallic filters that are used to cut the laser radiation.
- *Wavefront sensor* based on the Hartman type sensor to measure the beam pointing, its intensity distribution (divergence) and its wavefront.

All of those diagnostic tools will serve to optimize the source and characterize the beam in order to provide the best performance at the user end stations.

There will be two output arms of the harmonic beam

- A direct arm to be used for applications requiring high-flux VUV to soft X-ray beams primarily designated to atomic, molecular and optical sciences.
- A side arm to be used for ultrafast IR-VUV ellipsometry aimed at research in material science.

The source is designed to accommodate two driving laser beams split from the same laser system to be either used for driving two independent but mutually coherent VUV-soft X-ray beams that will allow XUV pump XUV probe experiments or to be used for two-colour driving scheme when one of the beams is converted to its second harmonic providing additional features in terms of conversion efficiency, accessible wavelengths or polarization states of the generated radiation.

The layout of the beamline in the two configurations is shown in Fig. 1.3 and the expected output parameters of the generated radiation are listed in Table 1.2.

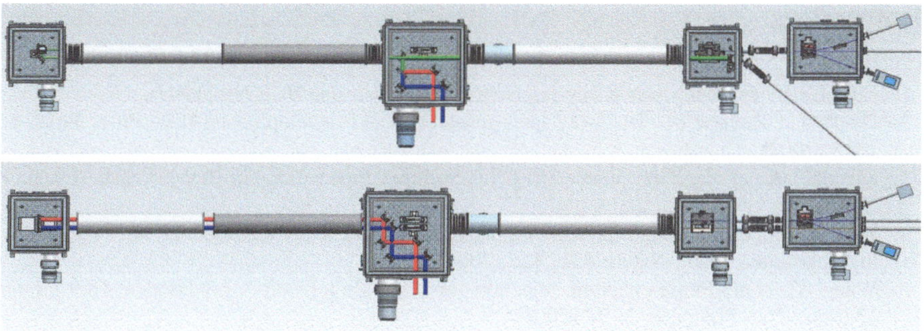

Fig. 1.3 Schematics of the two modes of operation of the HHG beamline with two driving beams: top: two-colour driving scheme with the two beams arriving from the bottom and recombining on a dichroic mirror and the XUV beam heading to the right. The straight and the side output arms are also shown. Bottom: each of the laser driver beams is reflected by a planar mirror with a central hole and focused onto the gas cell by a spherical mirror placed in the chamber on the left. The two independent XUV beams are generated and propagate along each other

Table 1.2 Expected output parameters of the HHG beamline driven by a single laser beam at ELI beamlines

Driver	1 kHz, 5 mJ, 35 fs	1 kHz, 100 mJ, 20 fs
Wavelength range (nm)	10–120	5–120
Photons/shot	10^7–10^9	few 10^9–10^{12}
$\Delta\lambda/\lambda$	10^{-2}	10^{-2}
Divergence (mrad)	<2	<1
Spatial profile	Gaussian-like	Gaussian-like
Wavefront RMS	$\lambda/10$	$\lambda/10$
Duration (fs)	<20	<20

1.4 Conclusions

The laser-driven sources of coherent XUV radiation are being extensively developed at the PALS research centre and this development will continue with the new ELI Beamlines facility, which is under construction in the Czech Republic. We believe that reliable operation of these sources based on stable and robust high-repetition-rate laser drivers will allow pump-probe experiments that are not possible elsewhere.

Acknowledgements We are deeply indebted to J. Hrebicek, T. Medrik and J. Golasowski for their technical support at PALS laboratory. This project was supported by the project ELI—Extreme Light Infrastructure—phase 2 (CZ.02.1.01/0.0/0.0/15_008/0000162) from European Regional Development Fund and Project LQ1606 of the Ministry of Education, Youth and Sports as part of targeted support from the National Programme of Sustainability II.

References

1. Rocca, J.J.: Table-top soft X-ray lasers. Rev. Sci. Instrum. **70**, 3799 (1999)
2. Daido, H.: Review of soft X-ray laser researches and developments. Rep. Prog. Phys. **65**, 1513 (2002)
3. Klisnick, A., et al.: X-ray lasers 2014. In: Springer Proceedings in Physics, vol. 169, p. 45 (2016)
4. Zeitoun, Ph., et al.: A high-intensity highly coherent soft X-ray femtosecond laser seeded by a high harmonic beam. Nature **431**, 426 (2004)
5. Wang, Y., et al.: Phase-coherent, injection-seeded, table-top soft-X-ray lasers at 18.9 nm and 13.9 nm. Nat. Photonics **2**, 94 (2008)
6. Corkum, P.B.: Plasma perspective on strong field multiphoton ionization. Phys. Rev. Lett. **71**, 1994 (1993)
7. Chang, Z.: Fundamentals of Attosecond Optics. CRC Press (2011)
8. Ruchon, T., et al.: Macroscopic effects in attosecond pulse generation. New J. Phys. **10**, 025027 (2008)
9. Balcou, Ph., et al.: Generalized phase-matching conditions for high harmonics: the role of field-gradient forces. Phys. Rev. A **55**, 3204 (1997)
10. Jungwirth, K., et al.: The prague asterix laser system. Phys. Plasmas **8**, 2495 (2001)
11. Hřebíček, J., et al.: 25TW Ti:sapphire laser chain at PALS. In: Proceeding of SPIE, vol. 8080 (2011)
12. Rus, B., et al.: Multimillijoule, highly coherent X-ray laser at 21 nm operating in deep saturation through double-pass amplification. Phys. Rev. A **66**, 063806 (2002)
13. Gartside, L.M.R., et al.: Extreme ultraviolet interferometry of warm dense matter in laser plasmas. Opt. Lett. **35**, 22 (2010)
14. Nejdl, J., et al.: Measuring the electron density gradients of dense plasmas by deflectometry using short-wavelength probe. Phys. Plasmas **17**, 122705 (2010)
15. Stehle, C., et al.: New probing techniques of radiative shocks. Opt. Commun. **285**, 64 (2011)
16. Chalupsky, J., et al.: Utilizing ablation of solids to characterize a focused soft X-ray laser beam. In: Proceeding of SPIE, vol. 6586, p. 65860S (2007)
17. Cassou, K., et al.: Study of damages induced in DNA by X-UV laser irradiation at 21.2 nm. J. Phys. IV Fr. **127**, 177 (2005)
18. Le Marec, A., et al.: Measurement of a subpicosecond coherence time in a quasi-steady-state XUV laser. Phys. Rev. A **92**, 033852 (2015)
19. Depresseux, A., et al.: Table-top femtosecond soft X-ray laser by collisional ionization gating. Nat. Photonics **9**, 817 (2015)
20. Kozlová, M., et al.: X-ray lasers 2014. In: Springer Proceedings in Physics, vol. 169, p. 35 (2016)
21. Albrecht, M., et al.: in preparation
22. Nefedova, V., et al.: Development of a high-flux XUV source based on high-order harmonic generation. J. Electron Spectrosc. **220**, 9 (2017)
23. Heyl, C.M., et al.: Scale-invariant nonlinear optics in gases. Optica **1**, 75 (2016)
24. Auguste, T., et al.: Quasi-phase-matching of high-order harmonics using a modulated atomic density. Phys. Rev. A **76**, 011802 (2007)
25. Shiner, A.D., et al.: Wavelength scaling of high harmonic generation efficiency. Phys. Rev. Lett. **103**, 073902 (2009)
26. Rus, B., et al.: Outline of the ELI-beamlines facility. In: Proceeding of SPIE, vol. 8080, p. 808010 (2011)

Chapter 2
High Average Power Table-Top Soft X-Ray Lasers Using Diode-Pumped Laser Drivers

J. J. Rocca, B. A. Reagan, C. Baumgarten, M. Pedicone, L. Yin, V. N. Shlyaptsev, Y. Wang, S. Wang, A. Rockwood, M. Berrill, M. C. Marconi and C. S. Menoni

Abstract Soft X-ray lasers (SXRLs) produce the highest energy pulses of coherent ultrashort wavelength radiation. Their large number of photons per pulse allow us to perform single-shot imaging of nanoscale objects, to develop material composition sensitive nanoprobes, and to conduct interferometric diagnostics of bright dense plasmas. However, until recently, with the exception of capillary discharge lasers at 46.9 nm, their average power was limited by the low repetition rate of the high energy optical pump lasers required to drive them and by the relatively low pumping efficiency. To overcome this limitation, we have developed a diode-pumped, picosecond Yb:YAG CPA laser driver that allowed us to demonstrate the first table-top SXRL capable of 100 Hz repetition rate gain saturated operation. These new pump lasers combined with efficient plasma heating techniques enable the operation of SXRLs at four orders of magnitude higher repetition rate than the first plasma-based collisional SXRLs (Fig. 2.1). Laser operation at 100 Hz repetition rate generated an average power of 0.2 mW at 18.9 nm (Ni-like Mo) and 0.1 mW at $\lambda = 13.9$ nm (Ni-like Ag). We have recently improved upon this diode-pumped laser technology, demonstrating a record 0.5 kHz repetition rate, 1 J, picosecond laser with good beam quality and high stability. We will discuss the results of the first demonstration of a compact table-top soft X-ray laser at repetition rates up to 400 Hz using this new pump laser.

J. J. Rocca (✉) · B. A. Reagan · C. Baumgarten · M. Pedicone · L. Yin · V. N. Shlyaptsev
Y. Wang · S. Wang · A. Rockwood · M. C. Marconi · C. S. Menoni
Colorado State University, Fort Collins, CO 80523, USA
e-mail: jorge.rocca@colostate.edu

M. Berrill
Oak Ridge National Laboratory, Oak Ridge, TN 37831, USA

T. Kawachi et al. (eds.), *X-Ray Lasers 2016*, Springer Proceedings in Physics 202, https://doi.org/10.1007/978-3-319-73025-7_2

2.1 Introduction

Bright and compact sources of coherent bright soft X-ray and extreme ultraviolet (EUV) radiation are of great interest for a large number of applications including high-resolution microscopy [1], compositional imaging [2], and the production of nanoscale features through coherent lithography [3] to name a few. Some of these applications, such as ultra-high resolution microscopy with an illumination wavelength of $\lambda = 13.2$ nm [4], have been demonstrated using table-top $\lambda = 10$–20 nm plasma-based soft X-ray lasers (SXRLs) driven by flashlamp-pumped chirped pulse amplification (CPA) driver systems [5–9]. However, the more photon-demanding applications have up to now only been demonstrated using the longer wavelength, $\lambda = 46.9$ nm capillary discharge lasers. The average power of plasma-based, $\lambda = 8$–20 nm SXRLs is usually limited by the average power of the optical-wavelength, picosecond-duration CPA laser driver. Using the grazing incidence [5, 10] transient pumping scheme, the required driver laser pulse energy ranges from a few hundred millijoules [11] to 10 J depending on the wavelength and SXRL pulse energy desired. For flashlamp-pumped CPA lasers based on either Ti:Al_2O_3 or Nd:glass, this energy requirement limits the repetition rate to a maximum of about 10 Hz, which in turn limits the average power of the SXRLs to a few tens of μW. In order to surpass this limitation, we have focused on developing directly diode-pumped CPA lasers producing Joule-level pulses at high repetition rates (Fig. 1).

Previously we have reported the development of an all-diode-pumped laser that produced 1 J pulses of ~5 ps duration at 100 Hz repetition rate [12]. We employed this laser to drive a $\lambda = 18.9$ nm SXRL operating on the $4d^1S_0 \rightarrow 4p^1P_1$ transition of Ni-like Mo [13]. At 100 Hz repetition rate, this laser produced >1.5 μJ pulses resulting in >0.15 mW average power, a record power for a compact coherent source at this wavelength. Using a high shot capacity, rotating disk target, we were able to operate this laser continuously at high repetition rates for long periods of time [14], and we made the first demonstration of printing nanometer-scale features through Talbot lithography at this wavelength. Furthermore, we were able to use this laser to drive SXRLs at a number of wavelengths down to $\lambda = 10.9$ nm [15], including the high repetition rate/high average power operation of $\lambda = 13.9$ nm

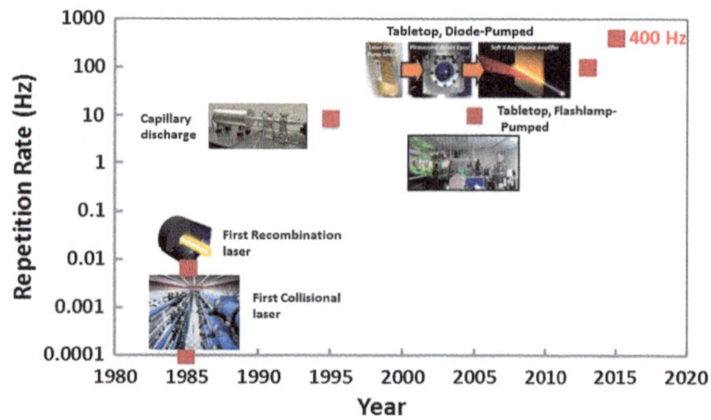

Fig. 2.1 Progress in the repetition rate of soft X-ray lasers from the first collisional and recombination soft X-ray lasers to the result presented here (400 Hz)

(>0.1 mW) and λ = 11.9 nm (20 μW) lasers from silver and tin targets, respectively. Very recently, we have further developed this diode-pumped laser technology to demonstrate a laser that produces 1 J, picosecond-duration pulses at 0.5 kHz repetition rate [16]. This laser was employed to make the first demonstration of a 400 Hz repetition rate SXRL. The characteristics of this new pump laser and the high repetition rate SXRL demonstration are discussed below.

2.2 Development of a 0.5 kHz, 1 J, Picosecond Laser

The all-diode-pumped 1 J, ps pump laser is shown in the diagram and photograph of Fig. 2.2a, b. Stretched λ = 1.03 μm pulses generated by a diode-pumped, mode-locked Yb:KYW oscillator seed a water-cooled, active mirror regenerative

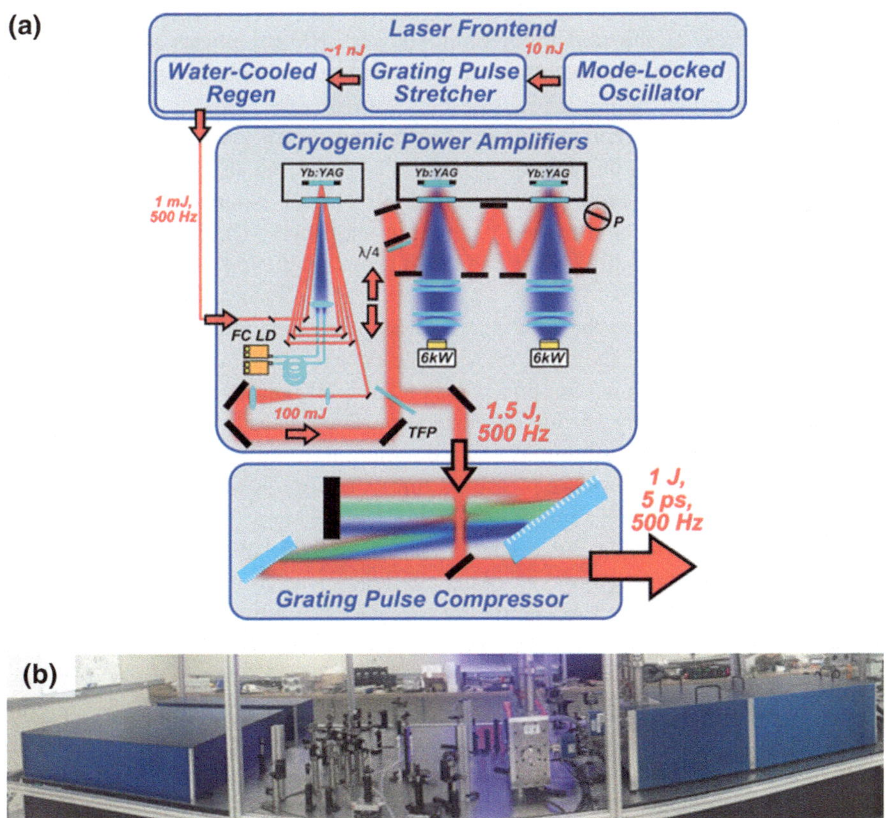

Fig. 2.2 **a** Schematic diagram and **b** photograph of the 1 J, 0.5 kHz CPA laser used to pump table-top soft X-ray lasers

amplifier. In turn, the ~1 mJ exiting the regenerative preamplifier seed a chain of two cryogenically cooled Yb:YAG multi-pass amplifiers. Cryo-cooling Yb:YAG improves its thermal properties [17], depopulates the laser lower level allowing efficient four-level operation, and increases the stimulated emission cross section [18]. These improved characteristics allow efficient, high energy, high average power operation in a compact, low pass number configuration. The first cryogenic amplifier features a Yb:YAG active mirror mounted in vacuum on a cryo-cooling head in which a subcooled cryogenic liquid flows in direct contact with the high reflecting face of the amplifier disk. This cooling scheme allows us to surpass the severe limitation in heat flux which can be dissipated by boiling heat transfers without the use of a heat spreader, and it results in largely longitudinal thermal gradients minimizing thermal lensing. The small amount of residual thermal lensing is compensated by adjusting the beam expanding telescopes at the output of both cryogenic amplifiers The amplifier is pumped in an approximately 4 mm diameter spot with up to 800 W total peak power in 400 μs pulses from two fiber-coupled laser diode modules. The millijoule-level pulses from the regenerative amplifier are directed to make four reflections from the active mirror. As can be seen from the plot of Fig. 2.2a, at the maximum pump power, 100 mJ pulses are obtained at 500 Hz repetition rate. This results in an optical-to-optical efficiency of 27% for this 50 W average power amplifier with good beam quality. This amplifier by itself represents an advancement in state-of-the-art picosecond laser technology and is envisioned to enable a number of applications including driving high peak and average power, single-cycle sources based on optical parametric chirped pulse amplification (OPCPA) and related parametric amplifiers. The pulses produced by this preamplifier are ideally suited for seeding our main, Joule-level amplification stage.

As shown in Fig. 2.3, the final stage of amplification includes two composite Yb:YAG active mirrors mounted on a single cryogenically cooled head. Each of these is pumped by 500 μs duration pulses from a 60 bar, $\lambda = 940$ nm laser diode stack in an approximately 16 mm diameter spot. The pulses from the previous stage of amplification are directed to make four double-passes through each active mirror. The plot of Fig. 2.3b shows the measured output energy as a function of combined pumped energy at 500 Hz repetition rate. As can be seen from this figure, at the maximum pump power, a pulse energy of 1.5 J (0.75 kW) is obtained with an optical-to-optical efficiency of 37%. This efficiency was computed using the pump power incident on the Yb:YAG active mirrors. Good beam quality was obtained as can be seen from the M^2 measurement shown in Fig. 2.3c at near full power operation. M^2 values of about 1.3 were obtained on both axes with an output pulse energy of 1.4 J at 500 Hz repetition rate. At this power, a slight amount of thermal lensing was observed, which we compensated by adjusting the beam expanding telescope at the output of the amplifier. Furthermore, no self-focusing was observed, which is compatible with an estimated total B-integral for the amplifier chain of approximately 0.8. After a short warmup period, the laser is quite stable. Figure 2.3d shows the measured pulse energy at the output of the final amplifier

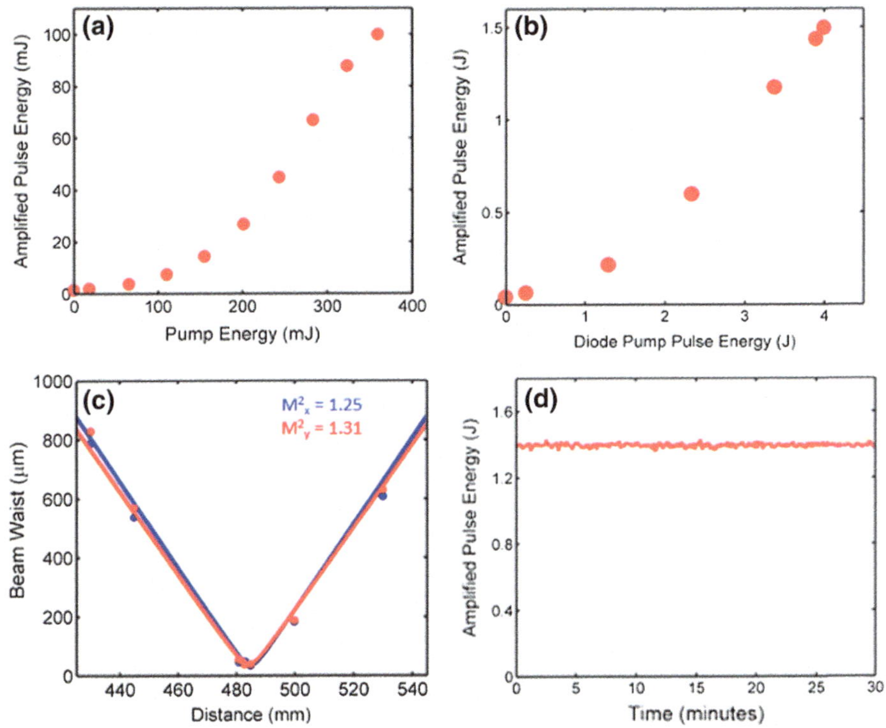

Fig. 2.3 **a** Measured pulse energy at the output of the first cryogenic Yb:YAG amplifier as a function of diode pump pulse energy at 0.5 kHz repetition rate. The amplifier was pumped with 400 μs duration, λ = 940 nm pulses from two fiber-coupled laser diode modules. With a total pump energy of 370 mJ, an output energy of 100 mJ is obtained (27% optical-to-optical efficiency). **b** Measured pulse energy at the output of the main amplifier as a function of pump pulse energy at 500 Hz repetition rate. **c** M^2 measurement of the 1.4 J pulses exiting the final amplifier at 500 Hz repetition rate. **d** Measured output energy at the full repetition rate over 30 min of continuous operation. A mean energy of 1.4 J was obtained with an RMS variation of 0.75%

over 30 min of continuous operation at 500 Hz repetition rate. A mean energy of 1.4 J was obtained with an RMS variation of 0.75%.

Following the final amplifier, we compress the pulses in vacuum using a pair of dielectric diffraction gratings with a groove density of 1740 mm^{-1}. Figure 2.4a shows the measured spectra of the amplified laser pulses after the 100 mJ-level and 1.5 J amplifiers. As can be seen from this figure, the amplified pulses maintain a bandwidth of about 0.36 nm FWHM, which supports sub-5 ps durations. Figure 2.4b shows a second harmonic generation autocorrelation measurement of

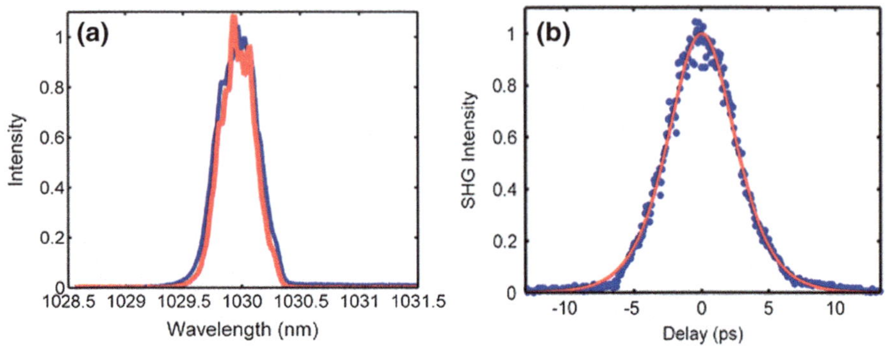

Fig. 2.4 **a** Measured amplified pulse spectra at the output of the first cryogenic amplifier (blue) and final amplifier (red). The pulses have an amplified bandwidth of 0.36 nm FWHM. **b** SHG autocorrelation of the compressed pulses. The line is sech^2 fit with a width corresponding to a pulse duration of 3.8 ps FWHM

the compressed pulse exiting the compressor. From this autocorrelation, we obtain a pulse duration of about 3.8 ps FWHM in an approximately sech^2 shape. The compressor has a measured overall efficiency of 72%, resulting in ~1 J, sub-5 ps pulses at 500 Hz repetition rate (0.5 kW average power).

2.3 First Demonstration of a 400 Hz Repetition Rate Soft X-Ray Laser

Using this new high repetition rate pump laser, we have made the first demonstration of an 18.9 nm soft X-ray laser operating at 400 Hz repetition rate. In this demonstration, ~1 J, temporally shaped laser pulses were focused at a grazing incidence angle of 29° into a 30 μm wide × 4 mm long line focus on a flat slab molybdenum target. A temporal pulse shape similar to what we used for our prior 100 Hz repetition rate demonstration was employed. The on-axis EUV plasma emission was detected using a flat-field spectrometer employing a grazing incidence 1200 mm^{-1} variable spaced diffraction grating and an EUV-sensitive CCD camera. Thin Al foils were used to reject visible plasma emission as well as to attenuate the SXRL. Figure 2.5a shows a single-shot EUV spectrum displaying strong lasing on the $\lambda = 18.9$ nm transition of Ni-like Mo. This spectrograph was obtained with the driver laser operating at 400 Hz repetition rate. To measure the SXRL operation at 400 Hz repetition rate, the CCD was replaced with an X-ray sensitive Si photodiode placed at the laser's spectral location in the image plane of the spectrometer.

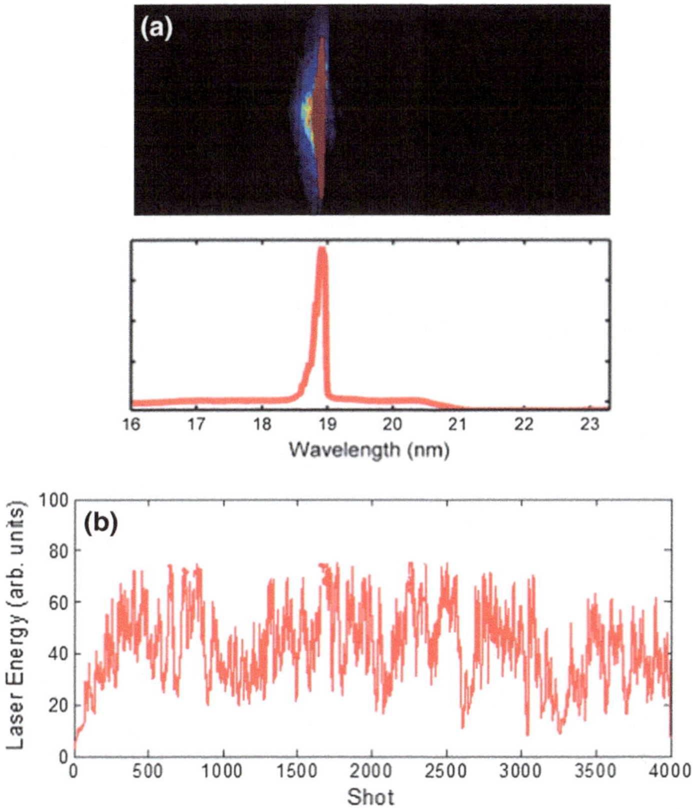

Fig. 2.5 **a** On-axis spectrum of the Mo plasma showing strong lasing on the λ = 18.9 nm Ni-like Mo transition. This single-shot spectrum was recorded with the driver laser operating at 400 Hz repetition rate. **b** Measured 18.9 nm SXRL pulse energy under 400 Hz operation

Figure 2.5b shows the measured 18.9 nm SXRL pulse energy at 400 Hz repetition rate over 4000 consecutive shots. Following this initial demonstration, the shot-to-shot energy variation is expected to be improved with further optimization.

2.4 Summary

In summary, we have developed an all-diode-pumped pump laser for soft X-ray lasers that produces 1 J pulses of picosecond duration at 0.5 kHz repetition rate. This laser is based on cryogenically cooled Yb:YAG amplifiers. Using this new pump laser, we have made the first demonstration of an 18.9 nm SXRL operating at record 400 Hz repetition rate. With future optimization, this approach provides a path to compact coherent sub-20 nm sources with milliwatt average powers.

Acknowledgements The pump laser development work was supported by the U.S. Department of Energy grant DE-SC0011375. The soft X-ray laser work was supported by the National Science Foundation under grant ECCS-1509925 and by the U.S. Department of Energy, Office of Science, Basic Energy Sciences, under Award# DE-FG02-04ER15592.

References

1. Vaschenko, G., Brewer, C., Brizuela, F., Wang, Y., Larotonda, M.A., Luther, B.M., Marconi, M.C., Rocca, J.J., Menoni, C.S., Anderson, E.H.: Sub-38 nm resolution tabletop microscopy with 13 nm wavelength laser light. Opt. Lett. **31**, 1214–1216 (2006)
2. Kuznetsov, I., Filevich, J., Dong, F., Woolston, M., Chao, W., Anderson, E.H., Bernstein, E.R., Crick, D.C., Rocca, J.J., Menoni, C.S.: Three-dimensional nanoscale molecular imaging by extreme ultraviolet laser ablation mass spectrometry. Nat. Commun. **6**, 6944 (2015)
3. Li, W., Urbanski, L., Marconi, M.C.: Invited article: progress in coherent lithography using table-top extreme ultraviolet lasers. Rev. Sci. Instrum. **86**, 121301 (2015)
4. Brizuela, F., Wang, Y., Brewer, C.A., Pedaci, F., Chao, W., Anderson, E.H., Liu, Y., Goldberg, K.A., Naulleau, P., Wachulak, P., Marconi, M.C., Attwood, D.T., Rocca, J.J., Menoni, C.S.: Microscopy of extreme ultraviolet lithography masks with 13.2 nm tabletop laser illumination. Opt. Lett. **34**, 271–273 (2009)
5. Luther, B.M., Wang, Y., Larotonda, M.A., Alessi, D., Berrill, M., Marconi, M.C., Rocca, J.J., Shlyaptsev, V.N.: Saturated high-repetition-rate 18.9-nm tabletop laser in nickellike molybdenum. Opt. Lett. **30**, 165–167 (2005)
6. Martz, D.H., Alessi, D., Luther, B.M., Wang, Y., Kemp, D., Berrill, M., Rocca, J.J.: High-energy 13.9 nm table-top soft-X-ray laser at 2.5 Hz repetition rate excited by a slab-pumped Ti:sapphire laser. Opt. Lett. **35**, 1632–1634 (2010)
7. Wang, Y., Larotonda, M.A., Luther, B.M., Alessi, D., Berrill, M., Shlyaptsev, V.N., Rocca, J.J.: Demonstration of high-repetition-rate tabletop soft-X-ray lasers with saturated output at wavelengths down to 13.9 nm and gain down to 10.9 nm. Phys. Rev. A **72**, 53807 (2005)
8. Alessi, D., Wang, Y., Luther, B.M., Yin, L., Martz, D.H., Woolston, M.R., Liu, Y., Berrill, M., Rocca, J.J.: Efficient excitation of gain-saturated sub-9-nm-wavelength tabletop soft-X-ray lasers and lasing down to 7.36 nm. Phys. Rev. X **1**, 2 (2011)
9. Zimmer, D., Zielbauer, B., Pittman, M., Guilbaud, O., Habib, J., Kazamias, S., Ros, D., Bagnoud, V., Kuehl, T.: Optimization of a tabletop high-repetition-rate soft X-ray laser pumped in double-pulse single-beam grazing incidence. Opt. Lett. **35**, 450–452 (2010)
10. Keenan, R., Dunn, J., Patel, P.K., Price, D.F., Smith, R.F., Shlyaptsev, V.N.: High-repetition-rate grazing-incidence pumped X-ray laser operating at 18.9 nm. Phys. Rev. Lett. **94**, 103901 (2005)
11. Banici, R.A., Cojocaru, G.V., Ungureanu, R.G., Dabu, R., Ursescu, D., Stiel, H.: Pump energy reduction for a high gain Ag X-ray laser using one long and two short pump pulses. Opt. Lett. **37**, 5130–5132 (2012)
12. Reagan, B.A., Baumgarten, C., Wernsing, K., Bravo, H., Woolston, M., Curtis, A., Furch, F. J., Luther, B., Patel, D., Menoni, C.S., Rocca, J.J.: 1 Joule, 100 Hz repetition rate, picosecond CPA laser for driving high average power soft X-ray lasers. In: CLEO: Science and Innovations, p. SM1F. 4 (2014)
13. Reagan, B.A., Wernsing, K.A., Curtis, A.H., Furch, F.J., Luther, B.M., Patel, D., Menoni, C.S., Rocca, J.J.: Demonstration of a 100 Hz repetition rate gain-saturated diode-pumped table-top soft X-ray laser. Opt. Lett. **37**, 3624–3626 (2012)
14. Reagan, B.A., Li, W., Urbanski, L., Wernsing, K.A., Salsbury, C., Baumgarten, C., Marconi, M.C., Menoni, C.S., Rocca, J.J.: Hour-long continuous operation of a tabletop soft X-ray laser at 50–100 Hz repetition rate. Opt. Express **21**, 28380–28386 (2013)

15. Reagan, B.A., Berrill, M., Wernsing, K.A., Baumgarten, C., Woolston, M., Rocca, J.J.: High-average-power, 100-Hz-repetition-rate, tabletop soft-X-ray lasers at sub-15-nm wavelengths. Phys. Rev. A **89**, 53820 (2014)
16. Baumgarten, C., Pedicone, M., Bravo, H., Wang, H., Yin, L., Menoni, C.S., Rocca, J.J., Reagan, B.A.: 1 J, 05 kHz repetition rate picosecond laser. Opt. Lett. **41**, 3339 (2016)
17. Aggarwal, R.L., Ripin, D.J., Ochoa, J.R., Fan, T.Y.: Measurement of thermo-optic properties of Y3Al5O12, Lu3Al5O12, YAIO, LiYF4, LiLuF4, BaY2F8, KGd(WO4)2, and KY(WO4)2 laser crystals in the 80–300 K temperature range. J. Appl. Phys. **98** (2005)
18. Dong, J., Bass, M., Mao, Y., Deng, P., Gan, F.: Dependence of the Yb 3^+ emission cross section and lifetime on temperature and concentration in yttrium aluminum garnet. J. Opt. Soc. Am. B **20**, 1975–1979 (2003)

Chapter 3
Progress and Prospects of X-Ray Laser Research in QST

M. Nishikino, N. Hasegawa, M. Ishino, T. Imazono, A. Sasaki, K. Mikami, T.-H. Dinh, T. Suemoto, S. Namba, A. Ya. Faenov, T. A. Pikuz, S. Ichimaru, M. Hatayama and T. Kawachi

Abstract This paper reviews recent progress made in the development of coherent x-ray sources and their applications in material science and laser processing in the National Institutes for Quantum and Radiological Science and Technology (QST). The upgrade of Ti:Sapphire laser for a grazing incidence pumping (GRIP) scheme was started, to further develop a coherent x-ray source. Investigating the applications of x-ray lasers, we observed temporal evolution of metal surface ablation using an x-ray laser probe, and the damage to an EUV mirror structure by the x-ray pulse irradiation.

3.1 Introduction

High-intensity short pulse X-ray laser (XRL) beam is expected for the application research of the material science using X-ray coherent imaging and X-ray laser ablation. The improvement of the X-ray source is also an important issue, and the development of coherent X-ray sources, such as XRL and high-order harmonic generation (HHG), is intensively studied. We have started the upgrade of the driving laser and the development of the 10-Hz GRIP system with HHG amplifi-

M. Nishikino (✉) · N. Hasegawa · M. Ishino · T. Imazono · A. Sasaki · K. Mikami
T.-H. Dinh · T. Kawachi
National Institutes for Quantum and Radiological Science and Technology, Kizugawa, Kyoto, Japan
e-mail: nishikino.masaharu@qst.go.jp

T. Suemoto
Toyota Physical and Chemical Research Institute, Aichi, Japan

S. Namba
Department of Engineering, Hiroshima University, Hiroshima, Japan

A. Ya. Faenov · T. A. Pikuz
Joint Institute of High temperature, Russian Academy of Science, Moscow, Russia

S. Ichimaru · M. Hatayama
NTT Advanced Technology Corporation, Kanagawa, Japan

T. Kawachi et al. (eds.), *X-Ray Lasers 2016*, Springer Proceedings in Physics 202, https://doi.org/10.1007/978-3-319-73025-7_3

cation [1, 2]. We are also starting the x-ray parametric amplification [3] by a collaborative effort. The high-quality and powerful soft XRL enables us to achieve quite high spatial-resolution imaging as a probe source and the nanoscale fabrication as a pump source by the high-repetition rate with over 10 Hz. The application will be expected the surface probe such as the mask inspection of the EUV lithography with 100 Hz repetition rate. As an application using the XRL probe, we have measured the morphology change of the spallation process in the femtosecond (fs) laser ablation. The dynamical processes of fs-laser-induced surface nanoscale modifications come to attract much attention for the micro/nano-fabrication. We have observed the surface dynamics of fs-laser ablation by the X-ray reflectometer and interferometer. The time- and spatial-resolved spallative ablation dynamics were measured by coherent X-ray imaging technique. As an application using the XRL pump, the damaged structures of the X-ray optics such as a multilayered (ML) mirror and resist material for the EUV lithography are evaluated after the irradiation of the high-intensity X-ray pulse.

3.2 Observation of fs-Laser Ablation Dynamics

X-ray images were observed by the fs-laser pump and XRL probe imaging system [4]. The laser-driven plasma XRL of 13.9 nm (89.2 eV) was used as the X-ray imaging probe. Because of the lowest attenuation length, the soft X-ray is suitable especially for the material surface measurement. The thin metal sample film (~100 nm) on the substrate is irradiated by the pumping fs-laser. The pumping fs-laser was a Ti:Sapphire laser with a duration of 80 fs (FWHM) at a central wavelength of 795 nm. The irradiated sample image was transferred to CCD camera and magnified by a spherical imaging mirror with a factor of about 40. In this X-ray imaging system, the spatial resolution is better than 1 μm in lateral direction and 2 nm in depth direction, respectively. The irradiation profile of the pump beam is nearly Gaussian distribution, and the FWHM of the focal spot was about 100 μm on the sample surface. The peak fluence of irradiance was around 1.0 J/cm^2. The time-evolution of the surface shape was obtained by varying the delay time between the pump fs laser beam and the probe XRL beam. The coherent X-ray interference imaging was done using a double Lloyd's mirror. The probe XRL beam passed through both ablated and non-ablated surface. The XRL beam was divided into two parts after the double Lloyd's mirror, and the X-ray interference pattern was generated by overlapping both parts of probe XRL beam. Figure 3.1 shows typical X-ray interference images of the Au surface at 95, 261, and 1983 ps after the fs-laser irradiation. The fringe shift in the left direction is indication of the surface expansion. The expansion surface profile is reconstructed by analyzing the fringe shifts in the interference pattern. From the measurement, we find the ablation surface dynamics. The timescale of fs-laser ablation of the Au is

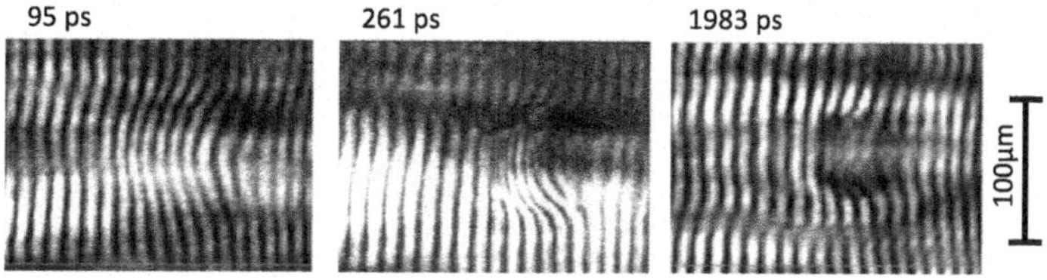

Fig. 3.1 X-ray laser interference patterns of Au at 95, 261, and 1983 ps after fs-laser irradiation

about ~ns region, and the morphology change can be measured by delay interval of every 100 ps. From this observation, we find that the ablation surface of Au expands, reaching up to ~60 nm above the original surface, at about 261 ps. The detailed results and discussion are described in [5].

3.3 Damage Structure on Multilayered Structure Sample Induced by Intense SXRL Irradiations

The damage test of EUV optics, such as Mo/Si and Nb/Si multilayered (ML) mirror, was performed by the irradiation of the high-intensity XRL pulse. The high-performance X-ray mirror has been used to support the high reflectivity and the long-life time at the same time in ideal. The detailed information of damage development is important for the practical use of Mo/Si ML surface and the future development of high-performance EUV optics. The focused X-ray laser pulse irradiated several types of ML structures. The experimental setup is similar to previous experiment [6]. The SXRL pulse was focused on a sample surface, and the focus size was about 20 μm (FWHM). The SXRL pulse was approximately about 50 nJ, corresponding to about 10–20 mJ/cm^2 for the sample surface. The irradiated ML samples were optimized for XRL at an incidence angle of 2°. The energy of XRL pulse is almost absorbed by the sample in the case of normal incidence. It is needed for the details of damaged structure to observe by a scanning electron microscope (SEM), a transmission electron microscope (TEM), and an atomic force microscope (AFM). The detailed results and discussion are described in this proceeding [7, 8]. The ablated damage holes are generated on the sample surface. The feature of the flat bottom structure is only observed in the case of ML damage. Figure 3.2 shows the comparison of the AFM images between (a) Mo/Si and (b) Nb/Si ML damaged sample. The damage size and depth are very small in the case of the Nb/Si ML sample. The damage depth is about half of the case of the Mo/Si ML sample. The mechanism of surface and inside damage generation induced by the high-intensity XRL pulses, such as the ablation threshold, ablated surface pattern, seems to be different from simple layer sample. Therefore, additional quantitative investigations of the observation of damage generation in EUV region would be required for deeper understanding.

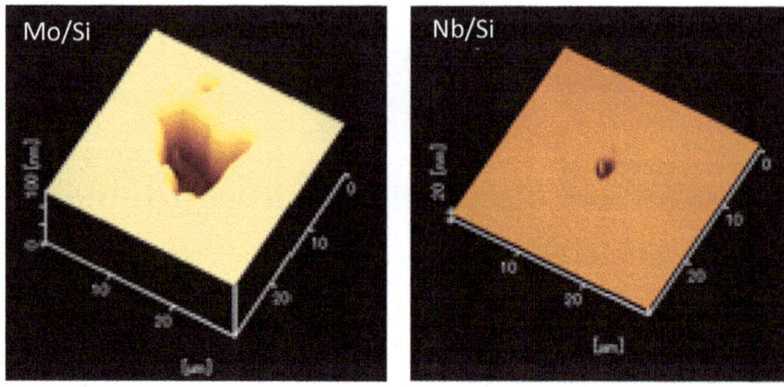

Fig. 3.2 The AFM image of damage structure comparison between Mo/Si and Nb/Si

3.4 Summary

We review the recent progress of SXRL application research in QST. Source developments of GRIP-XRL and/or HHG amplification are undergoing. Recent research highlights that by using the SXRL, observations of fs laser ablation and EUV ablation are possible. To summarize, we have shown experimental observation of the fs-laser ablation dynamics using soft X-ray laser interferometry. The similarities and differences shown in the experimental data from materials will provide important data for understanding correlations between the ablation properties and the material parameters. Mo/Si and Nb/Si ML samples of the EUV were irradiated with the high-intensity XRL pulse and observed the modified surface damage structures. The Nb/Si ML structure has the potential for higher durability and reflectivity, and may provide a good solution for the high power EUV light source in the near future.

References

1. Wang, Y., et al.: Phys. Rev. A **79**, 023810 (2009)
2. Hasegawa, N., et al.: Phys. Rev. A **76**, 043805 (2007)
3. Seres, J., et al.: Nat. Phys. **6**, 455–461 (2010)
4. Nishikino, M., et al.: Proc. SPIE **8849**, 88490E (2013)
5. Hasegawa, N., et al.: "The observation of transient thin film structures during the femto-second laser ablation process by using the soft x-ray laser probe" in this proceedings
6. Ishino, M., et al.: J. Appl. Phys. **109**, 013504 (2011)
7. Ichimaru, S., et al.: "Irradiation damage test of Mo/Si, Ru/Si and Nb/Si multilayeres using the soft x-ray laser built at QST" in this proceedings
8. Ishino, M., et al.: "Analysis of reflection signal from EUV multilayer mirror for irradiation induced famage study" in this proceedings

Chapter 4
DAGON: A 3D Maxwell–Bloch Code

E. Oliva, M. Cotelo, J. C. Escudero, S. Vicens, A. González and P. Velarde

Abstract High Order Harmonics are a powerful tool to unveil the inner dynamics of plasmas by studying their propagation. However, being the plasma an inhomogeneous medium, it is desirable to have a 3D description of the interaction of HOH with the ions. The 3D Maxwell-Bloch code DAGON, developed at the *Instituto de Fusión Nuclear* will allow to study the full spatio-temporal structure of amplified HOH.

4.1 Introduction

The amplification of High Order Harmonics (HOH) in inhomogeneous media such as plasmas is a powerful tool to access the inner evolution of plasma variables as ionic level populations [1] or the interaction of radiation with Zeeman sub-levels [2]. Moreover, amplified HOH are not only an excellent probe beam but also promise to be an ultra-intense and compact source of coherent soft X-rays [3, 4]. These applications require a strong effort both experimental and theoretical. Although these experiments have been successfully modelled using 1D [5] and 2D [6] Maxwell-Bloch codes the availability of a 3D description of the amplification would open the path towards exciting applications.

For example, Krypton amplifiers are a promising medium to amplify HOH and study its temporal structure, mainly to resolve Rabi oscillations. However, the strong inhomogeneity of the plasma medium, shown in Fig. 4.1, induces different features at different parts of the beam [7] making this problem a full tridimensional one.

Thus, at the *Instituto de Fusión Nuclear* we are developing a 3D Maxwell-Bloch code, DAGON, under the framework of an FP7 PEOPLE Marie Sklodowska-Curie grant.

E. Oliva (✉) · M. Cotelo · J. C. Escudero · S. Vicens · A. González · P. Velarde
Instituto de Fusión Nuclear, E.T.S.I. Industriales, Universidad Politécnica de Madrid, Madrid, Spain
e-mail: eduardo.oliva@upm.es

T. Kawachi et al. (eds.), *X-Ray Lasers 2016*, Springer Proceedings in Physics 202, https://doi.org/10.1007/978-3-319-73025-7_4

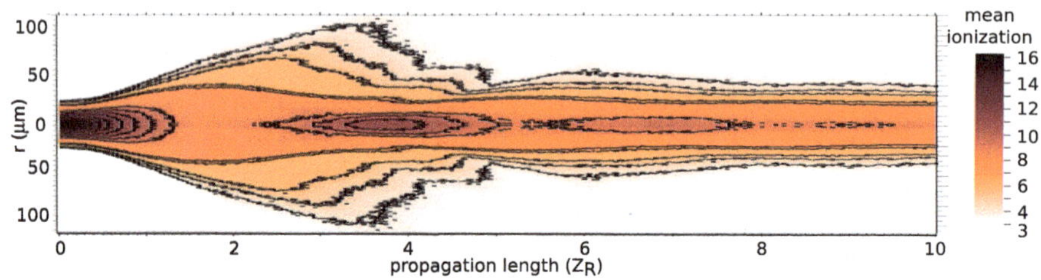

Fig. 4.1 Mean ionization of a plasma-based krypton amplifier. The details of the simulation are given in [8]

4.2 3D Maxwell-Bloch Simulations

The set of equations solved by our code are a wave equation in the slowly varying envelope approximation for the electric field E (4.1) of frecuency ω, enhanced with a constitutive relation for the polarization P (4.2). The populations $N_{1,2}$ are computed with standard rate equations coupled with the electric field (4.3). The electric field and the plasma are coupled by the plasma frequency ω_{pe}, the electron-ion collision frequency γ, the dipole element z_{21} and the spontaneous emission stochastic source term Γ. The coefficents C comprehend both collisional and radiative rates.

$$\frac{\partial E_\pm}{\partial t} \pm c\frac{\partial E_\pm}{\partial z} = i\frac{c^2}{2\omega}\nabla_\perp^2 E_\pm + \frac{i\omega}{2}\left[\mu_0 c^2 P_\pm - \left(\frac{\omega_{pe}}{\omega}\right)^2 E_\pm\right] \tag{4.1}$$

$$\frac{\partial P_\pm}{\partial t} = \Gamma - \gamma P_\pm - \frac{i z_{21}}{\hbar} E_\pm (N_2 - N_1) \tag{4.2}$$

$$\frac{\partial N_{1,2}}{\partial t} = \sum_k C_{k2,k1} N_k \mp \Im(E_\pm^* P_\pm)\frac{1}{2\hbar} \tag{4.3}$$

With this model it is possible to study the amplification of HOH through plasma amplifiers of several milimeters length and study its temporal structure, as shown in Fig. 4.2.

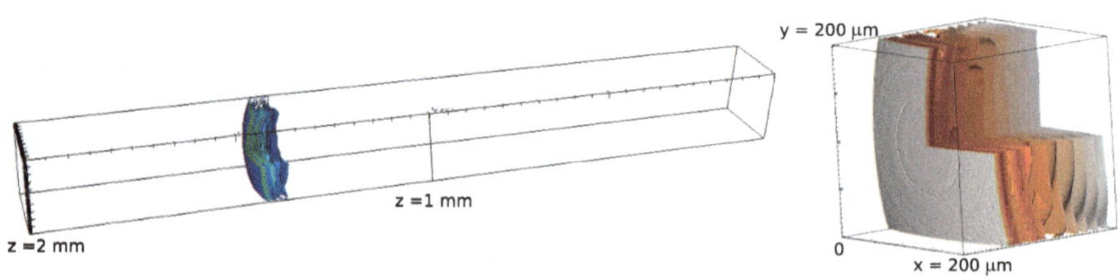

Fig. 4.2 (Left) 3D intensity profile of amplified high-order harmonic radiation inside the krypton amplifier. (Right) spatio-temporal structure of the HOH

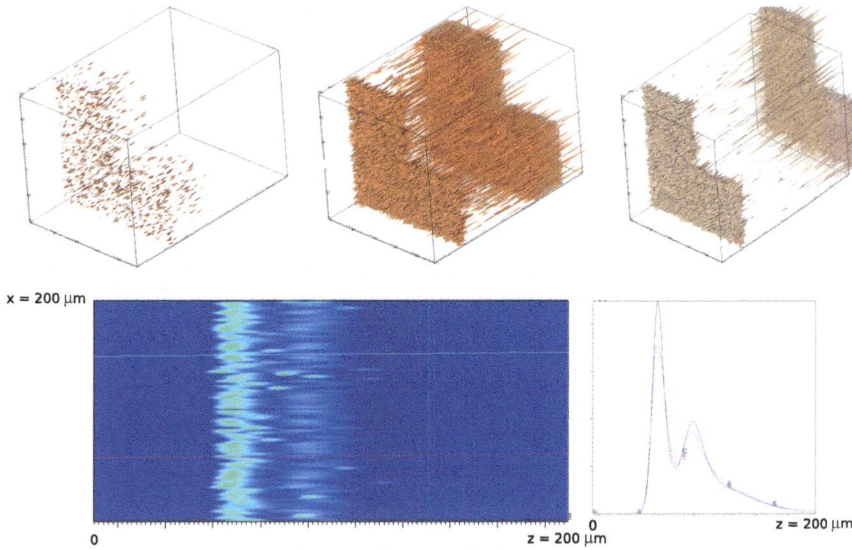

Fig. 4.3 3D structure of the amplified spontaneous emission in a 5 mm krypton amplifier. The upper panel shows different intensity isovolumes while the lower one shows the temporal structure in the propagation direction (from right to left)

The amplification of the injected HOH can be followed through all the length of the amplifier. Thus, the evolution of the different spatio-temporal structures that the HOH develops when it is amplified (wake, Rabi oscillations) can be studied at different positions.

In addition to this, DAGON can also resolve the 3D stochastic structure of the Amplified Spontaneous Emission, as shown in Fig. 4.3.

The upper panel of Fig. 4.3 shows isovolumes of ASE intensity. The stochastic nature of the spontaneous emission mixes the regions of high and low intensity in a seemingly chaotic pattern. However, if we study the structure of ASE in the propagation direction (Fig. 4.3 lower panel) we retrieve the expected structure of an intense spike followed by oscillations and decay, albeit with different maximum intensity, depending on the position in the amplifier.

In conclusion, the development of DAGON, a 3D Maxwell-Bloch code will extend the capabilities of current models. The full simulation of a 3D amplifier will allow us to study experiments where the inhomogeneities of the plasma play a fundamental role, as the development of spatio-temporal structures in amplified HOH or the statistical properties of ASE.

Acknowledgements The research leading to these results has received funding from the People Programme (Marie Curie Actions) of the European Union's Seventh Framework Programme (FP7/2007–2013) under REA grant agreement 627191 DAGON.

References

1. Wang, Y., et al.: Gain dynamics in a soft-X-ray laser amplifier perturbed by a strong injected X-ray field. Nat. Photonics **8**, 381–384 (2014)
2. Depresseux, A., et al.: Demonstration of a circularly polarized plasma-based soft-X-ray laser. Phys. Rev. Lett. **115**, 083901 (2015)
3. Oliva, E., et al.: A proposal for multi-tens of GW fully coherent femtosecond soft X-ray lasers. Nat. Photonics **6**, 764–767 (2012)
4. Depresseux, A., et al.: Table-top femtosecond soft X-ray laser by collisional ionization gating. Nat. Photonics **9**, 817–821 (2015)
5. Oliva, E., Zeitoun, P., Fajardo, M., Lambert, G., Ros, D., Sebban, S., Velarde, P.: Comparison of natural and forced amplification regimes in plasma-based soft-X-ray lasers seeded by high-order harmonics. Phys. Rev. A **84**, 013811 (2011)
6. Al'miev, I.R., Larroche, O., Benredjem, D., Dubau, J., Kazamias, S., Möller, C., Klisnick, A.: Dynamical description of transient X-ray lasers seeded with high-order harmonic radiation through Maxwell-Bloch numerical simulations. Phys. Rev. Lett. **99**, 123902 (2007)
7. Tissandier, F., Sebban, S., Gautier, J., Zeitoun, P., Oliva, E., Rousse, A., Maynard, G.: Three-dimensional Maxwell-Bloch calculation of the temporal profile of a seeded soft X-ray laser pulse. Appl. Phys. Lett. **101**, 251112 (2012)
8. Oliva, E., Depresseux, A., Tissandier, F., Gautier, J., Sebban, S., Maynard, G.: Self-regulated propagation of intense infrared pulses in elongated soft-X-ray plasma amplifiers. Phys. Rev. A **92**, 023848 (2015)

Chapter 5
Plasma-Source High-Resolution XUV Spectroscopy as Complementary to Beamlines Limitations

F. Barbato and D. Bleiner

Abstract Ultrashort pulses imply ultrabroad spectra, as a consequence of the transform limit. Hence, femtosecond time resolution is in trade-off with meV resolution spectroscopy. Further, beamline experiments will enormously profit from preparation, optimization and data replication at *home-lab* plasma-based sources, where beamtime limitations do not apply. We highlight the potential of such synergy among large and compact XUV sources, and how migrating advanced high-resolution spectroscopy to benchtop setups implies more efficient data acquisition and processing methods.

5.1 Introduction

There is a number of frontier applications in materials science and technology, and their progress depends on advanced analytical technologies. Advanced spectroscopies, e.g. based on beamlines, permit to probe *ultrafast* time scales in the femto- or even attosecond, or *ultrasmall* length scales down to a few nanometres, well before these fundamental studies become applications. Figure 5.1 summarizes the Fourier-transform trade-off between time resolution and spectral resolution and indicates parameter ranges for material science applications. Fourth generation light sources (XFEL) offer unprecedented fundamental capabilities, permitting detailed insights in the ultrafast, with intensities at and beyond the sample destruction limit.

On the flip side, two major *limitations* are to be reminded: (i) a ultrashort pulse, e.g. XFEL, is associated with a ultrabroad spectrum by the transform limit, i.e. excellent time resolution means poor spectroscopy resolution; (ii) centralized beamline facilities are so expensive that one can access them only in a few-shift beamtime, after strict proposal review. Such limited and discontinuous access makes all advanced beamline specifically dedicated to *proof-of-principle research.*

F. Barbato · D. Bleiner (✉)
Federal Laboratories for Materials Science & Technology (Empa),
Überlandstrasse 129, CH 8600 Dübendorf, Switzerland
e-mail: davide.bleiner@empa.ch

T. Kawachi et al. (eds.), *X-Ray Lasers 2016*, Springer Proceedings
in Physics 202, https://doi.org/10.1007/978-3-319-73025-7_5

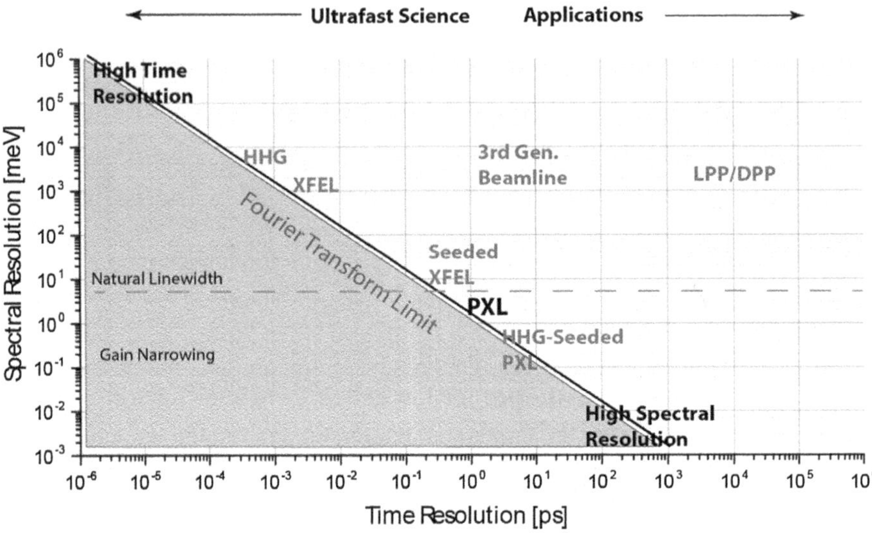

Fig. 5.1 The *time resolution* is in trade-off with the *spectral resolution* because of the well-known transform limit. The *spectral resolution* is in principle in trade-off with the *spatial resolution*, because for a given monochromator bandwidth (e.g. 0.1%) the reduction of the wavelength to access smaller structures means higher photon energies and thus larger absolute bandwidths. The approximate location of a few of state-of-the-art sources is also given, i.e. DPP = discharge-produced plasma, HHG = high-harmonic generation, LPP = laser-produced plasma, PXL = plasma X-ray Laser, XFEL = X-ray Free-Electron Laser

This means that high-risk research, long-term industrial R&D, and also young scientists' training are a few examples of '*excluded proposals*'.

In a synergic approach, large-scale centralized sources will benefit from home-lab *benchtop sources* for carefully preparing the beamtime and for performing 'beyond the proof-of-principle' applications. Furthermore, materials science applications, e.g. catalysis, photovoltaics, batteries, etc., are mainly on the '*slow* and *large* end' of Fig. 5.1 (see arrow 'Applications'), such as timescales of ns to ms and length scales of μm to mm are the requirement for most non-fundamental processes.

This makes the requirements for an *enabling source* substantially different than those for a proof-of-principle machine. Besides, advanced materials analysis, requires *high spectral resolution*, which is well matched with 'slow probes' (transform limit). In Fig. 5.1 one observes that plasma-based X-ray lasers (PXL) are substantially *complementary* to beamline sources, in the sense that the poorer time resolution is accompanied by an enhanced spectral resolution.

Aim of this paper is to highlight pro and cons of implementing plasma sources for home-lab high-resolution spectroscopic applications. Further, we suggest how the related *enabling character* is synergistic to the *proof-of-principle* character of accelerator beamlines.

5.2 Experiment

The experiments were conducted using either a laser (coherent) or a discharge (incoherent) plasma-based soft X-ray source. The pump laser was a 1054-nm Nd:glass chirped pulse amplification system, reconfigured to provide a maximum repetition rate with a short pulse (~1.5 ps) [1]. Alternatively, the incoherent XUV-radiation was generated on a system of hollow electrodes ('Pseudospark') [2]. The electrode gap is filled with the working gas and directly connected to a storage capacitor bank (C = 960 nF). The capacitor is charged until a discharge voltage ranging between 1.5 and 2.4 kV, depending on the working gas type, pressure (0.1–0.2 mbar for Ar) and electrode gap (d = 6 mm).

5.3 Results and Discussion

5.3.1 Unique PXL Characteristics

The plasma X-ray Laser (PXL), regardless all different pumping mechanisms [3], offers unique characteristics for spectroscopy, which are complementary to those available at accelerator sources. The laser line is *exceptionally narrow* thanks to gain narrowing across the plasma gain column. The effective linewidth is very difficult to measure by spectrometry, and was done by interferometry methods [3]. The reported linewidth ($\Delta\lambda$ = 0.07 pm or 0.4 meV), as well as the related 'discrete tunability' by choosing different targets, prove very appealing for high-resolution spectroscopic applications, e.g. X-ray Raman. However, given that the demonstrated PXL photon energy is here limited in the range 26–185 eV [4] the application in X-ray absorption spectroscopy is now only possible for a few L and M edges of low/mid Z elements. The extension of the PXL lines to shorter wavelength is thus very interesting. We demonstrated [5] that the state-of-the-art on saturated plasma lasing output is a fundamental limit for the collisional scheme, related to the maximum pumping depth into the plasma medium, such to reach for the optimum collisional density. For $\lambda < 6.7$ nm (Target Sm, Z = 60) the optimum electron density is 'buried' much below the pump turning point, and could not be accessed, save using a 2ω pump.

On the other hand, PXL application to *photoelectron spectroscopy* (PES) proved particularly powerful, as the photon energy of ~100 eV is well in excess for photoemission, and also considering the surface sensitivity. In a way, the use of XUV photons combines the advantages of electron beam methods with the advantages of X-ray. The strong matter sensitivity makes XUV very interesting for probing nanostructured samples and surface phenomena. Besides, high time coherence is another unique characteristic of the PXL, which permit interferometry and Fourier-transform spectroscopy in the home lab, whose development is underway in our laboratory. Narrow linewidth and time coherence characteristics of PXL are superior to what found at beamlines, where the focus is on ultrafast physics and chemistry.

5.3.2 Two-Colour X-Ray Laser for Pump-Probe Spectroscopy

In a recent work [6], we have investigated in detail the mechanisms, properties and dynamics of two-colour XUV lasers generated by means of plasma-driven amplified spontaneous emission. Two-colour pulses are a hot topic at beamlines, because allow pump-probe photochemistry with a single beam, but are complex to produce at XFELs, as discussed in our cited reference. Two-colour laser pulses can be produced in the PXL *either* by preparing multielemental targets, *or* by exploiting parallel radiative schemes, such as collisional and photo-pumping [4, 6]. Surprisingly, the 'geometrical characteristics' of the two lasing mechanisms, e.g. divergence, pointing stability, etc., were found to be similar.

However, it was found a major distinction in that the collisional ASE is almost one order of magnitude stronger than the photo-pumping. The weak λ_2 signal makes it less straightforward the utilization for pump-probe spectroscopy. Therefore, the '*target cocktail*' approach seems to be more robust. Nevertheless, studies in BaF_2 targets [1] have shown 'chemical dilution' and electron scavenging effects through fluorine injection in the plasma. Therefore, the target cocktail and plasma chemistry phenomena must be studied in more detail. The use of multicolour pulses is indeed very interesting for spectroscopy, and is shown below using discharge-produced plasma (DPP) spectra. Presently, the intensity and flexibility of DPP multicolour pulses are superior to that of two-colour PXL. Our experiment is a proof-of-principle, in order to demonstrate the advantage of multicolour pulses in spectroscopy.

5.3.3 Beamline Advanced Spectroscopy at Benchtop Plasma Sources

5.3.3.1 X-Ray Spectroscopy in the Home Lab

There are four modes of spectroscopy: (i) direct transmission mode, (ii) indirect mode via fluorescence yield, (iii) indirect mode via Auger yield and (iv) indirect via Raman scattering (energy loss). X-ray *absorption* spectroscopy (XAS), as either NEXAFS or EXAFS [7], can reveal elemental, oxidation and bonding information probing unoccupied states, i.e. the chemistry. XAS has one major requirement:

a tunable *multiwavelength* source, e.g. synchrotron. On the other hand, X-ray *emission* (XES) or *scattering* (XRS) spectroscopy can reveal the elemental and electronic information probing occupied states, i.e. the physics. XES has one major requirement: a highly *bright* source, e.g. XFEL. Although XAS, XES and XRS seems very distinguished spectroscopies, it is not really so, as shown below. For instance, a fluorescence event (XES) always follows to an ionizing absorption process (XAS and PES). The historical division between XAS and XES is due to the fact that the requirements imposed, mentioned above, are alternatively fulfilled at dedicated beamlines.

Indeed, the Kramers–Heisenberg (KH) dispersion formula links the XES, the XAS and XRS spectra. The KH is the expression for scattering cross section of a photon by an atomic electron in the quasi-resonant case [8], which is as follows:

$$XES(E_{out}) = \int \frac{E_{out}}{E_{in}} \frac{(E_o - E_f)(E_o + E_e)XAS(E)}{(E_o + E_e - E_{in})^2 + \Gamma^2_{initial}/4} \delta(E_{in} - E_f - E_e - E_{out})dE$$

where $E_{in,out}$ are the energies of the 'in' and 'out' photons, E_f and E_o those of the probed final and initial states, and E_e that of the photoelectron, while $\Gamma_{initial}$ is the initial state broadening. The KH formula is more simply the probability of emission of photons when a 3-level system is considered, made by an initial, a final and an intermediate (virtual) state.

The challenge to migrate advanced spectroscopies from dedicated beamlines to a flexible home-lab setup demands to reconsider the data acquisition and processing protocols in the light of the complete theory, to match with the available source capabilities. In this respect, the application of the KH formalism may help to demonstrate XAS without the *tunability* requirement.

5.3.3.2 Implementing Kramers–Heisenberg for Home-Lab Spectroscopy

A ground-breaking spectroscopic advance would be single-colour XAS, i.e. without tuning the wavelength across the absorption edge. Błachucki et al. [9] introduced high-energy resolution off-resonance spectroscopy (HEROS) as an extension of XAS, where one does not require a tunable source, but only a narrow linewidth source, in order to extract the XAS in post-processing from a XES run. In brief, if a probe photon is below the absorption edge, it is unable to ionize a core electron, and lead to either XAS or XES. However, as derived from the KH formula, absorption could anyway occur because of self-coupling of the probe photon with background fluorescence. The latter must fill the photon energy gap to get above threshold, i.e. $E_{probe} + E_{bkg} > E_{KLM}$. Tulkki et al. [8] reported that for photons below the K absorption edge (called *off-resonance* region), the shape of the emission spectrum is proportional to the unoccupied density of states of an atom.

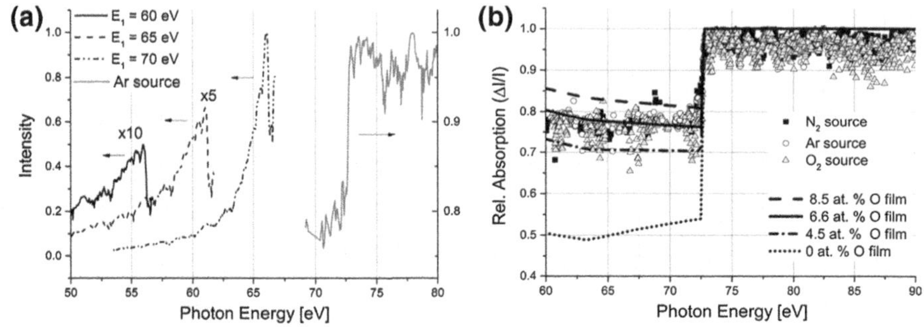

Fig. 5.2 **a** Experimental single-shot absorption spectrum for Al@Al_2O_3 sample using Ar plasma radiation, and corresponding emission spectrum, calculated for different incident photon energies (E_1) with the Kramers–Heisenberg formalism. See text for explanations. **b** Experimental single-shot absorption spectra for Al@Al_2O_3 sample using Ar, O_2 and N_2 plasma radiation. The surface oxidation layer is estimated by comparison with calculated absorption profiles

Figure 5.2a, b shows the experimentally obtained single-colour absorption spectrum (XAS, solid line) with Ar plasma home-lab source and Al@Al_2O_3 sample, where the L-edges at 72.5 and 72.9 eV [10] is evident. Figure 5.2a shows the corresponding XES, calculated for different off-resonant photon energies (E_1) via the KH formalism. The inelastic (inverse Raman) process involved in the absorption and emission mechanisms is responsible for the energy shift of the fluorescence. In Fig. 5.2a one observes that the farther the excitation from the edge, the smaller the XES. Figure 5.2b shows the experimental XAS for the Al@Al_2O_3 sample obtained with the home-lab source, using either Ar or N_2 or O_2 as the discharge gas. The three spectra did confirm the same results. The level of surface oxidation (ca. 20–40 nm oxidation film) was determined by estimating the level of deviation by a pure Al in the absorption profiles. Figure 5.2b shows calculated absorption profiles, i.e. the mean through the experimental spectra (6.6 at.% oxide), and higher (8.5 at.%) and lower (4.5 at.%) tolerances. Validation measurements with scanning electron microscopy (SEM)—Energy-Dispersive X-ray Spectrometry (EDX) confirmed an oxide content of 5.3–6.8%.

5.4 Conclusions

Resolution trade-offs (time spectral) have been discussed, highlighting the need for a *high spectral resolution* 'home lab' source. Preliminary results of plasma-source XAS were shown, demonstrating the potential.

References

1. Staub, F., Imesch, C., Bleiner, D., Balmer, J.E.: Soft-X-ray lasing in nickel-like barium at 9.2 nm using the grazing-incidence scheme. Opt. Commun. **285**(8), 2118–2121 (2012)
2. Arbelo, Y., Barbato, F., Bleiner, D.: He-doped pseudospark as a home-lab XUV source beyond the beamtime bottleneck. Plasma Sour. Sci. Technol. (2017) (in press)
3. Klisnick, A., Guilbaud, O., Ros, D., Cassou, K., Kazamias, S., Jamelot, G., Edwards, M., et al.: Experimental study of the temporal coherence and spectral profile of the 13.9 nm transient X-ray laser. J. Quant. Spectrosc. Radiat.Transf. **99**(1), 370–380 (2006)
4. Bleiner, D., Arbelo-Pena, Y., Masoudnia, L., Ruiz-Lopez, M.: Table-top X-ray lasers using a plasma gain-medium: limits and potentials. Phys. Scr. **2014**(T162), 014050 (2014)
5. Masoudnia, L., Bleiner, D.: Optimum electron temperature and density for short-wavelength plasma-lasing from nickel-like ions. Nucl. Instrum. Methods Phys. Res. Sect. B **323**, 59–70 (2014)
6. Masoudnia, L., Ruiz-Lopez, M., Bleiner, D.: Table-top two-color soft X-ray laser by means of Ni-like plasmas. Phys. Plasmas (1994-present) **23**(4), 043108 (2016)
7. Calvin, S.: XAFS for Everyone. CRC Press (2013)
8. Tulkki, J., Aberg, T.: Behaviour of Raman resonance scattering across the K X-ray absorption edge. J. Phys. B At. Mol. Phys. **15**(13), L435 (1982)
9. Błachucki, W., Szlachetko, J., Hoszowska, J., Dousse, J.C., Kayser, Y., Nachtegaal, M., Sá, J.: High energy resolution off-resonant spectroscopy for X-ray absorption spectra free of self-absorption effects. Phys. Rev. Lett. **112**(17), 173003 (2014)
10. McMaster, W.H., Del Grande, N.K., Mallett, J.H., Hubbell, J.H.: Compilation of X-ray cross sections. Lawrence Livermore National Laboratory Report UCRL-50174

Chapter 6
The Creation of Radiation Dominated Plasmas Using Laboratory X-Ray Lasers

G. J. Tallents, S. Wilson, V. Aslanyan, A. West, A. K. Rossall, C. S. Menoni and J. J. Rocca

Abstract When short wavelength extreme ultraviolet (EUV) and X-ray laser radiation is focused onto solid targets, narrow deep features are ablated and a dense, low-temperature plasma is formed. We examine the radiation dominated plasma formed by 46.9 nm laser radiation focused onto solids and show that ionisation can be significantly modified by electron degeneracy effects. Some experimental and theoretical considerations for investigating the interaction of capillary discharge lasers operating at 46.9 nm with solid and gas targets are presented.

6.1 Introduction

Extreme ultraviolet (EUV) and X-ray lasers can be used to generate strongly coupled plasmas and 'warm dense' matter [1]. Targets irradiated by EUV and X-ray lasers are heated by direct photoionisation as well as inverse bremsstrahlung. With EUV and X-ray lasers, lower temperature and higher particle density plasmas are produced than can be created by optical laser irradiation. With optical laser-produced plasmas, an expanding plume of plasma only allows absorption by inverse bremsstrahlung where the electron density drops below a critical value ($\simeq 10^{21}/\lambda^2$ μm^{-2} cm^{-3}, where λ is the laser wavelength in units of microns). By reducing the wavelength into the EUV to X-ray region, the critical electron density is greater than solid electron densities and the laser photon energy E_p becomes sufficient to directly photoionise elemental components (ionisation energy E_i), transferring energy ($E_p - E_i$) to the ejected electron. As the critical electron density is higher than solid, the laser is able to penetrate any expanding plasma plume and

G. J. Tallents (✉) · S. Wilson · V. Aslanyan · A. West · A. K. Rossall
York Plasma Institute Department of Physics, University of York, York YO10 5DD, UK
e-mail: greg.tallents@york.ac.uk

C. S. Menoni · J. J. Rocca
Center for Extreme Ultraviolet Science and Technology and Department of Electrical and Computer Engineering, Colorado State University, Fort Collins, CO 80523, USA

T. Kawachi et al. (eds.), *X-Ray Lasers 2016*, Springer Proceedings in Physics 202, https://doi.org/10.1007/978-3-319-73025-7_6

heat solid material directly throughout the duration of a laser pulse. A large depth of plasma is heated at high density to moderate (a few eV) temperatures.

We explore in this paper the potential for creating narrow and deep features by EUV laser ablation of solids [2], and also for using the plasma created by EUV lasers as sources for warm dense matter studies. Narrow feature can be ablated due to the short wavelength λ allowing focusing to a diameter of approximately $f\lambda$, where f is the f-number of the focusing optic. Our calculations show that the ionisation of plasma can be affected by free electron degeneracy effects with overlap to the plasma conditions found in inertial fusion, where degenerate plasma is created during X-ray driven compression of material. The use of a capillary laser [3, 4] operating at 46.9 nm in producing high-density degenerate plasma is explored.

6.2 Simple Model of EUV Interaction in Gases

It is possible to simply model EUV laser interaction with gas targets [5] as with low-density targets there is a large change in opacity after deposition of a defined quantity of energy per unit volume. The thickness of material heated by EUV or X-ray radiation incident on a target is controlled by the integral of the incident irradiance in time. A rapid drop of opacity associated with the ionisation of the target will progress through the target with time due to photoionisation, Auger ionisation and inverse bremsstrahlung heating (the last leading to collisional ionisation). If a thickness Δx of target material is ionised, we can equate the energy per unit area from the incident irradiance I in a time Δt with the energy per unit area which is ionised, i.e.

$$I\Delta t = N(E_{ion} + kT)\Delta x \tag{6.1}$$

where N represents the number of electrons per unit volume heated to temperature T and E_{ion} is the summation of the average ionisation energy per electron required to produce an ionisation stage with a sufficiently large ionisation energy that further photoionisation is not possible. It is assumed that the ion temperature remains small.

Rearranging gives an estimate of the velocity of an ionisation wave through the target:

$$v_{ion} = \frac{\Delta x}{\Delta t} = \frac{I}{N(E_{ion} + kT)}. \tag{6.2}$$

The plasma temperatures achieved can be estimated by relating the energy density absorbed $E_{ion} + kT$ to the temperature and by measuring the speed v_{ion} of the propagation of an ionisation wave (see 6.2). The necessary ionisation energy density E_{ion} can be estimated by calculating the energy density required to ionise

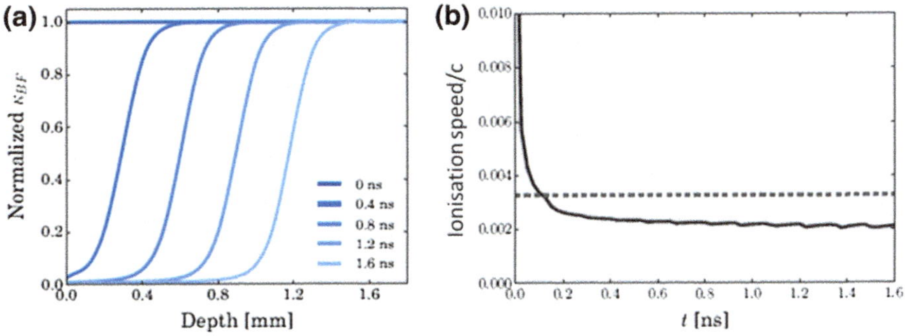

Fig. 6.1 **a** The normalised photoionisation absorption coefficient of helium at atmospheric pressure during irradiation by a laser pulse with a flat temporal profile and photon energy 26.4 eV at irradiance of 10^{10} W cm^{-2}. **b** Simulated speed of propagation of the ionisation wave *dx*/*dt* (solid), determined as the point where the normalised absorption coefficient takes a value of 0.5, compared to the value following (6.2) assuming zero particle temperature (dashed)

the material sufficiently so that the ionisation energy becomes greater than the photon energy. Equation (6.2) is remarkably accurate for EUV interactions in gases. In Fig. 6.1, we compare a detailed simulation of the ionisation wave propagation [6] with the analytic expression given by (6.2) for an EUV laser interaction with a gas at atmospheric density.

6.3 EUV Interactions with Solid Targets

With EUV interactions onto solid targets, the higher density plasma undergoes significant inverse bremsstrahlung absorption as well as photoionisation so that the model of 'bleaching' after deposition of energy $E_{ion} + kT$ per electron breaks down. Inverse bremsstrahlung causes additional heating raising the value of NkT without a change of the material opacity (see Fig. 6.2).

Experiments, where a capillary discharge laser is focused onto solid parylene targets, have been undertaken and simulated using a 2D fluid code [7]. The agreement between experiment and the simulations in the depth of material ablated in a single shot is good for experiments undertaken with a 1.3 ns duration capillary focused to different fluence values (see Fig. 6.3). A small fraction of the EUV laser output was focused with Fresnel zone plate operating at f/4. The depth of the ablated craters was measured post-shot using an electron microscope.

At irradiances higher than Fig. 6.3 (i.e. $>10^{10}$ W cm^{-2}), we found that shock wave or other effects cause additional target damage after the shot. An experiment utilising the full energy of the capillary laser was undertaken with spherical multilayer mirror focusing and collection optics to measure the transmission of the laser through thin targets of parylene or aluminium (see Fig. 6.4). Using spherical multilayer mirrors collecting the full laser beam profile enabled higher irradiances

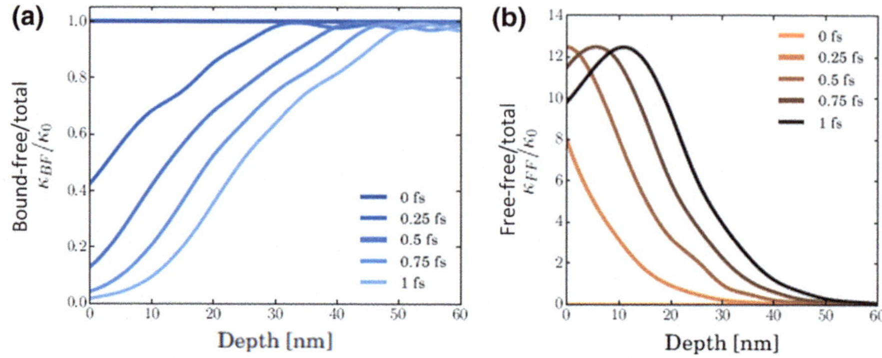

Fig. 6.2 The contribution to the absorption coefficient due to **a** photoionisation and **b** inverse bremsstrahlung for solid density carbon irradiated by photons at irradiance 10^{14} W cm^{-2} and photon energy 26.4 eV at times indicated. The absorption coefficients are normalised to the photoionisation coefficient for neutral carbon

Fig. 6.3 Comparison between experimental ablated depth measurements (squares) and simulations (solid line) as a function of EUV laser fluence on target. The dashed line indicate the resolution of the Eulerian mesh used in the simulations. Each data point represents the ablation depth for a single shot at the indicated fluence which was determined from the area of the ablated surface

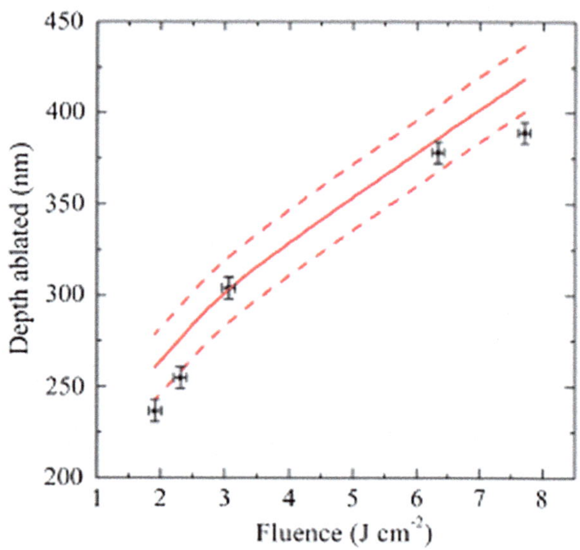

on target than was possible using Fresnel zone plate focusing optics. For more details of this experiments, see the paper by Aslanyan et al. [8]. Wave optic simulations of the focused laser beam profile including aberration effects due to a 4.7° tilt on the focusing and collection mirrors agreed with microscope images of the damaged targets. However, the transmission of the laser through the target was best modelled using our fluid code calculations of the shape of the ablated laser hole to evaluate the transmitted image including diffraction. Any cold target material is highly opaque to the EUV lasers, so the transmitted image is simply a sum of the diffracted images once ablation through the target thickness has occurred at some

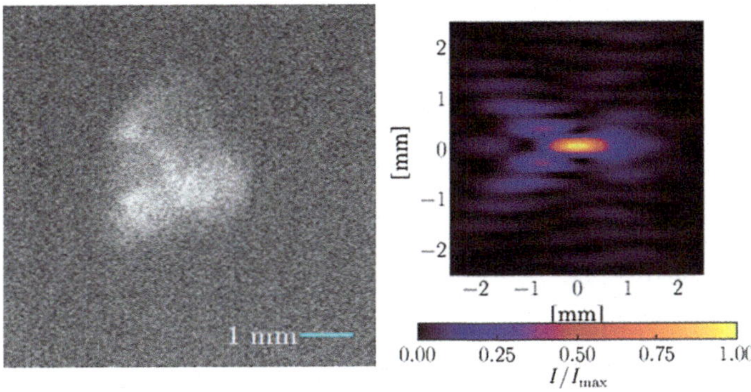

Fig. 6.4 A measured profile of the EUV laser (left) and a simulated profile (right) transmitted through a parylene target of thickness 1028 nm. The transmitted EUV laser light is collected with a multilayer mirror which is identically positioned as a multilayer mirror used to focus the EUV laser onto the target. Transmission of EUV light only occurs when the target has been ablated through the total target thickness

place. A sample transmission image and the simulation of this transmission is shown in Fig. 6.4. The agreement of the experimental and simulated images in Fig. 6.4 is better than if diffraction is assumed to occur through the post-shot hole produced by EUV laser irradiation confirming that additional post-shot damage to the target occurs due to shock waves and other effects.

6.4 Degeneracy Effects

Due to the penetration of EUV radiation into solid density material during EUV laser interaction with a solid target, plasma at modest temperatures (<10 eV) and solid density is produced. With only a small amount of ionisation, the free electrons can be affected by degeneracy effects. A measure of the degree of free electron degeneracy is given by the reduced chemical potential $\eta = \mu/kT_e$, where μ is the chemical potential and kT_e is the electron temperature [9]. Large negative chemical potential corresponds to conditions where degeneracy effects are unimportant. For densities $\rho > 1$ g cm^{-3} with temperatures $T_e < 3$ eV, degeneracy effects are important for the free electrons and $\eta > 0$.

Calculations show that electron degeneracy significantly reduces collisional rate coefficients for excitation and ionisation process due to the reduction of possible free electron quantum states for the colliding electrons [7]. These reductions can modify the ionisation and heat capacity of plasma as shown by Aslanyan and Tallents [10]. Steady state calculations of the degree of ionisation of solid density carbon show that in the presence of a significant irradiance (>10^{14} W cm^{-2}) of EUV radiation, the degree of ionisation is different assuming a Maxwellian distribution of electrons and assuming a Fermi–Dirac distribution (Fig. 6.5). At high

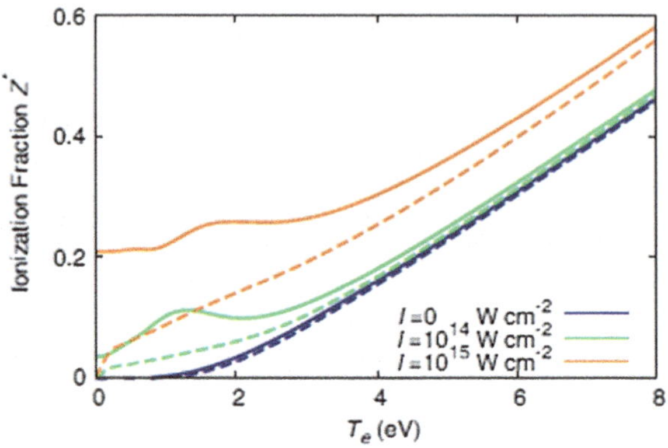

Fig. 6.5 The steady state ionisation Z^* of a carbon plasma at solid density irradiated by EUV photons of energy 50 eV at the indicated irradiance as a function of electron temperature. The free electron distribution is assumed to be Fermi–Dirac (solid curves) or Maxwellian (broken curves)

irradiance, photoionisation is in balance with three-body recombination. Due to the two free electrons involved in three-body recombination, the rate of this process is much more affected by degeneracy effects than photoionisation and so the steady state ionisation balance is changed due to the degeneracy effects.

6.5 Conclusion

We have shown that when short wavelength extreme ultraviolet (EUV) and X-ray laser radiation is focused onto solid targets, deep features are ablated. A dense, low temperature plasma is formed. We have examined the radiation dominated plasma formed by 46.9 nm laser radiation focused onto solids and have shown that ionisation can be significantly modified by electron degeneracy effects. The regimes where electron degeneracy effects are important have been quantified and sample simulations presented showing the effects of free electron degeneracy on plasma ionisation under conditions found in solid density plasmas irradiated by EUV lasers.

References

1. Berrill, M., Brizuela, Langdon, F.B., Bravo, H., Menoni, C.S., Rocca, J.J.: Warm photoionized plasmas created by soft X-ray laser irradiation of solid targets. J. Opt. Soc. Am. B **25**, p. 32 (2008)
2. Bravo, H., Szapiro, B.T., Wachulak, P.W., Marconi, M.C., Chao, W., Anderson, W.H., Menoni, C.S., Rocca, J.J.: Demonstration of nanomachining with focused extreme ultraviolet laser beams. IEEE J. Sel. Top. Quantum Electron. **18**, 443 (2012)
3. Rocca, J.J., Shlyaptsev, V.N., Tomasel, F.G., Cortazar, O.D., Hartshorn, D., Chilla, J.L.A.: Demonstration of a discharge pumped table-top soft X-ray lasers. Phys. Rev. Lett. **73**, 2192 (1994)

4. Benware, B.R., Macchietto, C.D., Moreno, C.H., Rocca, J.J.: Demonstration of a high average power tabletop soft X-ray laser. Phys. Rev. Lett. **81**, 5804 (1998)
5. Tallents, G.J., Whittaker, D.S., Wilson, L.A., Wagenaars, E.: Heating of high energy density plasmas using EUV and X-ray lasers. Proc. SPIE **8140**, 81400F (2011)
6. Aslanyan, V.: Extreme ultraviolet lasers and their interactions with matter. Ph.D. thesis (2016) (unpublished)
7. Rossall, A.K., Aslanyan, V., Tallents, G.J., Kuznetsov, I., Rocca, J.J., Menoni, C.S.: Ablation of submicrometer holes using an extreme ultra-violet laser. Phys. Rev. Appl. **3**, 064013 (2015)
8. Aslanyan, V., Kusznetsov, I., Bravo, H., Woolston, M.R., Rossall, A.K., Menoni, C.S., Rocca, J.J., Tallents, G.J.: Ablation and transmission of thin solid targets irradiated by intense extreme ultraviolet laser radiation. APL Photonics **1**, 066101 (2016)
9. Tallents, G.J.: Free electron degeneracy effects on collisional excitation, ionization, de-excitation and three-body recombination. HEDP **20**, 9 (2016)
10. Aslanyan, V., Tallents, G.J.: Ionization rate coefficients in warm dense plasmas. Phys. Rev. E **91**, 063106 (2015)

Chapter 7
Plasma Dynamics in Capillary Discharges

P. V. Sasorov

Abstract Capillary discharge plasma is used frequently as a source of UV radiation, as active media for EUV and soft X-ray lasers, for the formation of plasma wave-guides to transport high power laser beams, and as plasma lens to focus beams of accelerated charged particles. A brief review of main physical processes responsible for dynamics of plasma in capillary discharges is presented in this chapter. The review takes into account results of a lot of MHD simulations as well as their comparisons with different experiments. Two quite different types of capillary discharges that are used in many experiments are considered. Main physical processes that play an important role in plasma-wall interaction and determine wall material evaporation rate, are also considered.

7.1 Introduction

A through duct in a dielectric filled initially with a gas may conduct electric current that leads to the formation of plasma from the gas inside the duct. Such form of gas discharge is called usually as capillary discharge. Capillary discharge plasma has different applications: (a) as a source of UV radiation [1]; (b) as an active media for EUV and soft X-ray lasers [2, 3]; (c) as a plasma wave-guide for long enough transportation of power laser beams in laser particle accelerators [4]; (d) as a plasma lens to focus beams of accelerated charge particles [5, 6]; etc. This talk presents a review of physical processes that govern the main properties of capillary discharge plasma. This review is based on a lot of MHD computer simulations and their comparison with experimental data.

P. V. Sasorov (✉)
Keldysh Institute of Applied Mathematics, Russian Academy of Sciences, Moscow, Russia
e-mail: pavel.sasorov@gmail.com

T. Kawachi et al. (eds.), *X-Ray Lasers 2016*, Springer Proceedings in Physics 202, https://doi.org/10.1007/978-3-319-73025-7_7

7.2 General Properties of Capillary Discharges

We list here ranges of typical parameters of the capillary discharges:

- capillary length, 3–30 cm;
- its diameter, 0.2–2 mm;
- capillary materials: plastic, glass, ceramic, sapphire, etc.;
- peak electric current, 100 A–40 kA;
- electric current half period, 30–500 ns;
- filling: H_2, Ar,…;
- initial gas pressure, 0.3–50 mbar.

The main physical processes determining parameters of the capillary plasma and its dynamics can be sorted between the following types: mechanical, electromagnetic, thermal, and others. Two mechanical forces that move the plasma are gradient pressure force and Ampere's force ($\propto j \times$ B). The capillary discharge is distributed along its radius inductance and resistance. They determine electromagnetic properties of the discharge and, in particular, characteristic time of electric field penetration into the discharge. The main thermal processes that govern plasma temperature are Joule heating, electron thermal conduction and radiation of the relatively hot plasma. Other physical processes may play an important role in the dynamics of the capillary discharge plasma. We mention among such processes: ionization (recombination) and capillary wall material evaporation. The latter process is caused by energy deposition in the wall due to energy flux from the capillary plasma.

All of these physical processes for the capillary discharges are taken usually into account in the frame of magnetic hydrodynamics (MHD). A two-temperature one-dimensional version of such an MHD model is described for example in [7, 8]. It takes into account possible differences between electron and ion temperatures.

7.3 Two Types of Capillary Discharges

Electric current and the magnetic field caused by it leads to the following two effects that affected plasma dynamics. The first of them is the Ampere's force ($\propto j \times$ B), and the second one is Joule heating (j^2/σ) that heats the electron component of plasma. The relative importance of these two effects for plasma dynamics is determined by the following dimensionless parameter, magnetic Reynolds number:

$$\mathrm{Re}_m = \frac{av}{\nu_m}, \tag{7.1}$$

where a is typical size of the plasma (capillary diameter); $\nu_m = c^2/(4\pi\sigma)$ is the so-called magnetic viscosity, defined by the electric conductivity, σ; and v is typical velocity of plasma motion, that is usually of the order of Alfven velocity, c_A: $c_A^2 = B^2/(4\pi\rho)$. Here B is magnetic field, and ρ is plasma density.

If $\mathrm{Re}_m \gg 1$, then:

- Effects of the Ampere's force are much stronger than the effects caused by the Joule heating;
- Plasma pressure due to Joule heating cannot be prevented from magnetic compression;
- Plasma is detached from the capillary walls;
- There is a strong pinch effect;
- Plasma is heated mainly by shock waves; and
- Plasma is screened from electric and magnetic fields due to strong skin-effect.

In the opposite case, when $\mathrm{Re}_m \ll 1$, we have that:

- Effects caused by the Joule heating is more important than the effects caused by the Ampere force, tending to compress the plasma;
- Ampere's force is of relatively small effect;
- Plasma pressure tends to be constant across the discharge;
- Plasma is confined by the capillary walls;
- It is heated by the Joule heating, that is balanced by the thermal conduction towards relatively cold capillary walls; and
- There is no skin-effect, and electric current distribution is smooth across the discharge.

We will call the capillary discharges of the first type, when $\mathrm{Re}_m \gg 1$, as ***pinching*** capillary discharges, whereas the capillary discharges of the second type, when $\mathrm{Re}_m \ll 1$, as ***dissipative*** capillary discharges.

We may say very roughly that the type of capillary discharge is controlled mainly by the amplitude of electric current. We have exemplary dissipative capillary discharge at the peak current of 0.3 kA [4], while for 40 kA we have exemplary pinching capillary [2]. The transition between these two types takes place approximately at a few kilo-amps.

7.3.1 Capillary Discharges with Strong Pinch Effect

The typical capillary discharge of the pinching type is the capillary discharge used in [2] in the 46.9 nm Ar laser. It was used to compress strongly Ar plasma at the axis of a polyacetal capillary of 4 mm diameter and to get hot and dense Ar plasma with temperature of about 60 eV and electron density of about $4 \cdot 10^{19}$ cm^{-3}. These parameters are favourable for Ne-like lasing in Ar plasma. To obtain such Ar plasma a current pulse with 40 kA amplitude and 60 ns duration was applied to the polyacetal capillary of 4 mm diameter prefilled with Ar gas of 0.85 mbar initial pressure. The volume compression of the Ar gas-plasma was as high as about 200 times.

Many teams performed 1d MHD simulations of such discharges. See, for example [9–11]. The MHD codes take into account all physical processes mentioned above, accompanied by simulation of atomic levels population dynamics and by simulation of laser beam propagation. Such simulations show good agreement with experiments. Scaling laws describing plasma parameters at stagnation as function of initial Ar gas pressure, capillary diameter and amplitude and duration of the electric current pulse were obtained in [7] using such MHD simulations.

The capillary discharge used in [2] and simulated in [7, 9–11] leads to evaporation of polyacetal walls. This evaporation leads, in turn, to the formation of a plasma sheath. It begins capturing part of the total electric current, so that the ablated plasma takes part in Ar plasma compression. This effect leads to less effective compression of Ar plasma and hence demands higher electric current. Using more refractory capillary materials [3] such as ceramics allows applying considerably lower electric current, providing complete absence of wall material evaporation and 46.9 nm Ar lasing. MHD simulations of such discharges are significantly simpler.

Assuming absence of plasma ablation from the capillary walls, we may achieve in MHD simulations lasing for much shorter wavelength. See for example [12, 13] and the experiments [14].

7.3.2 Dissipative Capillary Discharges

Dissipative capillary discharges tend, as it was explained above, to mechanical and thermal equilibrium, with the Ampere's force being negligible. As a result, such discharges have a maximum of electron temperature at the axis and a minimum of electron density. Such long-lived radial electron density distribution is favourable for optical guiding. This property of dissipative capillary discharges was first demonstrated in [15]. They used initially evacuated plastic capillary. As a result, plasma of the discharge was formed from wall material, and hence its chemical composition was poorly controlled. Usage in [16] of ceramic capillary prefilled with hydrogen demonstrated almost complete suppression of capillary wall evaporation, that leads to the well-controlled chemical composition of plasma wave-guide. This situation is suitable for long transportation of power femtoseconds laser pulses, and hence to acceleration of electrons due to wake field effects. It was demonstrated in [4]. A 4.2 GeV electron beam of high quality was produced by propagation of 0.3 PW laser pulse in capillary discharge of 9 cm length and 0.5 mm diameter with 0.3 kA peak current of ~300 ns duration. The sapphire capillary was filled with hydrogen up to ~40 mbar initial pressure.

Owing to the existence of mechanical and thermal equilibrium, and tendency to its establishing for the dissipative capillary discharges, there are very simple scalings [8] that describe plasma parameters in the capillary discharges of this type as well as profile of refraction index in this optic wave-guide. They were checked with the MHD simulations as well as in part experimentally.

7.4 Evaporation of Wall Material

Possible wall material evaporation is an important feature of capillary discharges. The evaporation is caused by energy deposition in a relatively thin layer of the wall. The deposition is caused, in turn, by the contact of discharge plasma with the wall. Sometimes, it is an unavoidable effect (for discharges in initially evacuated capillaries, for example), and sometimes, it is a parasitic one.

It is useful to introduce the term "cost of evaporation", W. It can be measured for example in mJ/cm^2. It is important that it does not originate from specific heat of evaporation of wall material. The latter value can be usually neglected, because it is much lower than the specific energy of discharge plasma. Instead of this, the value of W says: how much energy should be deposited into the wall (per its unit area) to heat a certain thin layer of it to sufficiently high temperature, so the saturated vapour pressure becomes comparable with the discharge plasma pressure.

For reasonable evaluation of W, we should consider transport of heat into the capillary wall and determine a depth of this layer. This process was considered and simulated in [10, 11] simultaneously with the MHD simulations of the discharge plasma. The simulations show that a small fraction ($\sim 3 \times 10^{-4}$) of electrons from the valence band in the wall dielectric layer of about 0.5 μm thickness is exited into the conduction band, and that temperature of the electrons component becomes ~1.5–2 eV, that is much higher, than typical lattice temperature, several hundreds of K. The lattice temperature grows gradually in time due to energy exchange between electrons and lattice. The layer thickness is determined by this energy exchange and by transport of heat due to electron thermal conduction and radiation.

Typical value of W for fused materials like plastics is of the order of 10 mJ/cm^2, whereas for refractory ceramics or sapphire, it is of the order of 100 mJ/cm^2. This value depends on many factors like impurities and current pulse waveform. The above numbers are presented for orientation only.

It is important that time-integrated energy flux towards the capillary wall (F) during the important part of the electric current pulse is determined completely by the capillary plasma dynamics and not by processes in the capillary walls. It is the consequence of that typical temperature of the capillary plasma that is much higher than the temperature of the solid capillary wall.

The process of evaporation is easily simulated in two cases: $F \gg W$, and $F \ll W$. The former case corresponds to the situation when the evaporation is limited only by the energy flux from the discharge plasma [10], whereas the latter one corresponds to the complete absence of evaporation [8]. The case $F \sim W$ demands detailed simulation of heating of the thin layer of wall.

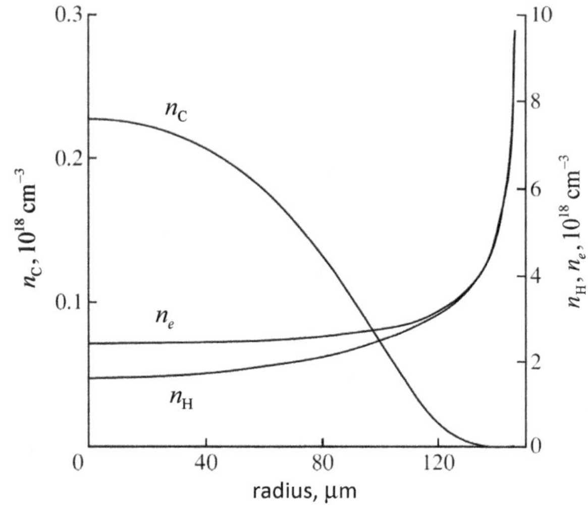

Fig. 7.1 Radial distributions of electron (n_e), hydrogen (n_H) and carbon (n_C) densities across the capillary discharge

The capillary wall evaporation supplies admixtures to the capillary plasma. Radial distribution of highly ionized heavy impurities in the discharge was considered in [17]. See Fig. 7.1.

7.5 Conclusions

We may conclude that theory of the capillary plasma dynamics is presently well developed. We mean here 1D radial MHD dynamics of the plasma in thin and long capillaries. The only unresolved problem is quantitative description of capillary wall evaporation, when time-integrated energy flux toward the wall is close to the threshold of the beginning of intensive evaporation: $F \sim W$. We have good description of the evaporation process in two limiting cases: $F \gg W$ and $F \ll W$.

It is now a good time for the beginning of 2D and 3D MHD simulation of capillary discharges to have an adequate description of electron density distribution in the vicinities of open ends of capillaries and gas supplying channels. Another 2D MHD problem, that could be interesting for applications, is discharges in capillaries of noncircular cross sections. See postal presentation by G. Bagdasarov et al. at this conference.

Acknowledgements The author is grateful to N. Bobrova, S. V. Bulavov, T. Razinkova, D. Farina, R. Pozzoli, V. Gasilov, O. Olkhovskaya, A. Boldarev and G. Bagdasarov for their permanent collaboration in the field considered in this paper. This work was supported in part by the Russian Foundation of Basic Researches, grant No. 15-01-06195.

References

1. Shevelko, A.P., Sobel'man, I.I., Yakushev, O.F., Knight, L.V., Turleyb, R.S.: A capillary discharge plasma source of intense VUV radiation. Quantum Electron. **33**(1), 3–6 (2003)
2. Rocca, J.J., Shlyaptsev, V., Tomasel, F.G., Gortazar, O.D., Hartshorn, D., Chilla, J.L.A.: Demonstration of a discharge pumped table-top soft-X-ray laser. Phys. Rev. Lett. **73**, 2192 (1994)
3. Heinbuch, S., Grisham, M., Martz, D., Rocca, J.J.: Demonstration of a desk-top size high repetition rate soft X-ray laser. Opt. Express **13**, 4050–4055 (2005)
4. Leemans, W.P., Gonsalves, A.J., Mao, H.-S., Nakamura, K., Benedetti, C., Schroeder, C.B., Toth, CS., Daniels, J., Mittelberger, D.E., Bulanov, S.S., Vay, J.-L., Geddes, C.G.R., Esarey, E.: Multi-GeV electron beams from capillary-discharge-guided subpetawatt laser pulses in the self-trapping regime. Phys. Rev. Lett. **113**, 245002 (2014)
5. Basko, M.M., Drozdovskii, A.A., Golubev, A.A., et al.: Plasma lens for the heavy ion accelerator at ITEP. Phys. Part. Nucl. Lett. **5**(7), 582–585 (2008)
6. van Tilborg, J., Steinke, S., Geddes, C.G.R., Matlis, N.H., Shaw, B.H., Gonsalves, A.J., Huijts, J.V., Nakamura, K., Daniels, J., Schroeder, C.B., Benedetti, C., Esarey, E., Bulanov, S. S., Bobrova, N.A., Sasorov, P.V., Leemans, W.P.: Active plasma lensing for relativistic laser-plasma-accelerated electron beams. Phys. Rev. Lett. **115**, 184802 (2015)
7. Bobrova, N.A., Bulanov, S.V., Farina, D., Pozzoli, R., Razinkova, T.L., Sakai, J.-I., Sasorov, P.V., Sokolov, I.V.: MHD simulations of plasma dynamics in pinch discharges in capillary plasmas. Laser Part. Beams **18**, 623–638 (2000)
8. Bobrova, N.A., Esaulov, A.A., Sakai, J.-I., Sasorov, P.V., Spence, D.J., Butler, A., Hooker, S. M., Bulanov, S.V.: Simulations of a hydrogen-filled capillary discharge waveguide. Phys. Rev. E, **65**, 016407 (2002)
9. Shlyaptsev, V.N., Gerusov, A.V., Vinogradov, A.V., Rocca, J.J., Cortazar, O.D., Tomasel, F. G., Szapiro, B.T.: Proceedings of the SPIE, 2012, vol. 99 (1993)
10. Bobrova, N.A., Bulanov, S.V., Razinkova, T.L., Sasorov, P.V.: Dynamics of a pinch discharge in capillarics. Plasma Phys. Rep. **22**(5), 349–362 (1996)
11. Moreno, C.H., Marconi, M.C., Shlyaptsev, V.N., et al.: Two-dimensional near-field and far-field imaging of a Ne-like Ar capillary discharge table-top soft-X-ray laser. Phys. Rev. A **58**, 1509 (1998)
12. Vrba, P., Vrbova, M., Bobrova, N.A., Sasorov, P.V.: Modelling of a nitrogen X-ray laser pumped by capillary discharge. Cent. Eur. J. Phys. **3**, 564–580 (2005)
13. Vrba, P., Bobrova, N.A., Sasorov, P.V., Vrbova, M., Hubner, J.: Modeling of capillary Z-pinch recombination pumping of boron extreme ultraviolet laser. Phys. Plasmas **16**, 073105 (2009)
14. Jancarek, A.: Capillary Discharge—A Way for Recombination XUV Laser (these proceedings)
15. Ehrlich, Y., Cohen, C., Zigler, A., Krall, J., Sprangle, P., Esarey, E.: Guiding of high intensity laser pulses in straight and curved plasma channel experiments. Phys. Rev. Lett. **77**, 4186 (1996)
16. Spence, D.J., Hooker, S.M.: Investigation of a hydrogen plasma waveguide. Phys. Rev. E **63**, 015401 (2000)
17. Kocharyan, A.E., Bobrova, N.A., Sasorov, P.V.: Nonuniformity of the chemical composition of a capillary discharge plasma. Plasma Phys. Rep. **32**, 887 (2006)

Chapter 8
MHD Simulation of Various Cross-Section Capillary Discharges

Gennadiy Bagdasarov, P. V. Sasorov, Alexey Boldarev, Olga Olkhovskaya, Anthony Gonsalves, Stepan Sergeevich Bulanov, C. B. Schroeder, E. Esarey, W. P. Leemans, Tadzio Levato, Daniele Margarone, Georg Korn, S. V. Bulanov and Vladimir Gasilov

Abstract Our paper concerns with simulation of plasma dynamics in hydrogen-filled capillary discharge with circular and square cross sections under the same conditions. External domain with moderate size which is formed by a passive insulator was taken into account to perform these simulations. The general dynamics of plasma inside these capillaries are found to be quite similar to each other.

8.1 Introduction

Presently capillary discharges are under intensive investigation due to various promising applications [1, 2]. A majority of the experiments implement the capillaries with circular cross section. Such capillary cross section significantly simplifies the theoretical and computer simulations studies, reducing the dimensionality of the problem and allowing the use of 1D computer codes. On the other hand, the square cross-section capillaries, which attracted significantly smaller attention, are advantageous from the point of view of transverse plasma diagnostics [3].

G. Bagdasarov (✉) · P. V. Sasorov · A. Boldarev · O. Olkhovskaya
Keldysh Institute of Applied Mathematics RAS, Moscow 125047, Russia
e-mail: gennadiy3.14@gmail.com

G. Bagdasarov · A. Boldarev · V. Gasilov
National Research Nuclear University MEPhI (Moscow Engineering Physics Institute), Moscow 115409, Russia

A. Gonsalves · S. S. Bulanov · C. B. Schroeder · E. Esarey · W. P. Leemans
Lawrence Berkeley National Laboratory, Berkeley, CA 94720, USA

T. Levato · D. Margarone · G. Korn
Institute of Physics ASCR, v.v.i. (FZU), ELI-Beamlines Project, 182 21 Prague, Czech Republic

S. V. Bulanov
Kansai Photon Science Institute, National Institutes for Quantum and Radiological Science and Technology (QST), 8-1-7 Umemidai, Kizugawa, Kyoto 619-0215, Japan

T. Kawachi et al. (eds.), *X-Ray Lasers 2016*, Springer Proceedings in Physics 202, https://doi.org/10.1007/978-3-319-73025-7_8

8.2 Formulation of the Problem

We use magnetohydrodynamic (MHD) code MARPLE [4] for the simulations of capillary discharges with arbitrary cross sections.

The physical model implemented in the code is similar to the model described in [5]: it is two-temperature (ion and electron), one-fluid MHD with the following dissipative processes taken into account: electron and ion thermal conductivities and electron-ion energy exchange, magnetic field diffusion, Joule heating and radiation losses. Similar to [5] both of state and dissipative coefficients incorporate degree of ionization.

The high symmetry of a circular discharge allows simulating plasma dynamics inside the capillary with the MHD code only. However, lines of magnetic field cross the interface between discharge plasma and insulator in the general case. Therefore we are forced to calculate the magnetic field distribution in the insulator together with the MHD simulation of the discharge plasma. In the simulations we consider both regions, so-called "plasma" and "insulator", in one computational domain. We solve the same equations in the "insulator" region, but the "insulator" is treated as immobile ($\vec{v} = 0$) fluid with zero conductivity ($\sigma = 0$).

Two section shapes of the capillary were considered in the discharge simulations—circular and square (Fig. 8.1). To minimize differences between the simulations and ease the comparison we used "plasma" and "insulator" regions for both capillaries.

In the simulations we consider the sapphire capillaries with diameter/side $d = 2r_o = 0.05\,\text{cm}$, prefilled with pure hydrogen of uniform density $\rho = 3.5 \cdot 10^{-6}\text{g/cm}^3$. The external passive insulator domain has radius $R = 0.05$ cm. The discharges are initiated by the same pulse of current with approximately sinusoidal profile, quarter period of 180 ns and a peak of 311 A. Model time is $t \in [0, 250]$ ns. These parameters are similar to those present in experiments [1].

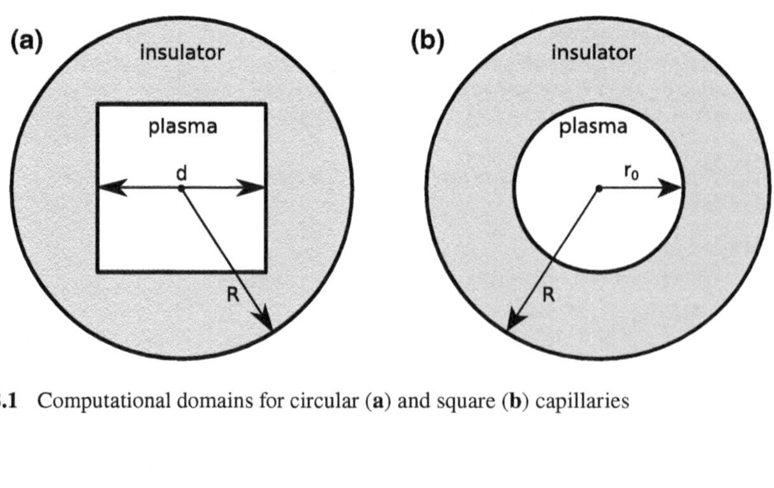

Fig. 8.1 Computational domains for circular (**a**) and square (**b**) capillaries

8.3 Results of the Simulations

On Fig. 8.2 distributions of magnetic field and current density are presented for both capillaries at the moment 200 ns of model time (slightly after current reaches its maximum). It can be noticed that the magnetic field lines are not perfect circumferences near the plasma–insulator interface for the square capillary while for the circular capillary they are perfect. However, near the center of square capillary and external boundary of the insulator, they again become perfect. These results indicate that the size of insulator region was chosen correctly.

Figure 8.3 presents information about the electron density and temperature for both circular and square capillaries at the moment 200 ns after discharge. Obtained results for the circular capillary show a good agreement with both experiment and other simulations [5]. For the square capillary values of both temperature (T_e = 5.34 eV) and density ($n_e = 1.2 \cdot 10^{18}\,\text{cm}^{-3}$) at the axis are lower than for the circular one (T_e = 5.73 eV and $n_e = 1.31 \cdot 10^{18}\,\text{cm}^{-3}$) because of difference in areas of the capillaries. However general dynamics of plasma and time dependence of axial plasma parameters for square capillary is quite similar to that in circular one.

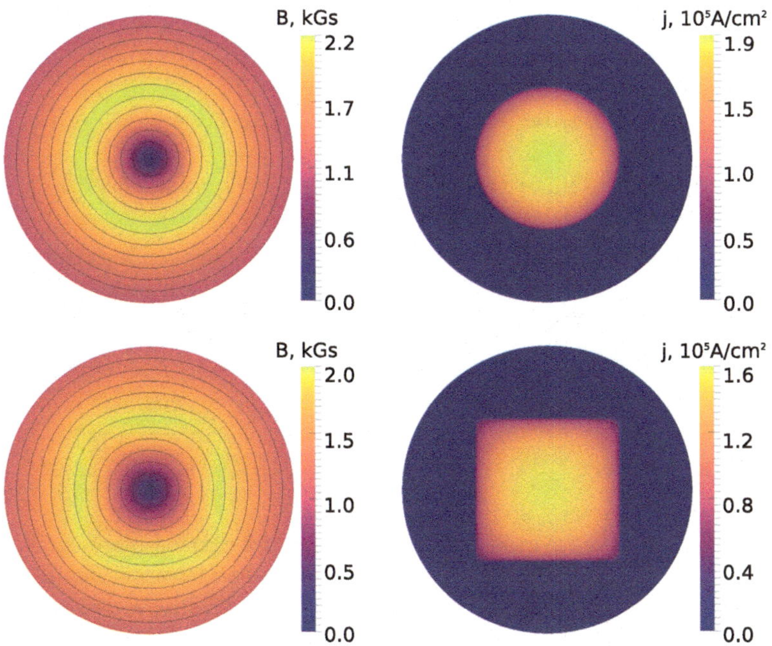

Fig. 8.2 Simulation results at 200 ns of model time. On the left pictures distributions of magnetic field together with its streamlines are shown for circular (top left) and square (bottom left) capillaries; right pictures presents distributions of current density for these capillaries respectively

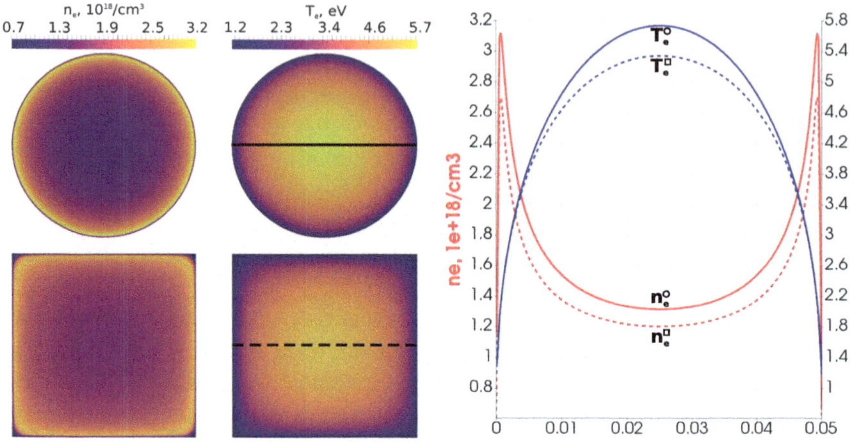

Fig. 8.3 Spatial distribution of electron density and temperature at $t = 200$ ns. Top row and solid lines correspond to circular capillary, bottom row and dashed lines—to the square one

Using simulation results, we obtained the following approximations for electron density distribution in the vicinity of the axes for both circular (○) and square (□) capillaries.

$$n_e^{\bigcirc}(r) = n_e(0) \cdot \left[1 + 0.33 \cdot (r/r_0)^2 + 0.4 \cdot (r/r_0)^4 + \cdots\right],$$
$$n_e^{\square}(r) = n_e(0) \cdot \left[1 + 0.29 \cdot (r/r_0)^2 + (0.24 + 0.075 \cdot \cos 4\varphi) \cdot (r/r_0)^4 + \cdots\right],$$

where φ is azimuthal angle. Obtained results for the circular capillary are in a good agreement with the results from [5]. Value of coefficient in front of $\cos 4\varphi$ shows that the azimuthal dependence of the solution in square capillary is rather weak and negligible near the center.

8.4 Conclusions

In conclusion, we have investigated the dynamics of hydrogen-filled capillary discharge with circular and square cross sections under the same conditions. We have found that the general dynamics of plasma inside these capillaries are quite similar to each other.

Acknowledgements The work was supported in part by the Russian Foundation for Basic Research (Grant No. 15-01-06195), by the Competitiveness Program of National Research Nuclear University MEPhI (Moscow Engineering Physics Institute), contract with the Ministry of Education and Science of the Russian Federation No. 02.A03.21.0005, 27.08.2013 and the basic research program of Russian Ac. Sci. Mathematical Branch, project 3-OMN RAS. The work at Lawrence Berkeley National Laboratory was supported by US DOE under contract No. DE-AC02-05CH11231. The work at KPSI-QST was funded by ImPACT Program of Council for Science, Technology and Innovation (Cabinet Oce, Government of Japan). At ELI-BL it has been supported by the project ELI—Extreme Light Infrastructure—phase 2 (CZ.02.1.01/0.0/0.0/15_008/0000162) from European Regional Development Fund, and by the Ministry of Education, Youth and Sports of the Czech Republic (project No. LQ1606) and by the project High Field Initiative (CZ.02.1.01/0.0/0.0/15_003/0000449) from European Regional Development Fund.

References

1. Leemans, W.P., et al.: Multi-GeV electron beams from capillary-discharge-guided subpetawatt laser pulses in the self-trapping regime. Phys. Rev. Lett. **113**, 245002 (2014)
2. Benware, B.R., et al.: Demonstration of a high average power tabletop soft X-ray laser. Phys. Rev. Lett. **81**, 5804–5807 (1998)
3. Gonsalves, A., et al.: Transverse interferometry of a hydrogen-filled capillary discharge waveguide. Phys. Rev. Lett. **98**, 025002 (2007)
4. Gasilov, V.A., et al.: Advances in Parallel Computing, vol. 22, IOS Press (2012)
5. Bobrova, N.A., et al.: Simulations of a hydrogen-filled capillary discharge waveguide. Phys. Rev. E **65**, 016407 (2001)

Chapter 9
Towards Generation of Sub-fs Pulses Using Lasing to Ground States of H-Like LiIII at 13.5 nm and He-Like CV at 4 nm

O. Kocharovskaya, T. R. Akhmedzhanov, V. A. Antonov, A. Morozov, A. Goltsov, M. O. Scully and S. Suckewer

Abstract An efficient method to transform a radiation of X-ray plasma lasers into a train of sub-fs pulses, as well as to form sub-fs pulses directly in an active medium of plasma lasers is suggested. This method is based on temporal and spatial modulation of the parameters of a resonant transition in absorbing or amplifying medium by a moderately strong IR field propagating in the same direction as an X-ray radiation.

9.1 Introduction

Sub-femtosecond (sub-fs) X-ray pulses provide a unique combination of high spatial and temporal resolution. Currently, they are produced only by high harmonic generation (HHG) of laser fields. The generated pulses are typically weak because of the low efficiency of HHG process (especially for given single harmonic). At the same time, modern table-top plasma lasers are able to generate soft X-ray pulses with much higher energy (from μJ up to several mJ), but with rather long picosecond duration.

A "water window", the range of wavelengths between K-absorption edges of oxygen and carbon (2.3–4.4 nm), is of particular interest for applications in biology and medicine due to the potentially high contrast and high-resolution dynamical imaging of proteins in live cells. High gain G ≈ 40/cm in He-like CV at $2^1P–1^1S$ transition of 4.03 nm wavelength was obtained recently at Princeton University via recombination process using very high intensity 800 nm pumping laser pulses and reported at this conference. Even higher gain G ≈ 150/cm is expected to be

O. Kocharovskaya · T. R. Akhmedzhanov · A. Goltsov · M. O. Scully
Texas A&M University, College Station, TX, USA

V. A. Antonov (✉)
Institute of Applied Physics RAS, Nizhny Novgorod, Russia
e-mail: antonov@appl.sci-nnov.ru

A. Morozov · A. Goltsov · M. O. Scully · S. Suckewer
Princeton University, Princeton, NJ, USA

T. Kawachi et al. (eds.), *X-Ray Lasers 2016*, Springer Proceedings in Physics 202, https://doi.org/10.1007/978-3-319-73025-7_9

achieved in H-like CVI at 2P–1S transition of 3.37 nm wavelength [1]. The duration of generated X-ray pulse in both cases is on the order of few hundred femtoseconds. Thus, a highly efficient method for transformation of an output radiation of X-ray plasma laser into the attosecond (as) pulse train is very desirable.

In this work, we study the potential to use recently proposed technique to produce ultrashort pulses from a quasi-monochromatic XUV radiation via the resonant interaction with atoms, dressed by a moderately strong IR laser field [2, 3], for an efficient transformation of the X-ray plasma laser radiation into the train of sub-fs pulses. We study also the possibility to produce sub-fs pulse directly in the active media of X-ray plasma lasers.

9.2 Efficient Transformation of X-Ray Laser Radiation into a Train of Sub-fs Pulses in the IR-Field-Dressed Resonantly Absorbing Medium

Let us consider joint propagation of a z-polarized X-ray and IR fields in x-direction in the plasma of H-like ions. The X-ray field is tuned to resonance with an atomic transition from the ground to the first excited energy level, split due to Stark effect into the sublevels, oscillating in time and space along with oscillations of the instantaneous IR field strength. The resonant scattering of X-ray field on the IR-field-dressed ions results in generation of Raman Stokes and anti-Stokes components [2]. Under the proper choice of intensity of the IR field, the sidebands are produced in-phase with each other and with the incident field. In this case, attenuation of the resonant component of the output X-ray field results in formation of ultrashort pulses. A possibility of an efficient transformation of radiation of H-like LiIII X-ray laser [4] into a train sub-fs pulses is illustrated by Fig. 9.1. Similarly,

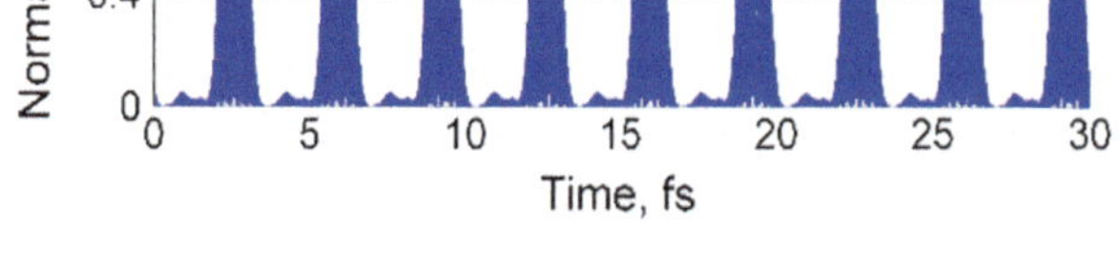

Fig. 9.1 Efficient transformation of the quasi-continues (ps) 13.5 nm radiation of H-like LiIII X-ray laser into a train of 0.9 fs pulses after passing of 1 mm long thermally equilibrium Li^{2+} plasma with ion density 10^{17} cm^{-3} in the presence of IR field with the wavelength 2 μm and intensity 3.6×10^{14} W/cm^2

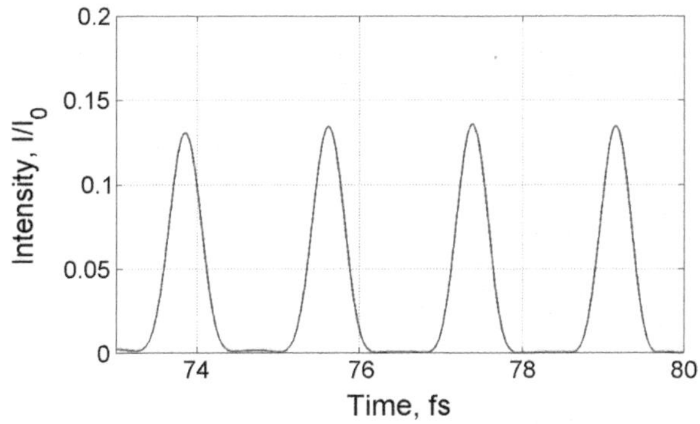

Fig. 9.2 Transformation of the quasi-continues (ps) 4 nm radiation of He-like CV X-ray laser into a train of 450 as pulses after passing 70 μm thermally equilibrium C^{4+} plasma with ion density 10^{18} cm^{-3} in the presence of IR field with the wavelength 1.058 μm and intensity 2×10^{15} W/cm^2

radiation of H-like CVI X-ray laser can be transformed into a train of 360 as pulses using modulating IR laser field with the wavelength 0.8 μm and intensity 9×10^{15} W/cm^2.

In the general case of non-hydrogen-like atoms application of the IR field also leads to generation of Stokes and anti-Stokes components of an incident XUV field [3]. Under the optimal choice of parameters of the IR and X-ray fields, the phases of all spectral components can be matched, resulting in formation of a train of sub-fs pulses. Thus, radiation of He-like CV X-ray laser can be efficiently transformed into a train of attosecond pulses, as illustrated in Fig. 9.2.

9.3 Formation of Amplified Trains of Sub-fs X-Ray Pulses in the IR-Field-Dressed Resonant Inverted Medium

Modulation of the atomic transition resonant to X-ray radiation by an IR field directly in an active medium of a plasma laser allows for a possibility to form strongly amplified sub-fs pulses.

Similarly to the case of a passive medium with optimal choice of parameters of the IR and X-ray fields the phases of all the sidebands will be the same. But in active medium, the sidebands are generated in antiphase to the incident X-ray field. It turns out that this phase difference can be compensated using plasma dispersion with a proper choice of the length of the medium, resulting in production of sub-fs pulses, as it is shown in Fig. 9.3. Application of scaling law for H-like ions [2] leads to conclusion that an amplified train of 380 as pulses of CVI X-ray radiation at 3.38 nm can be produced using an IR field with wavelength 0.8 μm and intensity 9×10^{15} W/cm^2.

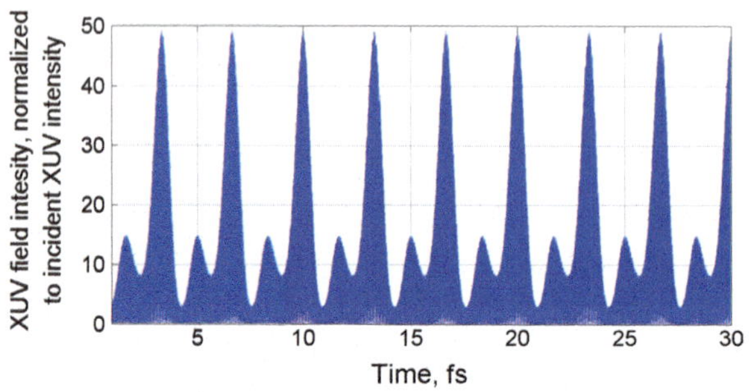

Fig. 9.3 Formation and amplification of 0.9 fs pulses at 13.5 nm carrier wavelength in the active medium of recombination LiIII X-ray laser with ion density 10^{17} cm^{-3} and total length 1 mm in the presence of IR field with the wavelength 2 μm and intensity 3.6×10^{14} W/cm^2

9.4 Conclusions

It is shown that picosecond radiation of the X-ray plasma lasers can be efficiently transformed into a train of sub-fs pulses by passing through the resonantly absorbing medium in the presence of an IR field, modulating the parameters of the resonant transition. Moreover, using an IR field of a proper wavelength and intensity directly in the inverted media of such recombination lasers may lead to formation of strongly amplified attosecond pulses.

Acknowledgements We acknowledge support by NSF under the grants No. PHY-130-73-46 and No. PHY-PHY-1506467. V. A. Antonov acknowledges support by RFBR under the grants No. 16-32-60173, No. 16-02-01034, and No. 16-02-00527, as well as the personal grant from the Grant Council of the President of Russian Federation. We are grateful to Jorge Rocca and Carmen Menoni for the stimulating discussions.

References

1. Luo, Y., Morozov, A., Gordon, D., Sprangle, P., Svidzinsky, A., Xia, H., Scully, M., Suckewer, S.: In: Rocca, J., Menoni, C., Marconi, M. (eds.) Proceedings of the 14th International Conference on X-ray Lasers. Springer Proceedings in Physics, vol. 169, pp. 21–32 (2014)
2. Antonov, V.A., Radeonychev, Y.V., Kocharovskaya, O.: Formation of ultrashort pulses via quantum interference between stark-split atomic transitions in a hydrogenlike medium. Phys. Rev. A **88**, 053849 (2013)
3. Akhmedzhanov, T.R., Antonov, V.A., Kocharovskaya, O.: Formation of ultrashort pulses from quasimonochromatic XUV radiation via infrared-field-controlled forward scattering. Phys. Rev. A **94**, 023821 (2016)
4. Korobkin, D.V., Nam, C.H., Suckewer, S., Goltsov, A.: Demonstration of soft X-ray lasing to ground state in Li III. Phys. Rev. Lett. **77**, 5206–5209 (1996)

Chapter 10
Numerical Calculation of Gain Coefficient for Recombination X-Ray Lasers in a Carbon Cluster Plasma

H. Kawazome and S. Namba

Abstract In order to optimize plasma parameters for a charge exchange recombination scheme X-ray laser, the gain coefficient has been calculated by solving a rate equation model. Fully ionized carbon ions are subject to the charge exchange between the helium atom with a large cross section, resulting in the generation of C^{5+} ions captured into a certain excited level via $C^{6+} + He \rightarrow C^{5+}(n = 3) + He^{+}$. The expected lasing transition in hydrogen-like carbon and its wavelength are C^{5+} $n = 2$–3 and $\lambda = 18.2$ nm, respectively. When the electron temperature ranges from 20 to 50 eV and the electron density is $\sim 10^{19}$ cm^{-3}, the calculations show an increase in the gain coefficient owing to the charge exchange process.

10.1 Introduction

As one of the candidates for high-intensity plasma X-ray lasers, a charge exchange recombination (CXR) is an attractive pumping process for realization of the efficient population inversion, since electrons are captured into selective energy states with a large cross section. Despite several proposals of the CXR X-ray lasers have been made [1–4], the gain coefficient that is large enough for the laser oscillation has not been obtained experimentally. In order to demonstrate the CXR X-ray lasers, the interaction between highly charged ions and neutral atoms is the most critical issue. In this study, we performed a numerical study to optimize plasma parameters for lasing of the CXR X-ray laser in hydrogen-like carbon (C^{5+} $n = 2$–3 transition: $\lambda = 18.2$ nm). The lasing medium is a cluster plasma, where a mixture of helium gas and C_{60} cluster is simultaneously irradiated with intense femtosecond

H. Kawazome (✉)
Department of Information Engineering, National Institute of Technology, Kagawa College, Mitoyo, Kagawa, Japan
e-mail: kawazome@di.kagawa-nct.ac.jp

S. Namba
Graduate School of Engineering, Hiroshima University, Higashihiroshima, Hiroshima, Japan

T. Kawachi et al. (eds.), *X-Ray Lasers 2016*, Springer Proceedings in Physics 202, https://doi.org/10.1007/978-3-319-73025-7_10

laser pulses. For C_{60} clusters, the ionization mechanism generating highly charged ions is electron impact ionization, because the solid-density cluster is efficiently heated by the inverse bremsstrahlung and resonance absorption. While for helium atoms, the ionization mechanism is, so-called, the optical field ionization (OFI). Here, we note that the laser intensity should be smaller than the OFI threshold to the He atom, so as to remain them neutral (threshold intensity: 2.4×10^{14} W/cm^2). Thus, slightly under this intensity, the He atoms do not significantly ionize, whereas the C_{60} clusters do. Therefore, we expect that the cluster plasma is subject to the CXR with the helium atoms, leading to the population inversion.

10.2 Coupled Rate Equations and the Plasma Model

As fundamental atomic processes in plasmas, the electron impact ionization, excitation and de-excitation, spontaneous transition, three-body recombination (TBR), and radiative recombination are considered. The energy levels up to $n = 20$ with l-levels unresolved are taken into account. The population densities of sublevels with the same principal quantum number are distributed according to their statistical weights. The rate coefficients for a given plasma temperature and density are calculated by the COLRAD code [5]. The CXR process (C^{6+} + He $\rightarrow$ $C^{5+}(n = 3)$ + He^+) resulting in the selective population flow into $n = 3$ level is also taken into account, whose cross section is $\sigma \sim 10^{-15}$ cm^{-2} at a low-energy collision [6]. It is assumed that the collision velocity is the same as the thermal velocity of the C^{6+} ion and the plasma is homogeneous.

As an initial plasma condition, the electron density n_e is set equal to six times of the C^{6+} ion density (fully ionized plasmas) and ion temperature T_i is equivalent to the electron T_e. Throughout the calculation, the temperatures do not vary, whereas the ion density changes due to the recombination processes. The He atomic density is assumed to be 10^{17} cm^{-3}. Coupled rate equations are solved by the fourth-order Runge–Kutta method with a time step of 10^{-19} s. Thus, we calculate the time evolution of the population densities of C^{5+} ion. The photoabsorption process is not considered. We also evaluate the gain coefficients, by assuming that the spectral profile attributed to the lasing transition has a Gaussian shape.

10.3 Results and Discussion

Figure 10.1 shows the typical time evolution of (a) the reduced population densities n/g of $n = 1$–5 with and (b) without the CXR process. The electron density and temperature are 10^{19} cm^{-3}, 30 eV, respectively. As clearly seen, the plasma including the CXR process shows that the population density of $n = 3$ level becomes larger than that without the process up to $\sim 10^{-12}$ s.

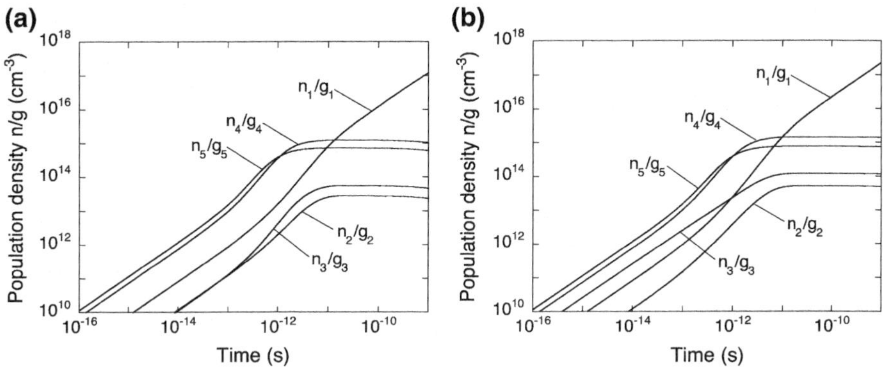

Fig. 10.1 Time evolutions of the population densities **a** without and **b** with the CXR process. The electron density is $n_e = 10^{19}$ cm^{-3} and the plasma temperature is $T_e = T_i = 30$ eV

Figure 10.2 shows the time evolutions of the gain coefficients for the lasing transition of $n = 2$–3 level with the CXR process. The electron densities are 10^{19}, 10^{20} and 10^{21} cm^{-3}. In this case, the TBR process plays an important role to generate the population inversion. Considering that the TBR rate coefficient depends on the temperature and strongly increases for low-temperature plasma, high gains obtained at the lower temperatures are reasonable. For high-plasma density, $>10^{19}$ cm^{-3}, the gain coefficients reach their maximum values around a few picosecond and then roll over due to the depletion of the C^{6+} ions via the TBR process.

On the other hand, Fig. 10.3 shows the temperature dependence of the gain coefficients with and without the CRX process. The electron densities are (a) 10^{19}, (b) 10^{20}, and (c) 10^{21} cm^{-3} and temperatures ranges from 10 to 50 eV. In (a), the increase in the gain coefficients above 20 eV is caused by the CXR process, whereas no distinct difference between them is observed for the densities of $>10^{19}$ cm^{-3}. It should be noted that the population inflows to C^{5+} $n = 3$ level by the CXR process are orders of 10^{27} –10^{29} cm^{-3} s^{-1}, while these values are smaller than those of the TBR process for the electron densities of 10^{20} and 10^{21} cm^{-3}. As a result, the increase in the gain coefficient due to the CXR process is only observed around the density of 10^{19} cm^{-3}. Also, note that too high temperature is not suitable for the presence of He atoms (ionization potential: 24.6 eV), so that lower temperature is preferable for the high-gain coefficient.

10.4 Summary

In order to optimize plasma parameters for the realization of charge exchange recombination scheme X-ray laser, we performed a numerical calculation of the gain coefficient in hydrogen-like carbon and He atom mixture plasmas.

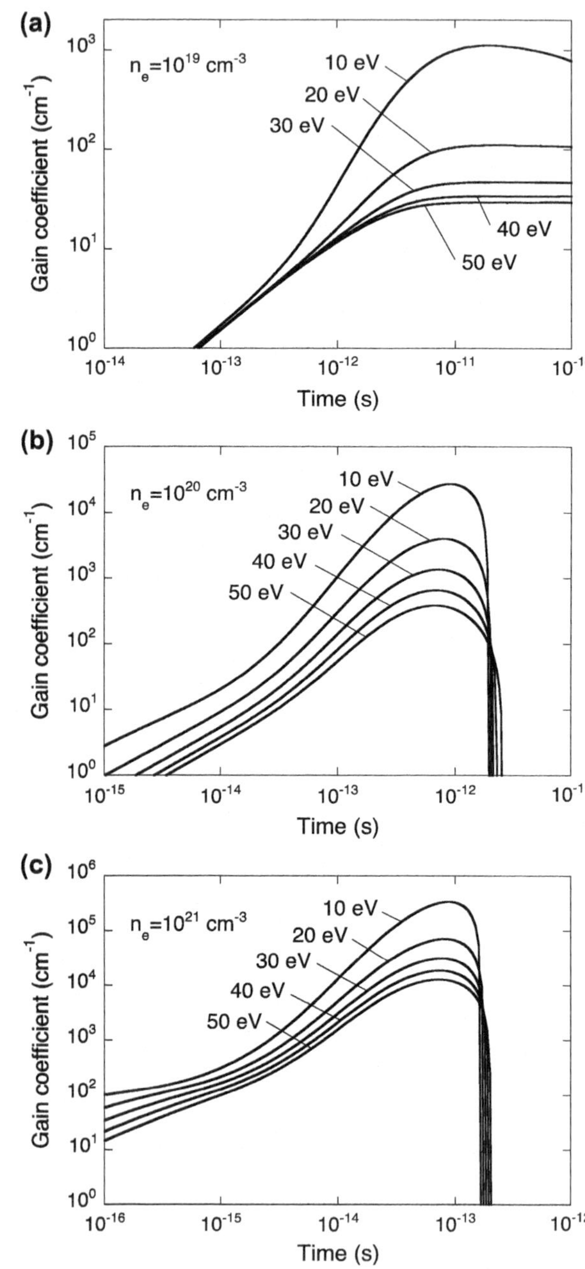

Fig. 10.2 Dependences of the gain coefficient on the temperature (with the CXR process). The electron densities are **a** 10^{19}, **b** 10^{20}, and **c** 10^{21} cm^{-3}

Considering that the temporal duration of population inversion necessary for the stable lasing is a few ps, a suitable condition for the high-gain medium is the electron density of $\sim 10^{19}$ cm^{-3} and temperature of less than 20 eV, where the gain coefficient reaches 10^3 cm^{-1}.

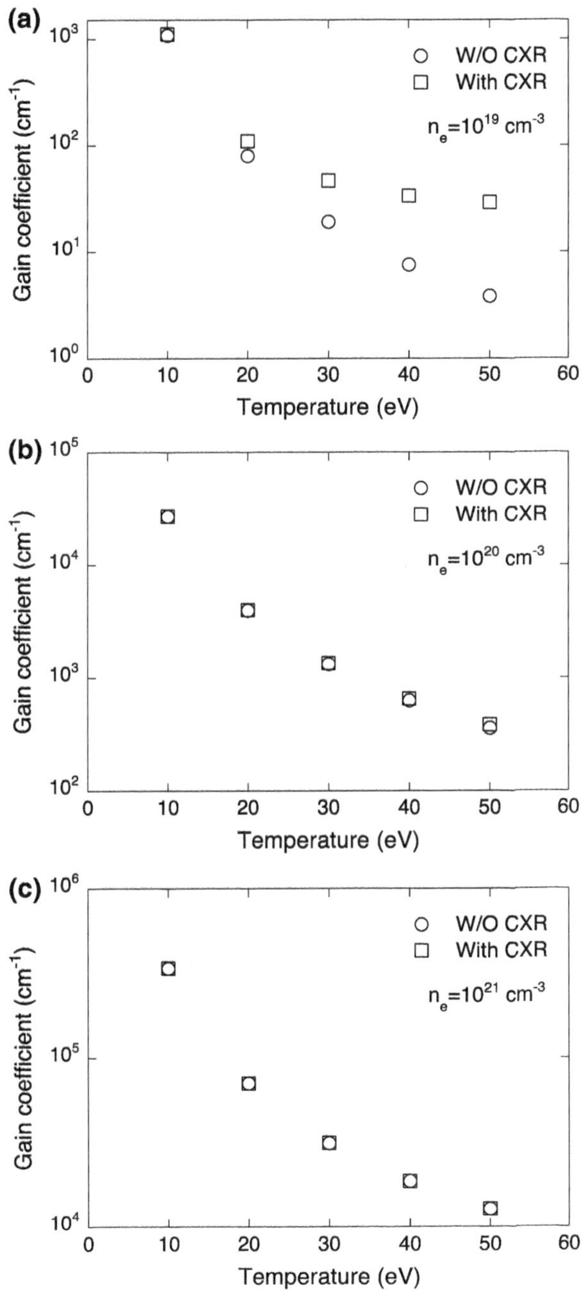

Fig. 10.3 Temperature dependence of the gain coefficient with and without the CXR process. The electron densities are **a** 10^{19}, **b** 10^{20}, and **c** 10^{21} cm^{-3}

References

1. Chichkov, B.N., Egbert, A., et al.: Soft-X-ray lasers with charge transfer pumping in a mixture of clusters and atomic gases. Jpn. J. Appl. Phys. **38**, 1975–1978 (1999)
2. Dixon, R.H., Elton, R.C.: Resonance charge transfer and population inversion following C^{5+} and C^{6+} interactions with carbon atoms in a laser-generated plasma. Phys. Rev. Lett. **38**, 1072 (1977)
3. Fuhl, F., Aschke, L., Kunze, H.-J.: Selective population of the $n = 3$ level of hydrogen-like carbon in two colliding laser-produced plasmas. Phys. Lett. A **225**, 107 (1997)
4. Namba, S., et al.: IOP conference proceedings. In: Zhang, J. (ed.) X-ray Lasers, vol. 481 (2004)
5. Ljepojevic, N.N., Hutcheon, R.J., Payne, J.: COLRAD—a program to calculate population densities of the excited atomic levels of hydrogen-like ions in a plasma. Comput. Phys. Commun. **44**, 157–176 (1987)
6. Kimura, M., Olson, R.E.: Electron capture to (nl) states in collisions of C^{4+} and C^{6+} with He. J. Phys. B **17**, L713 (1984)

Part II
Higher Harmonics and FEL Based X-Ray Lasers

Chapter 11
Investigations on Ultrafast Atomic and Molecular Dynamics with Harmonic Sources

Chang Hee Nam, Hyeok Yun, Kyung Taec Kim and Hyung Taek Kim

Abstract High harmonic light sources with attosecond duration in the EUV and soft X-ray spectral range can be applied to probe ultrafast dynamics of atoms and molecules. By tuning harmonic wavelength, He atoms can be excited to the $1s3p$ state that can be further ionized by a time-delayed infrared femtosecond laser pulse. The interference, between the directly photoionized electron and that from the excited state, was used to reconstruct the amplitude and phase of electron wavepacket from the $1s3p$ state, probing ultrafast photoionization dynamics of He. In addition, molecular high harmonic radiation, generated from multiple molecular orbitals, is investigated for molecular structural information. In order to resolve harmonic radiation from multiple orbitals of CO_2 molecules, two-dimensional high harmonic spectroscopy, using an orthogonally polarized two-color laser field, was employed. In this scheme odd and even harmonics carried the characteristics of the highest occupied molecular orbital (HOMO) and HOMO-1, respectively. With these versatile tools, high harmonic light sources will open a new route to explore ultrafast dynamics of atoms and molecules.

C. H. Nam (✉) · H. Yun · K. T. Kim · H. T. Kim
Center for Relativistic Laser Science, Institute for Basic Science, Gwangju 61005, Korea
e-mail: chnam@gist.ac.kr

C. H. Nam · K. T. Kim
Department of Physics and Photon Science, Gwangju Institute of Science and Technology, Gwangju 61005, Korea

H. T. Kim
Advanced Photonics Research Institute, Gwangju Institute of Science and Technology, Gwangju 61005, Korea

T. Kawachi et al. (eds.), *X-Ray Lasers 2016*, Springer Proceedings in Physics 202, https://doi.org/10.1007/978-3-319-73025-7_11

11.1 Introduction

High harmonic radiation emitted from gaseous atoms or molecules is a coherent light source with attosecond duration. High harmonics can be applied to probe ultrafast dynamics and electronic structures of atoms and molecules [1, 2]. Photoelectron spectra resulted by combining high harmonics and femtosecond infrared (IR) laser pulses have revealed inaccessible ultrafast phenomena in atoms and molecules [3–6]. In addition, molecular harmonic radiation has been used to probe molecular orbital structures with the help of the molecular alignment technique [7–9].

The ultrafast dynamics of photoionization process in He atoms can be fully described by measuring the amplitude and phase of electron wave packets (EWPs) in the continuum [10]. In the interferometric pump-probe technique [11], high harmonics can directly ionize He atoms (path 1), while simultaneously exciting He atoms to a certain state from the ground state. The EWP, directly ionized from the ground state, can be reconstructed by applying the complete reconstruction of attosecond burst (CRAB) method to the photoelectron spectra [12, 13]. Femtosecond IR laser pulses in the pump-probe technique ionize the excited He atom by adding IR photons (path 2). The EWPs from these two different paths interfere each other. The interference is used to reconstruct EWP from the path 2 and the reconstructed EWPs describe the excitation and ionization dynamics of He atoms.

Molecular high harmonic radiation is useful also for probing structural information of harmonic radiating molecules. Multi-orbital contributions to molecular high harmonic generation (HHG) can be resolved and separately probed by two-dimensional high harmonic spectroscopy [14]. The method is based on HHG in an orthogonally polarized two-color laser field, consisting of the fundamental (1ω) and its second harmonic (2ω). In the two-color laser field, odd and even harmonics are polarized along the polarization of the fundamental and the second harmonic field, respectively [15]. In addition, multiple orbitals possessing different structures induce strong dipole moments for harmonic radiation along different directions. By using two properties of the two-color driving field and structural characteristics of molecular orbitals, the characteristics attributed to the multiple orbitals is separately imprinted in odd and even harmonics. The two-dimensional harmonic spectroscopy of molecules, thus, offers the opportunity to simultaneously probe multiple molecular orbitals.

11.2 Photoionization Dynamics of Helium

Ionization of atoms or molecules is one of the most fundamental processes in physics. An intense laser field or a short wavelength light field can ionize an atom and this photoionization process usually occurs on a timescale shorter than one optical cycle of the light field. The development of high harmonic light sources with attosecond duration has enabled the experimental realization of ultrafast photoionization dynamics [4].

The ultrafast photoionization dynamics can be fully described by finding the amplitude and phase information of EWPs in the continuum. In order to find such information, an interferometric pump-probe technique is employed [11]. A schematic diagram of the technique, applied to He atom, is shown in Fig. 11.1a. A high harmonic pulse, as a pump, initiates a photoionization process, generating a reference EWP in the continuum, and an excitation process from the ground state to the $1s3p$ state. A subsequent laser pulse, as a probe, produces the signal EWP from the excited state. The EWPs from two different paths interfere with each other and generate an interference in the photoelectron spectra measured by scanning the time delay between the harmonic and the femtosecond laser pulses, as shown in Fig. 11.1b.

In order to reconstruct two EWPs, the reference EWP should be retrieved first. By using the strong field approximation model [12, 16], the transition amplitude and the temporal phase modulation of a free electron from the ground state to the continuum can be determined, and the amplitude and phase of EWP for each time delay τ can be completely characterized by applying the CRAB technique to the photoelectron spectra [12, 16, 17]. From the reference EWP completely characterized with the phase information, the signal EWP can be obtained by solving the interference equation. The reconstructed amplitude and phase are shown in Fig. 11.1b, c.

The reconstructed EWPs contain the information on the dynamics of electrons ionized through different pathways. The quadratic phase of the reference EWP, indicated by the thick solid line in the energy axis of the upper panel of Fig. 11.1c,

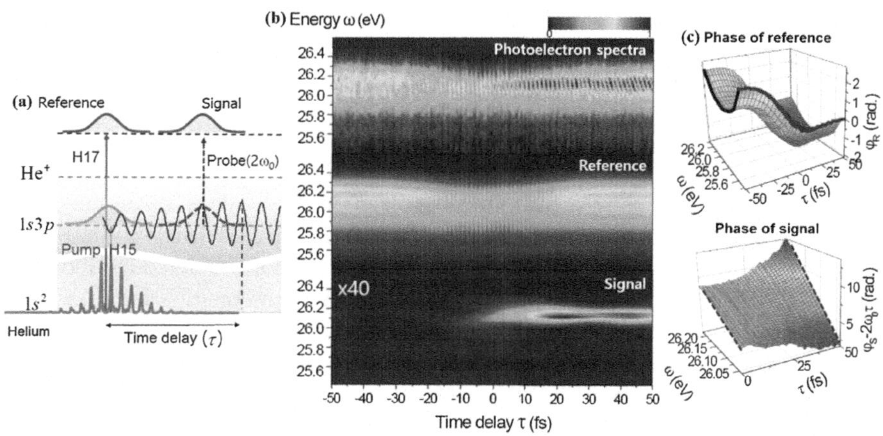

Fig. 11.1 **a** Schematic diagram of the excitation and the ionization processes of He by an attosecond high harmonic pulse superposed with a time-delayed IR femtosecond laser pulse. **b** Photoelectron spectra obtained in the experiment (upper panel), reference EWP (middle panel), and signal EWP (bottom panel). **c** Phase of the reference EWP (upper panel) and phase of the signal EWP (bottom panel)

shows the phase inherited from the positively chirped 17th harmonic. The slow change of the reference phase with respect to the time delay, indicated by the thick solid line in the time delay axis of the upper panel of Fig. 11.1c, corresponds to the ponderomotive shift of the laser electric field. The reconstructed phase of the signal EWP provides the timing information. The slope of the phase, shown in the lower panel of Fig. 11.1c, is the group delay of the signal wave packet, from which the time of ionization may be obtained. A more rigorous analysis with reconstructed amplitude and phase allows the detailed ionization dynamics of He atoms [10].

11.3 High Harmonic Generation from Multiple Molecular Orbitals

HHG from linear molecules has been intensively studied after the field-free molecular alignment technique was developed [18]. By aligning and controlling the molecular axis, the anisotropic characteristics of a molecule could be examined and, thus, enormous studies related to the ultrafast dynamics and electronic structures of molecules have been performed [6–9].

The multi-orbital contribution to molecular HHG is related to the molecular orbital structure. The highest occupied molecular orbital (HOMO) is mostly ionized in a strong laser field and dominantly reveals its characteristics in HHG. Energetically low-lying molecular orbitals referred to HOMO-1 and HOMO-2, however, can significantly contribute to the radiation when the orbital is aligned to induce strong dipole moment [8, 9, 19]. In this section, we present HHG from multiple orbitals of N_2 and CO_2 and a novel method to resolve multi-orbital contributions to separate harmonics from different orbitals.

11.3.1 HHG from Multiple Orbitals in N_2 Molecules

Harmonic intensity generated from linear molecules greatly depends on the degree of alignment. In the case of N_2 molecules, the harmonic intensity in the plateau increases as the degree of alignment increases along the polarization direction of the driving field for HHG. This is because a strong dipole moment is induced in the HOMO of N_2, as N_2 is aligned. The strong dipole moment is based on the orbital structure. As shown in Fig. 11.2a, the HOMO of N_2 has σ_g symmetry and relatively long electron distribution along the molecular axis. In this structure, the strong dipole moment is induced when the molecular axis is aligned along the polarization of the driving field.

The experimental result in Fig. 11.2c shows two different aspects with respect to the degree of alignment. In the plateau from the 21st to the 37th harmonics, harmonic intensities are maximized at the highest degree of alignment and this

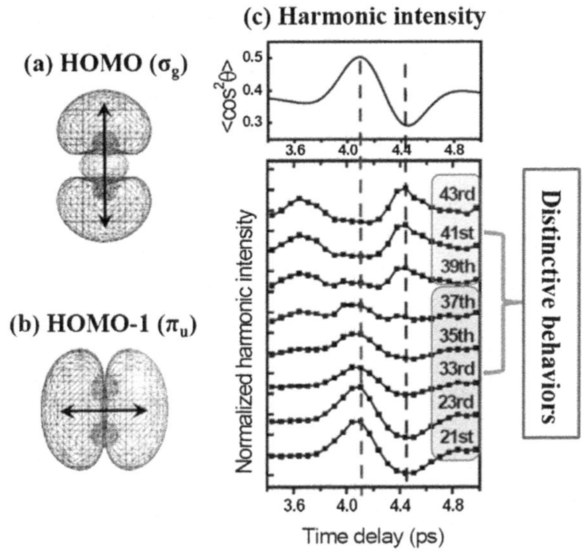

Fig. 11.2 **a** HOMO and **b** HOMO-1 of N_2. The black arrow in each orbital indicates a favorable direction for inducing strong dipole moment. **c** Calculated degree of alignment ‹$\cos^2\theta$› (upper panel), high harmonic intensities obtained from aligned N_2 molecules with respect to the time delay between aligning and harmonic generating pulses. The harmonic intensity was normalized with respect to that obtained without the aligning pulse

behavior is corresponded as discussed above. In the cutoff from 39th to 43rd harmonics, however, harmonics are maximized at the lowest degree of alignment, in other words, at the time when N_2 is perpendicularly aligned to the driving field.

The harmonic behavior in the cutoff can be understood by considering the contribution from the HOMO-1. The HOMO-1 of N_2 has π_u symmetry, as shown in Fig. 11.2b. The HOMO-1 does not have dipole moment along the molecular axis because two electron distributions around the molecular axis have different signs and cancel each contribution. In the perpendicular direction, however, the HOMO-1 can have much stronger dipole moment than the HOMO and enhance harmonic intensity in the cutoff. The higher ionization potential of an electron in the HOMO-1 than that in the HOMO gives rise to the higher order cutoff harmonics contributed only from the HOMO-1. As a consequence, the weak harmonics from the HOMO-1 can be distinguished from those from the HOMO.

11.3.2 Two-Dimensional High Harmonic Spectroscopy of CO_2 Molecules

Two-dimensional high harmonic spectroscopy (HHS) provides the orbital information of molecules on a two-dimensional space via high harmonic generation (HHG). Simultaneous two-dimensional probing can be possible by applying an orthogonally polarized two-color laser field consisting of the fundamental (1ω) and its second harmonic (2ω) [15]. When a molecule is aligned in the lab frame, for example, with two-dimensional HHS one can simultaneously extract both information along the molecular axis and along its perpendicular direction.

Two-dimensional HHS can be exploited to resolve harmonic radiation from multiple molecular orbitals by taking into account the structural characteristics of molecular orbitals. Figure 11.3b shows the HOMO(π_g) and HOMO-1(π_u) of CO_2 molecule. The black arrows in the orbitals represent the relative strength of dipole moment induced during recombination in a two-color laser field, as obtained by considering the orbital symmetry. The HOMO and the HOMO-1 have a larger dipole moment along the x- and along the y-axis, respectively, than along the other direction. Since two-dimensional HHS can probe along two orthogonal directions, the HOMO in the x-axis and the HOMO-1 in the y-axis can be projected to odd and even harmonics, respectively.

The experimental result in Fig. 11.3d shows a distinctive behavior between odd and even harmonics. The intensity of odd harmonics (25th and 27th harmonics) is inversely proportional to the degree of alignment, whereas the evolution of even harmonics (24th and 26th harmonics) follows the degree of alignment. At the time of alignment, the decrease in odd harmonic intensity can be understood by considering the destructive two-center interference occurring in the recombination of the HOMO [20]. The increased even harmonic intensity reveals the strong HOMO-1 contribution due to the induced strong dipole moment along the y-axis. The distinctive behavior in odd and even harmonics shows that the HOMO and the HOMO-1 can be separately characterized in the scheme of two-dimensional HHS.

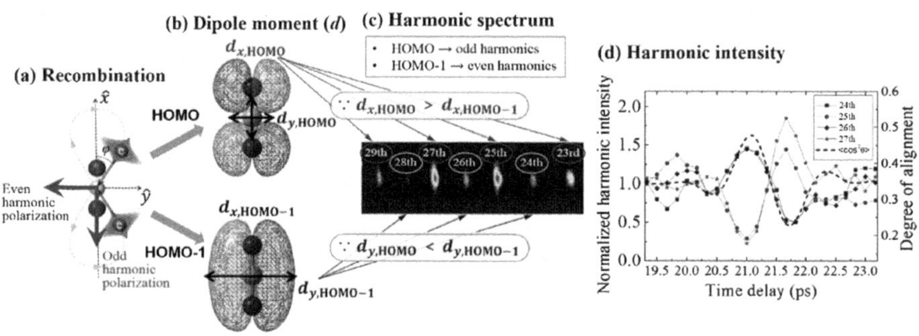

Fig. 11.3 Pictorial concept and experimental result of two-dimensional high harmonic spectroscopy for resolving high harmonics from multiple orbitals. **a** Recombination of aligned molecule in an orthogonally polarized two-color laser field. Note that odd and even harmonics are polarized x- and y-axes, respectively **b** Dipole moments induced in the HOMO and HOMO-1 of CO_2 during recombination. Black arrows in the orbitals represent relative strength of induced dipole moment. **c** Harmonic spectrum generated from CO_2 in the two-color laser field. **d** Harmonic intensities generated from CO_2 with respect the time delay between aligning and harmonic generating pulses. Right axis: degree of alignment, $\langle\cos^2\theta\rangle$ (calculation, dashed line). Left axis: normalized harmonic intensity (experiment, line and symbol)

11.4 Conclusion

High harmonic radiation is a unique light source for investigating ultrafast dynamics and electronic structures of atoms and molecules with a timescale up to attosecond. We have shown the photoionization dynamics of He atom by reconstructing the EWPs from the $1s3p$ state, and presented the high harmonics generated from multiple orbitals of molecules. EWPs in the continuum, including the phase information, were successfully reconstructed for He. In the case of CO_2 molecules, two-dimensional HHS has been investigated by applying a two-color laser field consisting of the fundamental and second harmonic field. The characteristics of the HOMO and HOMO-1 were separately imprinted in odd and even harmonics, respectively, offering a tool to investigate ultrafast molecular dynamics involving multiple orbitals. These techniques can be conveniently used for understanding attosecond atomic and molecular dynamics including multiple excitation and ionization, and chemical reactions.

References

1. Corkum, P.B., Krausz, F.: Attosecond science. Nat. Phys. **3**, 381 (2007)
2. Krausz, F., Ivanov, M.: Attosecond physics. Rev. Mod. Phys. **81**, 163 (2009)
3. Drescher, M., et al.: Time-resolved atomic inner-shell spectroscopy. Nature **419**, 803 (2002)
4. Uiberacker, M., et al.: Attosecond real-time observation of electron tunneling in atoms. Nature **446**, 627 (2007)
5. Schultze, M., et al.: Delay in photoemission. Science **328**(5986), 1658 (2010)
6. Calegari, F., et al.: Ultrafast electron dynamics in phenylalanine initiated by attosecond pulses. Science **346**(6207), 336 (2014)
7. Itatani, J., et al.: Tomographic imaging of molecular orbitals. Nature **432**, 867 (2004)
8. McFarland, B.K., Farrell, J.P., Bucksbaum, P.H., Gühr, M.: High harmonic generation from multiple orbitals in N_2. Science **322**, 1232 (2008)
9. Smirnova, O., et al.: High harmonic interferometry of multi-electron dynamics in molecules. Nature **460**, 972 (2009)
10. Kim, K.T., et al.: Amplitude and phase reconstruction of electron wave packets for probing ultrafast photoionization dynamics. Phys. Rev. Lett. **108**, 093001 (2012)
11. Mauritsson, J., et al.: Attosecond electron spectroscopy using a novel interferometric pump-probe technique. Phys. Rev. Lett. **105**, 053001 (2010)
12. Mairesse, Y., Quere, F.: Frequency-resolved optical gating for complete reconstruction of attosecond bursts. Phys. Rev. A **71**, 011401(R) (2005)
13. Yakovlev, V.S., Gagnon, J., Karpowicz, N., Krausz, F.: Attosecond streaking enables the measurement of quantum phase. Phys. Rev. Lett. **105**, 073001 (2010)
14. Yun, H., et al.: Resolving multiple molecular orbitals using two-dimensional high-harmonic spectroscopy. Phys. Rev. Lett. **114**, 153901 (2015)
15. Shafir, D., et al.: Atomic wavefunctions probed through strong-field light-matter interaction. Nat. Phys. **5**, 412 (2009)
16. Kitzler, M., et al.: Quantum theory of attosecond XUV pulse measurement by laser dressed photoionization. Phys. Rev. Lett. **88**, 173904 (2002)
17. Kim, K.T., et al.: Complete temporal reconstruction of attosecond high-harmonic pulse trains. New J. Phys. **12**, 083019 (2010)

18. Stapelfeldt, H., Seideman, T.: Colloquium: aligning molecules with strong laser pulses. Rev. Mod. Phys. **75**, 543 (2003)
19. Lee, G.H., et al.: Alignment dependence of high harmonics contributed from HOMO and HOMO-1 orbitals of N_2 molecules. J. Phys. B Atom. Mol. Opt. Phys. **43**, 205602 (2010)
20. Kanai, T., Minemoto, S., Sakai, H.: Quantum interference during high-order harmonic generation from aligned molecules. Nature **435**, 470 (2005)

Chapter 12
Characterization of Partially Coherent Ultrashort XUV Pulses

Charles Bourassin-Bouchet

Abstract Ultrafast metrology often considers that the measured light pulse is fully coherent, or equivalently that its properties are stable enough to enable its description by a given time-dependent electric field. However, these conditions are never entirely satisfied in practice, especially in the domain of ultrashort extreme-ultraviolet (XUV) sources. To sort out this issue, we have adapted Frequency-Resolved Optical Gating (FROG), the well-known technique for pulse characterization, to enable the measurement of partially coherent XUV pulses in the femtosecond and attosecond range. Especially, this technique allows one to overcome the sources of decoherence that normally prevent pulse measurement, such as the spectrometer resolution or the presence of XUV/laser arrival time jitter.

12.1 Introduction

Temporal metrology, that is the process of measuring the temporal structure of ultrashort pulses of light, is crucial for emerging ultrashort extreme ultraviolet and X-ray (XUV) sources, such as high-harmonic based attosecond (1 as $= 10^{-18}$ s) sources [1, 2] or free electron lasers (FEL) [3, 4]. In attosecond physics, the flagship approach for temporal metrology is known as the FROG-CRAB technique (Frequency-Resolved Optical Gating for Complete Reconstruction of Attosecond Bursts) [5], adapted from the FROG technique used in conventional near-visible ultrafast laser metrology [6]. By sending the XUV pulse through a gas jet in the presence of an infrared (IR) laser field, two-color XUV+IR photoionisation is induced. Measuring the spectrum of the produced photoelectrons while varying the IR/XUV delay gives a two-dimensional trace called a spectrogram. Such a spectrogram can then be processed with a phase retrieval algorithm in order to obtain the temporal profile of the XUV pulse.

C. Bourassin-Bouchet (✉)
Institut des Sciences Moléculaires d'Orsay, UMR 8214, CNRS, Université Paris-Sud, 91405 Orsay Cedex, France
e-mail: charles.bourassin-bouchet@institutoptique.fr

Synchrotron SOLEIL, Université Paris-Saclay, BP 34, 91192 Saint-Aubin, Gif-sur-Yvette, France

T. Kawachi et al. (eds.), *X-Ray Lasers 2016*, Springer Proceedings in Physics 202, https://doi.org/10.1007/978-3-319-73025-7_12

One fundamental condition for this measurement to succeed is the absolute stability of the pulse during the process [7]. Equivalently, the pulse must be fully coherent, i.e., any phenomenon that can make its physical properties inhomogeneous (e.g., shot-to-shot fluctuations, spatio-temporal distortions, etc.) must be avoided, a condition that is very difficult to satisfy. In order to overcome this problem, we propose a new technique called Mixed FROG [8] which enables the characterization of ultrashort pulses in the presence of decoherence.

12.2 Metrology of Partially Coherent Ultrashort Pulses

The measurement of an ultrashort pulse can be seen as a statistical process where an ensemble of pulses takes part to the process [7]. Given the chosen detection apparatus, the signal will be resolved in some degrees of freedom (e.g., in wavelength or photon energy for a spectrometer), and will be averaged out in other hidden degrees of freedom (e.g., the shot-to-shot fluctuations of the pulse if the detector cannot perform single-shot measurements). In regular ultrafast metrology, it is assumed that all the detected pulses are identical in the statistical ensemble, that is that the ensemble is fully coherent. In these conditions, it is relevant to describe a light pulse by a well-determined electric field. This is the underlying hypothesis of all femto/attosecond pulse characterization techniques, such as FROG, FROG-CRAB, or SPIDER (Spectral Phase Interferometry for Direct E-field Reconstruction) [9].

When this assumption is no longer fulfilled as, for example, when a shot-averaged measurement is performed on a source fluctuating from shot-to-shot, the retrieved electric field becomes meaningless as it cannot give a trustworthy image of the pulse ensemble. The pulse obtained in this case is known as the coherent artifact [10].

In these conditions, one needs to describe the ensemble of pulses with the help of statistical quantities such as the two-time correlation function

$$C\left(t, t'\right) = \left\langle E\left(t\right) E^{*}\left(t'\right)\right\rangle, \tag{12.1}$$

where $E(t)$ stands for the time-dependent electric field of the waveforms contained in the ensemble and where $\langle\rangle$ denotes the ensemble averaging. Characterizing a partially coherent ultrashort pulse thus requires measuring $C\left(t, t'\right)$.

12.3 Principle of Mixed FROG

In FROG, one measures the spectrum of the cross-correlation signal between two pulses referred to as the pulse $P(t)$ and gate $G(t)$. In some variants of FROG, the gate can be related to the pulse (e.g., Second Harmonic Generation or Polarization Gating FROG), or it can be completely independent (e.g., XFROG or Blind FROG) [6]. By sending the two pulses into a suitable nonlinear medium, one obtains a signal proportional to the product $P(t) \cdot G(t - \tau)$ where τ is the variable delay between the two pulses. The spectrum of this signal is then measured with a regular optical spectrometer, which provides a spectrogram $S(\omega, \tau)$ representing the evolution of this spectrum with respect to τ

$$S(\omega, \tau) = |\mathscr{F}[P(t) G(t - \tau)]|^2 \tag{12.2}$$

Here $\mathscr{F}$ represents the Fourier transform and ω is the angular frequency. A numerical algorithm is then used to reconstruct the pulse and gate from this spectrogram. In the FROG-CRAB variant, the pulse corresponds to an ultrashort XUV pulse [5] and the gate to an infrared laser pulse which are both focused into a gas jet. In these conditions, an electron wave packet is freed by the XUV pulse, and its kinetic energy is then modulated by the laser pulse. It has been shown that recording the kinetic energy spectrum of the released photoelectrons with an electron spectrometer with respect to the laser/XUV delay τ provides a signal that reads as a spectrogram. Unless otherwise mentioned we will now focus on the FROG-CRAB technique and will leave aside the other variants of FROG.

In the presence of partial coherence, the measured spectrum $\langle S \rangle$ corresponds to the incoherent average of the spectra of all the pulses contained in the statistical ensemble. Without loss of generality, the ensemble averaging $\langle\rangle$ is rewritten as the sum over a discrete number of spectra $\sum_k S_k$, so that the partially coherent spectrogram reads

$$\langle S \rangle \propto \sum_k S_k(\omega, \tau) \propto \sum_k \left| \mathscr{F}\left[P_k(t) G(t - \tau)\right] \right|^2 . \tag{12.3}$$

whereas the conventional FROG aims at recovering a given pulse electric field $P(t)$ from the spectrogram, measuring partially coherent pulses with FROG implies the extraction of the two-time correlation function $\propto \sum_k P_k(t) P_k^*(t')$ describing the statistical ensemble of pulses from the averaged spectrogram $\langle S \rangle$. We have christened this new technique Mixed FROG. Note that in the above equation the k indexing does not apply to $G(t)$. In other words, the gate is assumed to remain fully coherent. In practice this means that the infrared pulse must not experience shot-to-shot fluctuations, space-time coupling or any phenomenon that would alter its degree of coherence. This hypothesis ensures that one will be able to successfully separate numerically the contributions of the pulse from that of the gate in the spectrogram. For that reason, it is probable that the transposition of Mixed FROG in the near-visible range will be restricted to the versions of FROG relying on a gate independent of the pulse (e.g., XFROG, Blind FROG).

In practice, we have modified one of the most widespread FROG algorithms, namely the PCGPA (Principal Components Generalized Projections Algorithm) to make it able to synthesize incoherent sums of spectrograms $S_k(\omega, \tau)$. This new capability is obtained by reconstructing in parallel several pulses that are characteristic of the pulse ensemble. A detailed description of the algorithm is available in the supplementary information of [8].

12.4 A Numerical Example

We now illustrate the capabilities of Mixed FROG with a numerical example where decoherence of the ensemble comes from the presence of arrival time jitter between the IR and XUV pulses.

In the simulation depicted in Fig. 12.1, an XUV pulse with a duration of 270 as full width at half maximum (FWHM) is combined with a laser pulse (5 fs FWHM) centered at 800 nm with an intensity at focus of 8 TW/cm^2. In the first case in

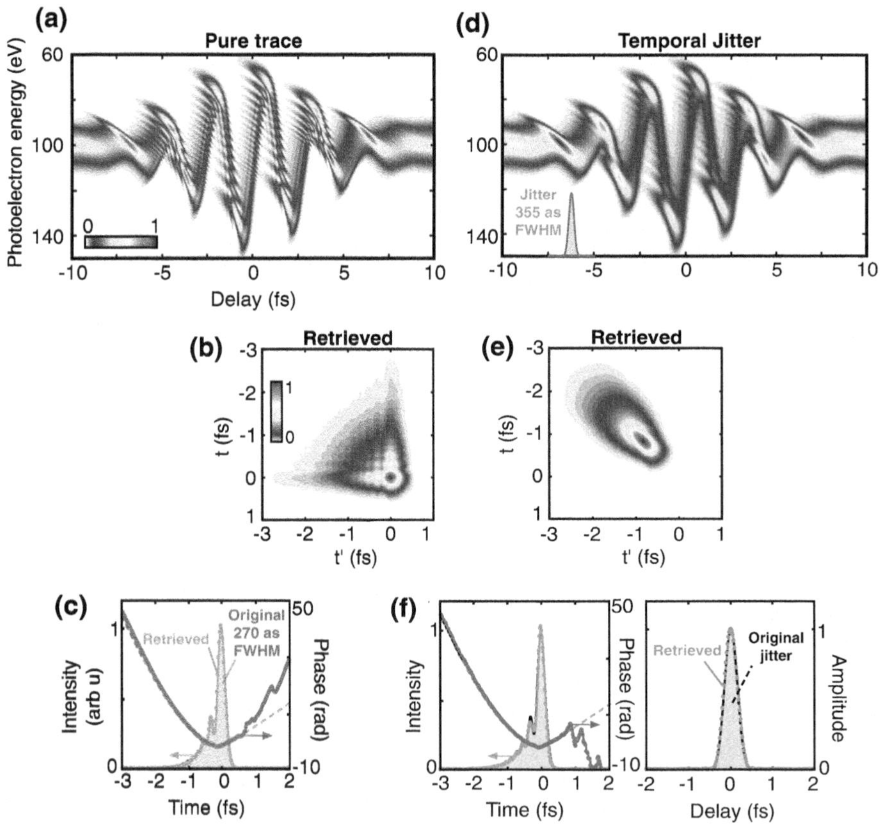

Fig. 12.1 Examples of Mixed FROG reconstructions. In the first case (left column), the XUV pulse is fully coherent. The corresponding spectrogram is shown (**a**) along with the modulus of the correlation function $\left|C\left(t,t'\right)\right|$ (**b**) retrieved with Mixed FROG. Being in the fully coherent case, the pulse amplitude and phase can be easily extracted from C (**c**). In the second case (right column), a laser/XUV arrival time jitter is accounted for. The spectrogram (**d**) and the reconstructed correlation function (**e**) exhibit clear differences with the fully coherent case. (**f**) Fitting C with a suitable decoherence model enables the retrieval of the XUV pulse (left panel) along with the jitter envelope (right panel)

Fig. 12.1a–c, no jitter is present and the obtained spectrogram shows a detailed structure. Processing the latter spectrogram with the Mixed FROG algorithm, the two-time correlation function depicted in Fig. 12.1b is retrieved. Note that $C(t, t')$ is a complex quantity but we only show its modulus in Fig. 12.1 for clarity. The retrieved correlation function exhibits a clear symmetry along the diagonal (the $t = t'$ axis) which indicates the full coherence of the pulse ensemble. Indeed in the present case, no arrival time jitter is present. By isolating a line out of $C(t, t')$, one can extract the amplitude and phase of the XUV pulse, which are properly retrieved, see Fig. 12.1c.

In Fig. 12.1d, we take into account the influence of random arrival time fluctuations (jitter) of the XUV pulse with respect to the laser pulse. The jitter envelope is gaussian with a FWHM of 355 as. This parasitic phenomenon erases the fine structure in the spectrogram, and notably affects the retrieved correlation function, see Fig. 12.1e. Although apparently not related to the jitter-free correlation function in Fig. 12.1b, the retrieved correlation function still contains enough information about the original XUV pulse to enable its reconstruction. To do so, we perform a least squares fit of $C(t, t')$ with a model of mixture [8]. In this model, we consider that the pulse ensemble is composed of identical waveforms $P(t)$ (the XUV pulse to retrieve) randomly delayed in the jitter envelope $J(t)$. Despite its simplicity, this assumption strongly constrains the shape of $C(t, t')$. Fitting the correlation function with this model enables the retrieval of both the XUV pulse $P(t)$ and the jitter envelope $J(t)$, see Fig. 12.1f.

By its ability to dissociate coherent and incoherent contributions in the spectrogram, Mixed FROG enables the XUV pulse reconstruction even in the presence of arrival time jitter. The limited resolution of the electron spectrometer is another common limiting factor in FROG-CRAB measurements. As in the previous example, a simple modeling of the mixture can enable the simultaneous reconstruction of the spectrometer response and the XUV pulse. Many more examples are described in [8] and illustrate the possibilities opened by Mixed FROG.

12.5 Experimental Demonstration

An experimental demonstration would constitute the last step to validate the Mixed FROG technique. In that scope, a measurement campaign has been performed on a source of attosecond pulse trains in the Lasers, Interactions and Dynamics Laboratory (LIDyL) at CEA Saclay, France. The obtained results confirm the potential of the technique and will be thoroughly described in a forthcoming publication.

12.6 Conclusions

Ultrafast metrology has been restricted so far to the characterization of fully coherent pulses, which has been very limiting in numerous applications. In this work, we have shown that in case of partial coherence the relevant quantity becomes the two-time correlation function, and that it is possible to reconstruct such a quantity from a regular FROG spectrogram through a minimal modification of the processing algorithm. We have named Mixed FROG this novel variant of FROG, and we have illustrated the possibilities opened by this new approach. Especially, the scheme proposed to suppress the influence of arrival time jitter or of the spectrometer response should dramatically relax the synchronization and spectral resolution requirements in XUV pulse metrology.

Acknowledgements We acknowledge V. Gruson, L. Barreau, F. Quéré, P. Salières, T. Ruchon and A. Klisnick for their continuous support to this work.

References

1. Paul, P.M., et al.: Observation of a train of attosecond pulses from high harmonic generation. Science **292**, 1689–1692 (2001)
2. Hentschel, M., et al.: Attosecond metrology. Nature **414**, 509–513 (2001)
3. Emma, P., et al.: First lasing and operation of an ångstrom-wavelength free-electron laser. Nat. Photonics **4**, 641–647 (2010)
4. Allaria, E., et al.: Highly coherent and stable pulses from the FERMI seeded free-electron laser in the extreme ultraviolet. Nat. Photonics **6**, 699–704 (2012)
5. Mairesse, Y., Quéré, F.: Frequency-resolved optical gating for complete reconstruction of attosecond bursts. Phys. Rev. A **71**, 011401 (2005)
6. Trebino, R., et al.: Measuring ultrashort laser pulses in the time-frequency domain using frequency-resolved optical gating. Rev. Sci. Instrum. **68**, 3277–3295 (1997)
7. Walmsley, I.A., Dorrer, C.: Characterization of ultrashort electromagnetic pulses. Adv. Opt. Photonics **1**, 308–437 (2009)
8. Bourassin-Bouchet, C., Couprie, M.E.: Partially coherent ultrafast spectrography. Nat. Commun. **6**, 6465 (2015)
9. Iaconis, C., Walmsley, I.A.: Spectral phase interferometry for direct electric-field reconstruction of ultrashort optical pulses. Opt. Lett. **23**, 792–794 (1998)
10. Ratner, J., et al.: Coherent artifact in modern pulse measurements. Opt. Lett. **37**, 2874–2876 (2012)

Chapter 13
High-Order Harmonic Generation by Relativistic Plasma Singularities: The Driving Laser Requirements

A. S. Pirozhkov, T. Zh. Esirkepov, T. A. Pikuz, A. Ya. Faenov, K. Ogura, Y. Hayashi, H. Kotaki, E. N. Ragozin, D. Neely, H. Kiriyama, J. K. Koga, Y. Fukuda, A. Sagisaka, M. Nishikino, T. Imazono, N. Hasegawa, T. Kawachi, H. Daido, Y. Kato, S. V. Bulanov, K. Kondo and M. Kando

Abstract We discuss the new regime of high-order harmonic generation by relativistic-irradiance multi-terawatt femtosecond lasers focused onto gas jet targets [*PRL*, 108, 135004, 2012; *NJP*, 16, 093003, 2014]. The laser induces multi-stream

---

A. S. Pirozhkov (✉) · T. Zh. Esirkepov · K. Ogura · Y. Hayashi · H. Kotaki · H. Kiriyama
J. K. Koga · Y. Fukuda · A. Sagisaka · M. Nishikino · T. Imazono · N. Hasegawa
T. Kawachi · S. V. Bulanov · K. Kondo · M. Kando
Kansai Photon Science Institute, National Institutes for Quantum and Radiological Science and Technology (QST), 8-1-7 Umemidai, Kizugawa, Kyoto 619-0215, Japan
e-mail: pirozhkov.alexander@qst.go.jp

T. A. Pikuz
Graduate School of Engineering, Osaka University, 2-1 Yamadaoka, Suita, Osaka 565-0871, Japan

T. A. Pikuz · A. Ya. Faenov
Joint Institute for High Temperatures of the Russian Academy of Sciences, 13/19 Izhorskaja Street, Moscow 125412, Russia

A. Ya. Faenov
Open and Transdisciplinary Research Initiatives, Osaka University, Suita, Osaka 565-0871, Japan

E. N. Ragozin
P. N. Lebedev Physical Institute of the Russian Academy of Sciences, 53 Leninsky Prospekt, Moscow 119991, Russia

E. N. Ragozin · S. V. Bulanov
Moscow Institute of Physics and Technology (State University), 9 Institutskii pereulok, Dolgoprudnyi, Moscow Region 141700, Russia

D. Neely
Central Laser Facility, Rutherford Appleton Laboratory, STFC, Chilton, Didcot, Oxon OX11 0QX, UK

D. Neely
Department of Physics, SUPA, University of Strathclyde, Glasgow G4 0NG, UK

T. Kawachi et al. (eds.), *X-Ray Lasers 2016*, Springer Proceedings in Physics 202, https://doi.org/10.1007/978-3-319-73025-7_13

relativistic plasma flow resulting in the formation of density singularities: structurally stable, oscillating electron spikes coherently emitting high-frequency radiation. Here we analyse the dependence of the harmonic yield on the focal spot quality and derive the required laser parameters for efficient harmonics generation. We show the status of the J-KAREN-P laser [*IEEE J. Sel. Topics Quantum Electron.*, 21, 1601118, 2015] and report on the progress towards satisfying these requirements.

13.1 Introduction

In the new regime of high-order harmonic generation, intense (>10^{18} W/cm^2) high-power (multi-TW) femtosecond (30–50 fs) laser pulses focused onto gas targets induce multi-stream relativistic flows in an underdense plasma ($n_e \sim 10^{19}$ cm^{-3}). This results in the formation of density singularities, which are structurally stable, oscillating electron spikes coherently emitting high-frequency radiation with spectra comprising high-order harmonics [1, 2]. The experiments with the J-KAREN laser [3] demonstrated strong dependence of the harmonics yield on the laser pulse energy and focal spot quality.

Here we report on the dependence on the focal spot quality characterized by the Strehl ratio, analyse the obtained dependence, and derive laser parameters required for efficient harmonics generation. We show that the focal spot quality should approach the diffraction limit, i.e. the Strehl ratio should exceed 0.5. The focal spots of high-power lasers typically suffer from wavefront distortions and angular dispersion. For noise-like wavefront distortions, the above stated requirement corresponds to an rms wavefront error of below 100 nm. The angular dispersion should be kept smaller than a fraction of the diffraction divergence, i.e. μrad level for 100–300 mm beam diameters. The corresponding angular chirp should be <10^{-2} μrad/nm for 50 nm bandwidth. We show the status of the J-KAREN-P laser and report on the progress towards satisfying these requirements.

H. Daido
Naraha Remote Technology Development Center, JAEA, 1-22 Nakamaru, Yamadaoka, Naraha Futaba, Fukushima 979-0513, Japan

Y. Kato
The Graduate School for the Creation of New Photonics Industries, 1955-1 Kurematsu-cho, Nishiku, Hamamatsu, Shizuoka 431-1202, Japan

S. V. Bulanov
A. M. Prokhorov Institute of General Physics of the Russian Academy of Sciences, 38 Vavilov Street, Moscow 119991, Russia

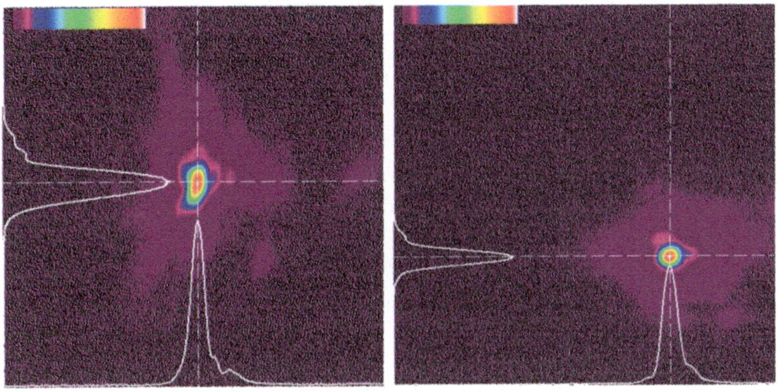

Fig. 13.1 Focal spot shapes corresponding to the minimum (0.16) and maximum (0.33) Strehl ratios used in the experiments. The full frame size is 210 μm

13.2 Experiments

We performed several experiments with the J-KAREN laser [3], where the laser parameters were approximately the same, except notable variation of the focal spot quality. The laser pulse energy varied from 0.4 to 0.8 J, the effective pulse duration was approximately 40 fs. The pulses were focused with an *f*/9 off-axis parabolic mirror into a supersonic helium gas jet. The peak irradiance (in vacuum) varied due to the pulse energy and spot quality difference from 4×10^{18} to 1.3×10^{19} W/cm^2. Figure 13.1 shows the focal spot shapes corresponding to the lowest and highest Strehl ratios. The data described here were obtained with linearly polarized pulses, which generated harmonics more efficiently than the circularly polarized ones.

The harmonics were recorded in all experiments in the forward direction with the flat-field grazing-incidence spectrograph comprised of a toroidal mirror, slit, optical blocking filters, flat-field spherical grating, and back-illuminated CCD. The spectrograph is described in more detail in [2].

13.3 Dependence on the Spot Quality

In each experiment, we optimized the harmonic generation by varying the He backing pressure and gas jet position with respect to the focus. Despite similar laser parameters, the obtained optimum yields were rather different.

The first hypothesis we had was dependence on the peak irradiance. Indeed, the optimized harmonic yield tended to increase with the irradiance. However, the lowest yield was not obtained with the lowest-irradiance pulses. We found that the harmonic yield increased monotonically with the Strehl ratio, Fig. 13.2. To facilitate comparison of data, we show here the yield normalized by pulse energy squared, in accordance with the scaling suggested in [1, 2].

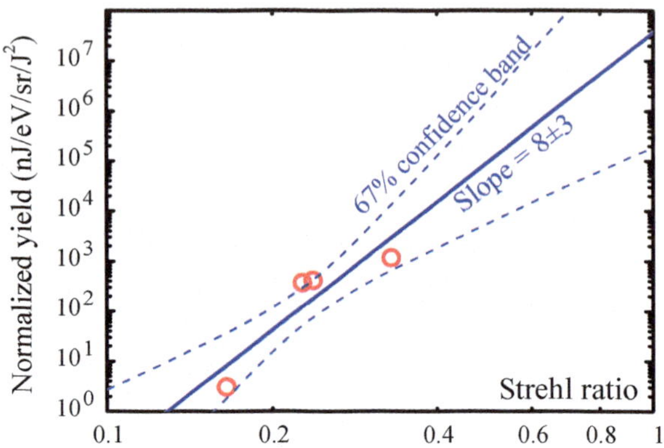

Fig. 13.2 Normalized harmonic yield at the photon energy of 200 eV versus Strehl ratio

Although the dependence on the Strehl ratio might be rather complicated, here we fitted the data with the power-law dependence, resulting in the exponent of 8 ± 3.

13.4 Driving Laser Requirements

The steep growth in Fig. 13.2 looks rather promising; it is intriguing to see whether the harmonic yield continues to grow rapidly for the Strehl ratios above 0.4. However, for this a high-power laser is needed with a focal spot quality approaching the diffraction limit. In practice, many effects reduce the focal spot quality. Here we discuss the requirements on the wavefront and angular chirp, which are two imperfections very typical for high-power lasers.

For noise-like high-frequency wavefront distortions, the Strehl ratio S can be estimated as

$$S = \exp\left[-\left(\frac{\eta\pi\sigma}{\lambda}\right)^2\right], \tag{13.1}$$

where σ is the wavefront distortion rms, λ is the laser wavelength, and the factor $\eta \approx 2$ [4]. We found that for our wavefront sampling and wavefront distortion statistics, the value $\eta \approx 1.85$ gave a good fit to the data, Fig. 13.3. We note that if the wavefront exhibited low-order aberrations, e.g. astigmatism, the relation (13.1) did not hold.

Thus, for efficient harmonic generation, the wavefront distortion should be kept smaller, or even much smaller, than 100 nm rms. Lower-order aberrations should be minimized as well.

Angular dispersion, i.e. the dependence of the propagation direction on wavelength, is a kind of spatiotemporal coupling [5–7] which distorts the focal spot and elongates the pulse duration. Most commonly, the angular dispersion is caused by

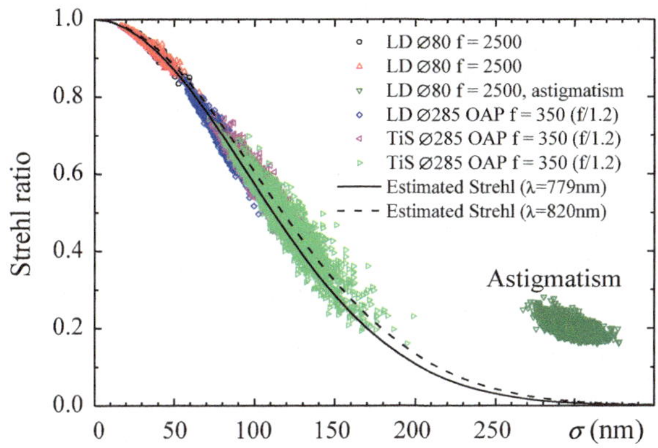

Fig. 13.3 Dependences of the Strehl ratio calculated from the measured wavefront on the rms wavefront distortion σ obtained with two lasers (LD, $\lambda = 779$ nm, and Ti:Sapphire, $\lambda = 820$ nm), beam diameters of 80 and 285 mm, focused with an $f = 2500$ mm lens and $f = 350$ mm (*f*/1.2) off-axis parabolic (OAP) mirror. The solid and dashed lines show the Strehl ratio estimated using (13.1) with $\eta = 1.85$

the stretcher or vacuum compressor misalignment. To keep the spot nearly diffraction-limited, the angular dispersion should be much smaller than the diffraction divergence. For linear angular chirp C [μrad/nm], this gives the requirement

$$C\Delta\lambda \ll \frac{\lambda}{\varnothing}, \tag{13.2}$$

where $\Delta\lambda$ is the bandwidth and $\varnothing$ is the beam diameter. Thus, this requirement is particularly severe for the high-power femtosecond lasers, which have simultaneously large bandwidth and beam diameter. For the J-KAREN-P laser (the final

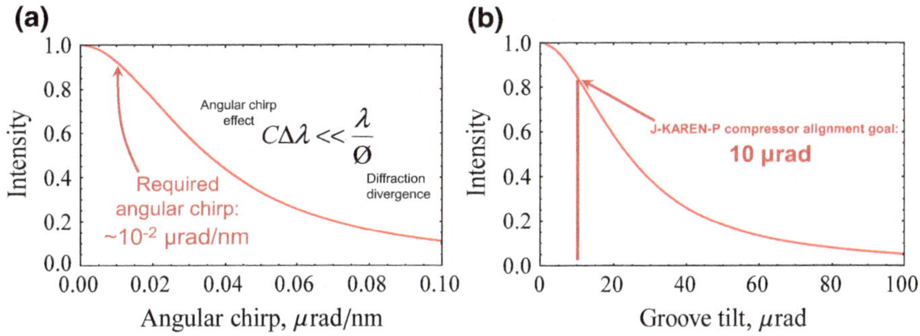

Fig. 13.4 Influence of the angular dispersion and compressor misalignment on the intensity (estimates based on [5] for the J-KAREN-P compressor with 1480 lines/mm gratings). **a** Estimated intensity versus amount of angular chirp. **b** Estimated intensity versus compressor misalignment

beam diameter of 250 mm, bandwidth of 40 nm) this gives $C \ll 0.08$ μrad/nm, Fig. 13.4a. In order to satisfy this requirement, all angles of all compressor gratings should be aligned with the accuracy of the order of 10 μrad, Fig. 13.4b. We note that the intensity estimates [5] shown in Fig. 13.4 include both the effect of spot increase and pulse elongation.

13.5 J-KAREN-P Laser Preparation

We have recently upgraded the J-KAREN-P laser which can now supply petawatt pulses at the repetition rate of 0.1 Hz [8]. We are presently performing preparations for the first experiments, which will include harmonic generation by relativistic electron spikes. To ensure efficient generation, we are taking measures to achieve good laser parameter stability and high spot quality.

Figure 13.5 shows the Power Amplifier energy stability, the data is of 558 shots taken during 4 h operation. The energy fluctuation was 1.8% rms.

Apart from energy, rapid wavefront distortions due to air fluctuations in the long beamline can also affect the shot-to-shot variations in the experiment. To suppress this, we have enclosed nearly all the beamline in plastic boxes. The effect of the beamline covering is demonstrated in Fig. 13.6, which shows distributions of the Strehl ratio calculated from the measured 80 mm diameter LD beam wavefront for the cases of open and enclosed beamline.

In order to achieve high quality wavefront, we set up two deformable mirrors, one after the final amplifier (beam diameter of up to 90 mm) and another before the final expander (beam diameter of up to 135 mm). The wavefront distortion of the Ø285 mm alignment LD beam after the compressor focused with an *f*/7 OAP was reduced down to 90 nm rms. The corresponding Strehl ratio calculated from the measured wavefront and near field distribution was 0.63 ± 0.06 (2000 frames average ± standard deviation), Fig. 13.7 (right red distribution). Work is now in progress to achieve similar or better wavefront with the Ti:Sapphire beam.

The focal spot after the *f*/7 OAP is shown in Fig. 13.7 (left inset). The spot had FWHM just ~10% wider than the diffraction limit, and full width at $1/e^2$ level just

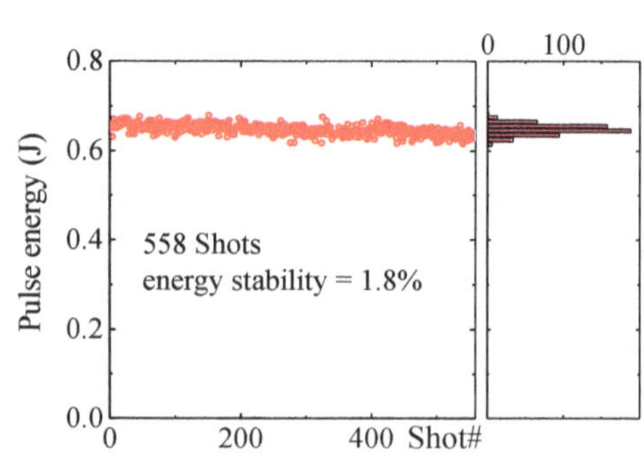

Fig. 13.5 J-KAREN laser pulse energy stability (Power Amplifier, 4 h operation)

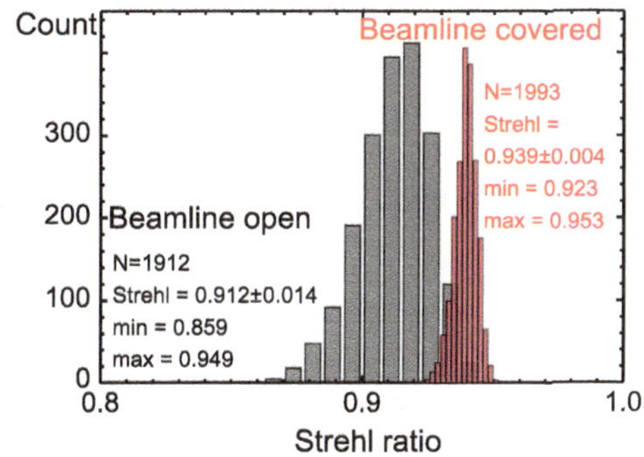

Fig. 13.6 Strehl ratio distributions demonstrating the effect of the beamline covering on the wavefront fluctuations; the Strehl ratios are calculated from the wavefront data

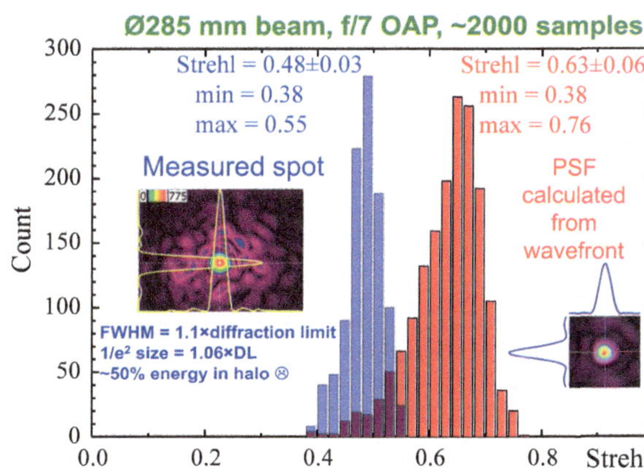

Fig. 13.7 Strehl ratios from the measured spot and PSF calculated from the measured wavefront

~6% wider. Despite this, the Strehl ratio estimated from the measured spot was 0.48 ± 0.03, Fig. 13.7 (left blue distribution), i.e. somewhat lower than predicted from the wavefront data (PSF is shown in Fig. 13.7 right inset). This was probably due to high-frequency wavefront distortions (missed by the wavefront sensor) which produced a low-intensity halo containing up to 50% of the energy around the main spot.

In order to satisfy the strict compressor alignment accuracy requirement, we developed an alignment system based on a combination of autocollimation with an absolute vertical reference. The alignment system exhibited sensitivity of better than 10 μrad for all twelve angles (four gratings, three angles each). The alignment system will be described in more detail elsewhere.

13.6 Conclusion

We have presented the dependence of the harmonic generation by relativistic plasma singularities on the driving laser quality, in particular, the Strehl ratio. We found that the harmonic yield increased dramatically when the Strehl ratio increased from 0.16 up to 0.33, which suggested that higher Strehl ratios may be even more favourable. It also suggested that increasing the Strehl ratio twice may be more advantageous than increasing the laser power twice, although there is a natural limit of the Strehl ratio increase. It would be interesting to test this hypothesis experimentally.

We derived the requirements on the wavefront quality and compressor alignment accuracy to achieve the Strehl ratio approaching 1. We started preparation of the recently upgraded J-KAREN-P laser for the first experiments, and reported on the present status of its stability and wavefront quality.

Acknowledgements We acknowledge financial support from Japan Society for the Promotion of Science (JSPS) KAKENHI JP 26707031 and Ministry of Education, Culture, Sports, Science and Technology (MEXT).

References

1. Pirozhkov, A.S., Kando, M., Esirkepov, T.Zh., Gallegos, P., Ahmed, H., Ragozin, E.N., Faenov, A.Ya., Pikuz, T.A., Kawachi, T., Sagisaka, A., Koga, J.K., Coury, M., Green, J., Foster, P., Brenner, C., Dromey, B., Symes, D.R., Mori, M., Kawase, K., Kameshima, T., Fukuda, Y., Chen, L., Daito, I., Ogura, K., Hayashi, Y., Kotaki, H., Kiriyama, H., Okada, H., Nishimori, N., Imazono, T., Kondo, K., Kimura, T., Tajima, T., Daido, H., Rajeev, P., McKenna, P., Borghesi, M., Neely, D., Kato, Y., Bulanov, S.V.: Soft-X-ray harmonic comb from relativistic electron spikes. Phys. Rev. Lett. **108**, 135004 (2012)
2. Pirozhkov, A.S., Kando, M., Esirkepov, T.Zh., Gallegos, P., Ahmed, H., Ragozin, E.N., Faenov, A.Ya., Pikuz, T.A., Kawachi, T., Sagisaka, A., Koga, J.K., Coury, M., Green, J., Foster, P., Brenner, C., Dromey, B., Symes, D.R., Mori, M., Kawase, K., Kameshima, T., Fukuda, Y., Chen, L., Daito, I., Ogura, K., Hayashi, Y., Kotaki, H., Kiriyama, H., Okada, H., Nishimori, N., Imazono, T., Kondo, K., Kimura, T., Tajima, T., Daido, H., Rajeev, P., McKenna, P., Borghesi, M., Neely, D., Kato, Y., Bulanov, S.V.: High order harmonics from relativistic electron spikes. New J. Phys. **16**, 093003 (2014)
3. Kiriyama, H., Shimomura, T., Sasao, H., Nakai, Y., Tanoue, M., Kondo, S., Kanazawa, S., Pirozhkov, A.S., Mori, M., Fukuda, Y., Nishiuchi, M., Kando, M., Bulanov, S.V., Nagashima, K., Yamagiwa, M., Kondo, K., Sugiyama, A., Bolton, P.R., Tajima, T., Miyanaga, N.: Temporal contrast enhancement of petawatt-class laser pulses. Opt. Lett. **37**, 3363–3365 (2012)
4. Ruze, J.: The effect of aperture errors on the antenna radiation pattern. Il Nuovo Cimento **1943–1954**(9), 364–380 (1952)
5. Pretzler, G., Kasper, A., Witte, K.J.: Angular chirp and tilted light pulses in CPA lasers. Appl. Phys. B **70**, 1–9 (2000)
6. Akturk, S., Gu, X., Gabolde, P., Trebino, R.: The general theory of first-order spatio-temporal distortions of Gaussian pulses and beams. Opt. Express **13**, 8642–8661 (2005)
7. Pariente, G., Gallet, V., Borot, A., Gobert, O., Quéré, F.: Space–time characterization of ultra-intense femtosecond laser beams. Nat. Photonics **10**, 547–553 (2016)
8. Kiriyama, H., Mori, M., Pirozhkov, A.S., Ogura, K., Sagisaka, A., Kon, A., Esirkepov, T.Zh., Hayashi, Y., Kotaki, H., Kanasaki, M., Sakaki, H., Fukuda, Y., Koga, J., Nishiuchi, M., Kando, M., Bulanov, S.V., Kondo, K., Bolton, P.R., Slezak, O., Vojna, D., Sawicka-Chyla, M., Jambunathan, V., Lucianetti, A., Mocek, T.: High-contrast, high-intensity petawatt-class laser and applications. IEEE J. Sel. Top. Quantum Electron. **21**, 1601118 (2015)

Chapter 14
HHG Beam Wavefront Characterization at 30 nm

P. Homer, B. Rus, J. Hrebicek and J. Nejdl

Abstract For the purpose of characterization and improvement of the HHG (High-order Harmonics Generation) eXtreme Ultra-Violet (XUV) coherent beam at the wavelength 30 nm, we developed a unique wavefront sensor based on PDI (Point Diffraction Interferometer) technique. A simple self-referencing monolithic device produces interferometric pattern with encoded information about the wavefront profile of the measured beam. We describe the development, fabrication, and alignment of the sensor, as well as the obtained results and their interpretation.

14.1 Introduction

We designed and fabricated an XUV PDI sensor for HHG beam wavefront measurements. The main motivation was obtaining a device able to analyze the OPD (Optical Path Difference) phase map of the HHG beam which has been used in various experiments at the PALS (Prague Asterix Laser System) laboratory [1]. This information indicates aberrations in the beam wavefront and also provides valuable data about the generating source. With implementation of a suitable phase corrector (a deformable mirror) such setup will bring new features to the investigated HHG beam (e.g., smaller focal spot size, better properties for interferometric experiments, etc.) and will make it possible many new applications.

14.2 Measured HHG Beam Specifications

The High-order Harmonics (HHG) are generated by the Titan:sapphire ultrashort laser pulses (central wavelength 810 nm, energy 1.2 mJ, duration 35 fs, repetition 1 kHz) focused into a gas cell filled with Argon. Nonlinear effects involved in HHG

P. Homer (✉) · B. Rus · J. Hrebicek · J. Nejdl
ELI-Beamlines, Institute of Physics v.v.i., Academy of Sciences of the Czech Republic, 182 21 Prague 8, Czech Republic
e-mail: homer@fzu.cz

T. Kawachi et al. (eds.), *X-Ray Lasers 2016*, Springer Proceedings in Physics 202, https://doi.org/10.1007/978-3-319-73025-7_14

generation produce a narrowly collimated XUV beam, which is in the given experimental conditions the 27th harmonics of the Titan:sapphire laser [2], with wavelength of 30 nm, pulse energy 1–10 nJ and with duration 35 fs. The beam with divergence <1 mrad is generated at frequency 1 kHz.

14.3 PDI: XUV Wavefront Sensing Technique

At present there exist many wavefront profile sensing techniques available in visible and IR domain of the spectrum. However, these techniques are not easily applicable for the XUV region, with calibration being one of the most difficult issues. With currently available XUV optics, calibration is not a straightforward task due to their parameters such as unavailability of efficient and high-quality beam splitters. On these grounds, we decided to design a PDI (Point Diffraction Interferometer) wavefront sensor as this constitutes a self-referencing device that does not require calibration.

The sensor itself consists of a miniature pinhole and a thin foil (as presented in Fig. 14.1). The pinhole is positioned in the axis of the XUV focal spot and it works as a diffraction aperture that generates a reference spherical wave from the part of the XUV beam. The rest of the beam is transmitted through the foil (semitransparent for the selected XUV radiation) that acts as a “density filter”. The essential condition for obtaining well-contrasted interferometric fringes is that the signal level generated by the pinhole must be equal to the signal contained in the transmitted beam. This condition is ensured by selection of an appropriate thickness of the foil, which determines the transmitted signal level. An XUV detector located behind the PDI sensor captures the interferogram, which arises as an interference between the reference spherical wave and the original (measured) wave. By using a

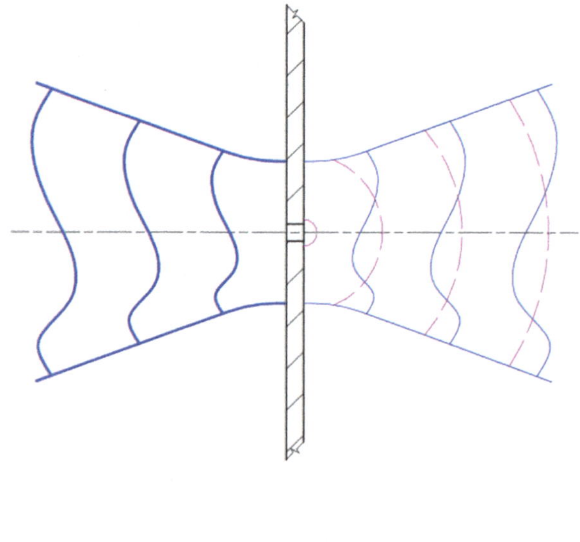

Fig. 14.1 Principle of the PDI wavefront sensor: the original measured wavefront is represented by full blue line, the spherical reference wavefront by dashed red line

deconvolution software it is possible to reconstruct the incident beam wavefront profile from the information encoded in the interferometric pattern.

The theoretical design and calculation of the PDI sensor parameters has been thoroughly described in the [3]. The resulting parameters are following: (1) Filter material Al, (2) Filter thickness 1.5 μm, (3) Pinhole diameter 3.5 μm, (4) Maximum allowed energy in incident beam to avoid foil ionization 0.11 μJ, (5) Phase difference between reference and signal due to phase shift in the material of the foil 2.5 lambda.

14.4 Fabrication of PDI Sensor

The fabrication process involved production of the ultra-thin aluminum foil and the micro pinhole. The high-quality homogeneous foil was commercially obtained in a plastic frame (Ø 28.4 mm). In this foil, we produced a set of twelve circular pinholes with various diameters ranging from 1 to 10 μm (distance between holes 0.5 mm) which provide a device for testing different configurations of the pinhole and/or the focus. The pinholes were made by a custom micropattering procedure utilizing the focused ion beam, resulting in highly precise pinholes with high circularity and with high-quality edges. Additionally, for alignment of the sensor to the optical axis a rectangular aperture (30 × 10 μm), serving to navigation and alignment, was produced along with the array of the individual pinholes.

14.5 Measurement

The experimental setup of the HHG wavefront profile measurement is represented in Fig. 14.2. It contains an XUV detector (X-ray camera Andor DX-440), the PDI sensor and an XUV off-axis parabolic mirror (f = 254 mm, angle between the incident and reflected beam 14°). During the experiment, we temporarily replaced the parabolic mirror with a spherical mirror (f = 250 mm), in order to deform deliberately the measured wavefront with astigmatism aberration and thereby to demonstrate the PDI sensor detection ability.

The alignment procedure of the PDI sensor is illustrated in Fig. 14.3. For rough adjustment to the beam focus, an optical telemicroscope (with 8 × magnification) has been used. By acquiring the IR Titan:sapphire laser focal point (the IR and HHG beams propagate coaxially) on the sensor foil by the telemicroscope we were able bringing the navigation rectangular aperture to this focal point. The fine alignment has been done using a spherical mirror (f = 125 mm) and another XUV detector (PI Roper Pixis), forming an optical imaging system with the magnification factor ten times. The object plane was the back surface of the PDI foil, therefore, we were able observing simultaneously the selected pinhole (backlight illuminated by a LED diode) and the HHG focus spot. This enabled us bringing the hole to the HHG

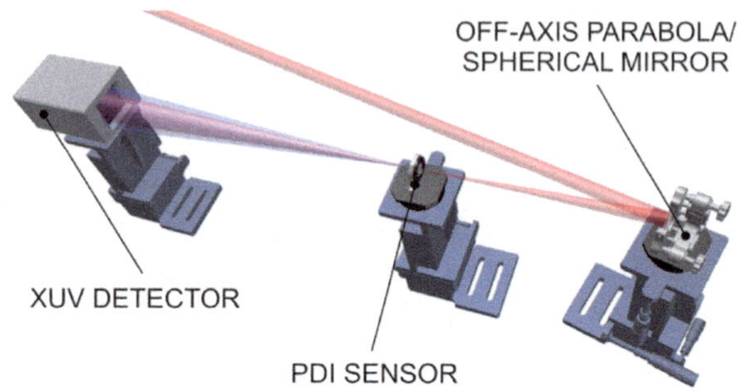

Fig. 14.2 Experimental setup of the HHG wavefront measurement

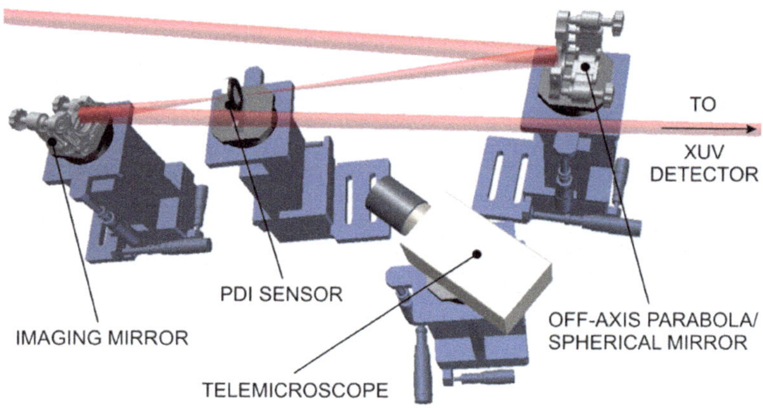

Fig. 14.3 PDI sensor alignment setup. The distance between the focusing mirror and PDI sensor is 254 (250) mm according to the used mirror, distance from PDI to imaging mirror is 137.5 mm, and distance from imaging mirror to the detector is 1375 mm

focus, utilizing a precise motorized stage (adjustable in *x*, *y*, *z* directions) under vacuum. Afterward, the imaging mirror and the LED illuminator were remotely removed out of the beam trajectory and the PDI sensor became ready for the XUV wavefront measurement as shown in the scheme in Fig. 14.2.

14.6 Simulations and Acquired Experimental Data

The acquired data are presented in Figs. 14.4, 14.5 and 14.6, along with the corresponding computer simulations created in the Matlab software. In the experiment, the distance between the focusing parabolic mirror and the PDI sensor was 254 mm, and the distance between the PDI sensor and the detector was 295 mm.

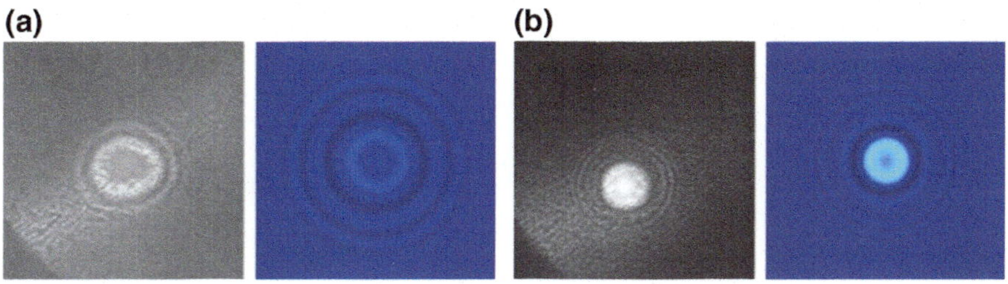

Fig. 14.4 Acquired data and computer simulation: **a** off-axis parabola, 5 μm pinhole, detected 4 mm behind the focus; **b** off-axis parabola, 10 μm pinhole, detected 8.3 mm behind the focus

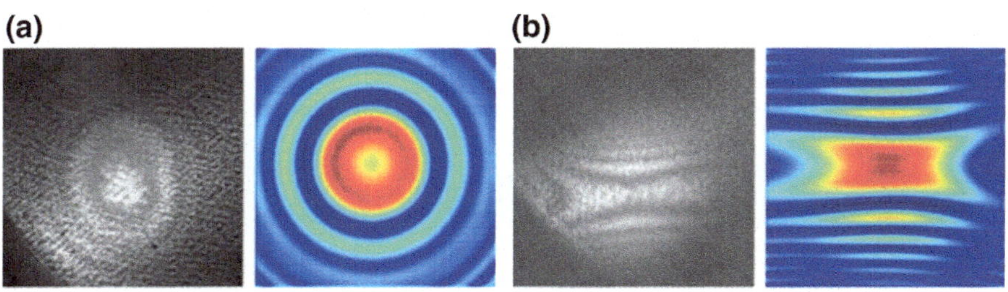

Fig. 14.5 Acquired data and computer simulation: **a** off-axis parabola, 3.5 μm pinhole, detected 2 mm in front of the focus; **b** spherical mirror, 3.5 μm pinhole, detected 2 mm in front of the focus

When the off-axis parabola was replaced by the spherical mirror (in order to deliberately introduce astigmatism), the distance between the focusing spherical mirror and the detector was decreased to 250 mm.

The simulations have been calculated concerning the Gaussian beam propagation and they take into the account only aberrations introduced by the focusing optics (astigmatism in Figs. 14.5a, b and 14.6a, b) and does not include other aberrations in the real measured beam. A decent correlation between the simulated and measured results indicates high-HHG beam wavefront quality. The measured data confirm the operation ability of the sensor.

14.7 Wavefront Data Analysis

Using the measured data accurate wavefront analysis was carried out by using a commercially available software. The results indicate presence of higher orders (equal to or higher than 3) of optical aberrations in the measured XUV beam, as illustrated in Fig. 14.7. The analyzed interferograms suggest wavefront distortions with rms of typically 0.8 (varying between 0.4 and 0.9 on day-to-day basis).

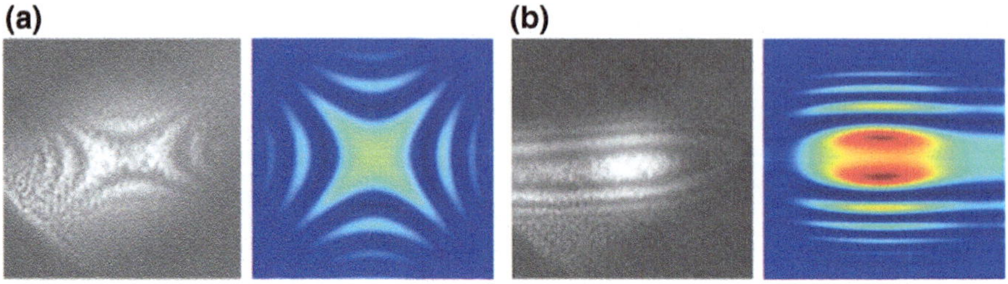

Fig. 14.6 Acquired data and computer simulation: **a** spherical mirror, 3.5 μm pinhole, detected in the focus; **b** acquired data and computer simulation: spherical mirror, 5 μm pinhole, detected 3 mm in front of the focus

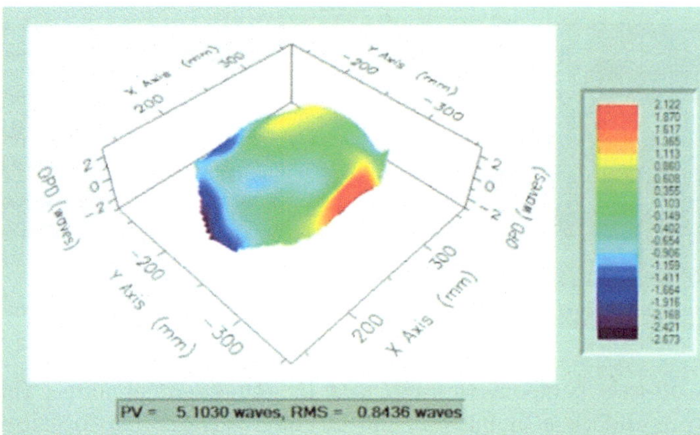

Fig. 14.7 Typical example of 3D wavefront profile of the incident XUV beam, corresponding to the interferogram shown in Fig. 14.4a

14.8 Conclusion

The obtained data represent, to our best knowledge, the first authentic interferometric wavefront sensing in the XUV region. The developed sensor demonstrated capability of PDI technique to provide data exploitable by accurate characterization of the XUV beam wavefront up to 5th order of aberrations. The sensor with the used wavefront analysis technique can be employed to provide data for phase correction in a closed loop, using a deformable mirror with appropriate XUV coating.

Acknowledgements Authors acknowledge the support from grants "CZ.1.05/1.1.00/02.006", "CZ.1.07/2.3.00/20.0091". We thank Dr. Ales Jäger from the Institute of Physics ASCR, v.v.i., for pinhole micropatterning and acknowledge support of MEYS LM2015087.

References

1. Rus, B., et al.: Development and applications of multimillijoule soft X-ray lasers. J. Mod. Opt. **54**, 2571–2583 (2007)
2. Jakubczak, K., et al.: Beam properties of fully optimized, table-top, coherent source at 30 nm. Opto-Electron. Rev. **19**(2), 169–175 (2011)
3. Homer, P., et al.: Measurements of X-ray laser wavefront profile using PDI technique. Proc. SPIE **6702**, 670211 (2007)

Chapter 15
Using the XFEL to Drive Gain in K-Shell and L-Shell Systems Using Photoionization and Photoexcitation of Inner Shell Transitions

Joseph Nilsen

Abstract Six years ago, X-ray lasing on an inner shell transition was demonstrated at 849 eV (1.46 nm) in singly ionized neon gas using the XFEL at 960 eV to photoionize the 1s electron in neutral neon followed by lasing on the 2p-1s transition in singly ionized neon. That research was done at the SLAC Linac Coherent Light Source (LCLS) by a multi-laboratory team led by Nina Rohringer and published in the January 26, 2012, issue of Nature. It took many decades to demonstrate this scheme because it required a very strong X-ray source that could photoionize the 1s (K-shell) electrons in neon on a timescale comparable to the intrinsic auger lifetime in neon, which is typically 2 fs. In this chapter, we have shown how the XFEL could be used to photoionize L-shell electrons to drive gain on n = 3–2 transitions in singly ionized Ar and Cu plasmas. These bright, coherent and monochromatic X-ray lasers may prove to be very useful for doing high-resolution spectroscopy and for studying nonlinear process in the X-ray regime.

15.1 Introduction

Scientists have proposed schemes to achieve lasing at shorter wavelengths since the invention of the laser. In the 1960s, Duguay and Rentzepis proposed using photoionization to create an X-ray laser on the inner-shell K-α line in sodium vapour [1]. In the 1970s, Ray Elton [2] discussed the challenges of making quasi–steady-state inner shell K-α lasers in Si, Ca and Cu. The dream of demonstrating an inner shell X-ray laser was realized at the SLAC Linac Coherent Light Source (LCLS) in 2011 when the X-ray free electron laser (XFEL) at 960 eV was used to photoionize the K-shell of neutral neon gas and create lasing at 849 eV in singly ionized neon gas [3].

J. Nilsen (✉)
Lawrence Livermore National Laboratory, Livermore, CA 94551, USA
e-mail: nilsen1@llnl.gov

T. Kawachi et al. (eds.), *X-Ray Lasers 2016*, Springer Proceedings in Physics 202, https://doi.org/10.1007/978-3-319-73025-7_15

An alternative approach for creating X-ray lasers was the idea of a resonantly photo-pumped laser, where a strong emission line in one material could be used to photoexcite a transition in another material and create lasing. A classic example is the Na-pumped Ne X-ray laser scheme proposed 40 years ago by Vinogradov et al. [4] and modeled by Nilsen et al. [5] that used the strong Na He-α line at 1127 eV to resonantly photo-pump the Ne He-γ line and lase on the 4f-3d transition at 23.1 nm in He-like Ne. This scheme was studied extensively and weak gain [6] was inferred in several experiments. The difficulty with this type of scheme was creating a sufficiently strong pump line. With the availability of strong XFEL sources, the pump line in the traditional photo-pumped schemes can be replaced with an XFEL that is tuned to the appropriate resonance. Since the resonant photo-pumped scheme selectively pumps a transition, it offers the potential for higher gain and lower drive intensity than the photoionization pumping.

Recently, we looked at the advantages and challenges of using the XFEL to resonantly photo-pump the 1s-3p line in neutral neon as a mechanism for creating gain on the K-α line in Ne and compare this with the photoionization pumping that has already been demonstrated [7]. We showed that with the use of a sufficiently short XFEL pulse (1-fs) the resonant photoexcitation could reduce the XFEL flux requirements by two orders of magnitude. In this chapter, we look at how the inner shell X-ray laser can be extended to lasing on L-shell transitions in Ar and Cu. For Ar, we consider an XFEL pulse that photoionizes the 2p or 2s electrons and creates lasing on the 3s-2p or 3p-2s transitions. In the case of Cu, we consider an XFEL pulse that photoionizes the 2p electron and creates lasing on the strong 3d-2p transitions near 1 keV.

15.2 Modelling Ar and Cu L-Shell X-Ray Lasers

Starting with the neon X-ray laser based on photoionization of the K-shell, we look at extending inner shell X-ray lasers to the other principal shells beginning with the L and M shells. Neutral argon gas is the first promising candidate we investigate. Figure 15.1 shows the energy level diagram that uses XFEL above the L-shell edge of neutral Ar I to create an L-shell hole in singly ionized Ar II. Tuning an XFEL between the two L-edges at 250 and 326 eV, one could create a 2p hole that would lase on the 3s-2p transitions at 219 and 221 eV. If the XFEL drive was tuned above the L-edge at 326.3 eV then, one would create holes in the 2s and 2p shells simultaneously that would create lasing on the 3p-2s transitions at 310.4 and 310.6 eV in addition to the 3s-2p transitions. By tuning the XFEL from low to high energy, one could watch the 3s-2p lasing turn on followed by lasing on both sets of lines.

To model these photoionization schemes, we created a simple atomic model of the levels shown in Fig. 15.1. We used the Cretin code [8] to model the kinetics and gain of the system under various conditions. For the baseline XFEL beam, we assume that the XFEL beam has 10^{12} photons in a 0.1% linewidth focused to a

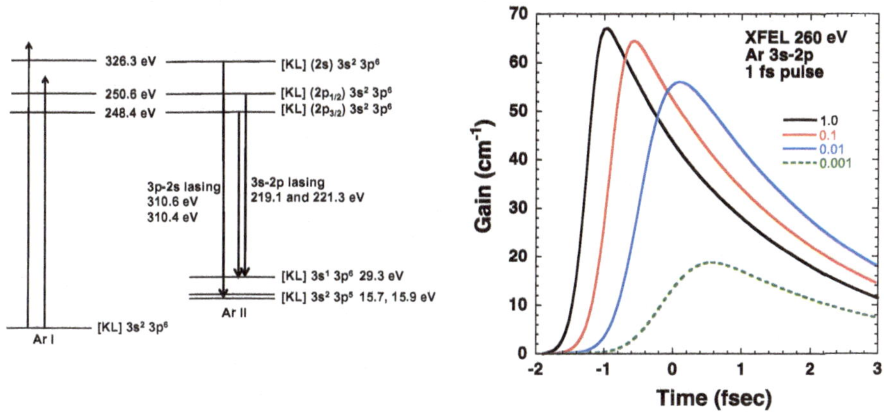

Fig. 15.1 Energy level diagram (left) for the photoionization-driven inner shell argon X-ray laser. Gain versus time (right) for the 3s-2p line at 219 eV in the argon X-ray laser driven by the photoionization mechanism using a 1-fs duration 260 eV XFEL drive pulse. The XFEL intensity is varied using a multiplier between 1.0 (nominal) and 0.001. The peak of the XFEL pulse is defined as time = 0

1-μm diameter. In our previous work, we had studied the nominal LCLS conditions that used a 100-fs full-width half-maximum (FWHM) Gaussian pulse [7]. In this work, we examine using a shorter 1-fs FWHM pulse, which produces much higher gain than the 100-fs pulse used in the LCLS experiments [3, 7].

Figure 15.1 shows the gain versus time for the Argon 3s-2p X-ray laser line at 219 eV driven by a 260 eV XFEL with 10^{12} photons in a 1-fs pulse focused to 1-μm diameter spot. The 219 eV line has twice the gain of the 221 eV line. One observes that the peak gain of 67 cm^{-1} falls very slowly to 56 cm^{-1} as the XFEL flux is reduced by a factor of 100. By comparison, using a 330 eV XFEL pulse, the 3p-2s line at 310 eV has a peak gain of about 30 cm^{-1}.

Figure 15.2 shows the energy level diagram for using a 1-keV XFEL whose energy is above the L-shell edge of neutral Cu I to photoionize an L-shell hole in singly ionized Cu II that results in lasing on the strong 3d-2p lines at 928 and 948 eV. The gain at 928 eV is predicted to be about twice the gain at 948 eV. Fig. 15.2 shows the gain versus time for the Copper 3d-2p X-ray laser line at 928 eV driven by a 1000 eV XFEL with 10^{12} photons in a 10-fs pulse focused to 1-μm diameter spot. The gain peaks at 136 cm^{-1} and falls slowly as the XFEL intensity is reduced. As an alternative, photoexcitation of the 2p-4d transition in Cu I would also create lasing on the 3d-2p line in Cu I and might require an even lower XFEL intensity.

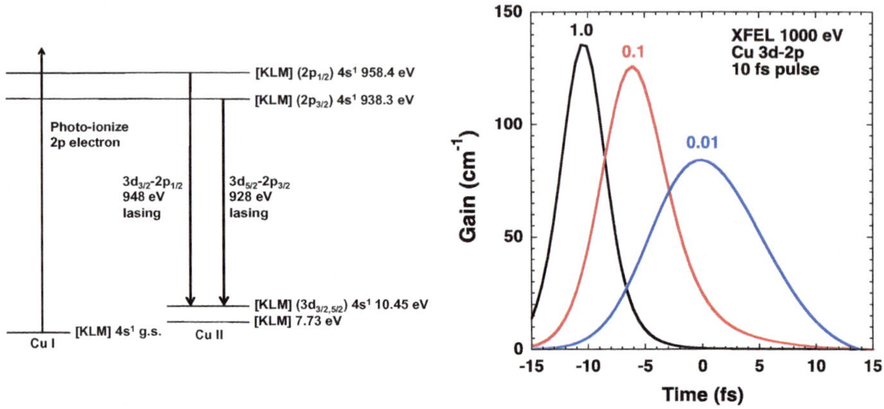

Fig. 15.2 Energy level diagram (left) for the photoionization-driven inner shell copper X-ray laser. Gain versus time (right) for the 3d-2p line at 928 eV in the copper X-ray laser driven by the photoionization mechanism using a 10-fs duration 1000 eV XFEL drive pulse. The XFEL intensity is varied using a multiplier between 1.0 (nominal) and 0.01. The peak of the XFEL pulse is defined as time = 0

15.3 Conclusions

In this chapter, we have shown how the XFEL could be used to photoionize L-shell electrons to drive gain on n = 3–2 transitions in singly ionized Ar and Cu plasmas. We model those systems and estimate peak gain of 67 cm^{-1} on the 3s-2p line at 219 eV in Ar and peak gain of 136 cm^{-1} on the 3d-2p line at 928 eV in Cu. Substantial gain is predicted in other n = 3–2 lines in Ar at 221 and 310 eV as well as 948 eV in Cu. These bright, coherent and monochromatic X-ray lasers may prove to be very useful for doing high-resolution spectroscopy and for studying nonlinear process in the X-ray regime.

Acknowledgements This work was performed under the auspices of the U.S. Department of Energy by Lawrence Livermore National Laboratory under Contract DE-AC52-07NA27344.

References

1. Duguay, M.A., Rentzepis, P.M.: Appl. Phys. Lett. **10**, 350–352 (1967)
2. Elton, R.C.: Appl. Opt. **14**, 2243–2249 (1975)
3. Rohringer, N., Ryan, D., London, R.A., Purvis, M., Albert, F., Dunn, J., Bozek, J.D., Bostedt, C., Graf, A., Hill, R., Hau-Riege, S.P., Rocca, J.J.: Nature **481**, 488–491 (2012)
4. Vinogradov, A.V., Sobelman, I.I., Yukov, E.A.: Sov. J. Quantum Electron. **5**, 59–63 (1975)
5. Nilsen, J., Scofield, J.H., Chandler, E.A.: Appl. Opt. **31**, 4950–4956 (1992)
6. Nilsen, J., Chandler, E.A.: Phys. Rev. A **44**, 4591–4598 (1991)
7. Nilsen, J.: Matter Radiat. Extremes **1**, 76–81 (2016)
8. Scott, H.A.: JQSRT **71**, 689–701 (2001)

Chapter 16
Superfluorescence/Superradiance in Helium Following Free-Electron Laser Excitation

J. R. Harries, S. Kuma, H. Iwayama and E. Shigemasa

Abstract We discuss the observation of superfluorescence following the excitation of helium atoms with pulses from a free-electron laser. From semi-classical simulations and consideration of transition parameters, we predict that it should be possible to generate pulses of EUV superfluorescence using two-photon excitation.

16.1 Introduction

Superradiance is a fundamental effect which can occur when ultrafast, intense radiation is used to excite dense atomic samples. Its development is sensitive to the spectral and coherence properties of the incident radiation, and as such is an excellent test of our understanding of these properties. This is particularly relevant with the increasing availability of coherent light sources operating at EUV and X-ray wavelengths. First discussed by Dicke [1], superradiance occurs following the creation of a macroscopic polarization by coherent excitation. With incoherent excitation the related process superfluorescence can occur, where the macroscopic polarization emerges spontaneously, leading to a characteristic and stochastic delay [2–4]. Superfluorescence has been observed in gaseous and solid media, Bose–Einstein condensates, and nanometre-sized structures [5], but never (to our knowledge) at wavelengths significantly shorter than the visible, although an effect related to superradiance has been observed at X-ray wavelengths [6]. Recently, we reported the observation of superfluorescence at visible wavelengths following the excitation of helium atoms using intense pulses of radiation from a free-electron laser (FEL),

J. R. Harries (✉)
National Institutes for Quantum and Radiological Science and Technology,
SPring-8, 1-1-1 Kouto, Sayo, Hyogo 679-5148, Japan
e-mail: harries@spring8.or.jp

S. Kuma
Atomic, Molecular, and Optical Physics Laboratory, RIKEN,
Wako, Saitama 351-0198, Japan

H. Iwayama · E. Shigemasa
UVSOR, IMS, Nishigo-Naka 38, Myodaiji, Okazaki, Aichi 444-8585, Japan

T. Kawachi et al. (eds.), *X-Ray Lasers 2016*, Springer Proceedings
in Physics 202, https://doi.org/10.1007/978-3-319-73025-7_16

resonant to 1s3p (53.7 nm) and 1s4p (52.2 nm) excitation [7–10]. Simulations (unpublished) have also shown that the partial coherence of the FEL pulses may also lead to superradiant decay of the 1snp states back to the ground state, at the same EUV wavelengths. However, the experimental verification of this is difficult since the emitted radiation is at the same wavelength as the intense excitation pulse. Here we propose making use of the partial coherence of the excitation and using two-photon excitation to induce yoked superfluorescence [11–14], leading to the emission of superfluorescence at wavelengths in the EUV region and at shorter wavelengths than the excitation pulse.

16.2 Yoked Superfluorescence

To generate superfluorescence at EUV wavelengths we propose using two-photon excitation to excite either the 1s3s/1s3d or 1s4s/1s4d states. To demonstrate that this is feasible using pulses typical of a free-electron laser, Fig. 16.1 shows the results of a simulation of the interaction of a 4 μJ FEL pulse with a duration of around 50 fs and a central wavelength of 107.4 nm. To simulate the partially coherent nature of FEL pulses the partial coherence method of Pfeifer et al. was used [15]. For details of the simulation methods see [16, 17]. For this particular pulse, it is clear that significant excitation to 1s3d can be expected, along with some excitation of 1s3s, and also 1s4s, 1s4d. Different realizations of the stochastic pulses lead to different final populations, but at this central wavelength the majority of atoms are either excited to 1s3d or remain in the ground state. While the results are not shown here, similar simulations with an excitation wavelength of 104.4 nm suggest that the most likely state to be excited is 1s4d, with 1s4s excitation weaker, but stronger than is 1s3s excitation at 107.4 nm.

Having established that it is feasible to create high densities of helium atoms in the 1s3d or 1s4d states, we discuss whether yoked superfluorescence can be expected

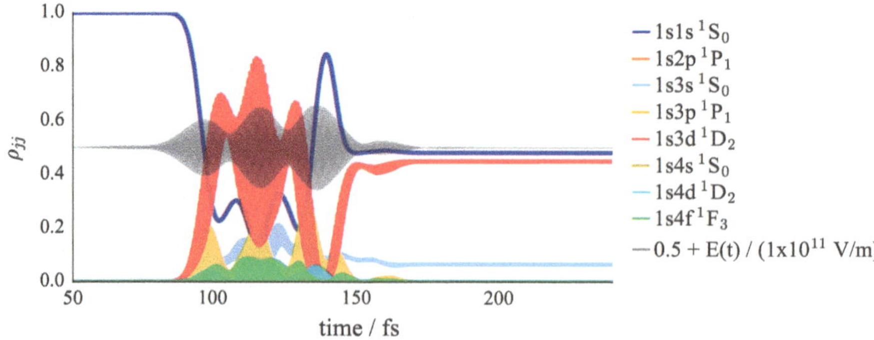

Fig. 16.1 Simulation of the two-photon excitation process with a partially coherent FEL pulse (see text), showing the evolution of the populations of selected levels. The black trace shows the electric field, centred and scaled

Table 16.1 Wavelengths, characteristic superfluorescence times and threshold column densities for the relevant transitions

Upper	Lower	λ/nm	$\sigma\tau_{SF}$/(ps nm^{-2})	σ_{th}/nm^{-2}
1s3s	1s2p	728.3	0.86	1.2
1s3d	1s2p	668.0	0.29	0.37
1s4s	1s2p	504.9	4.9	4.6
1s4s	1s3p	2114	0.41	1.6
1s4d	1s2p	492.3	1.7	1.6
1s4d	1s3p	1909	0.32	1.2
1s2p	1s1s	58.4	1.4	0.15
1s2p	1s2s	2059	1.0	3.9
1s3p	1s1s	53.7	5.1	0.52
1s3p	1s2s	501.7	2.5	2.4
1s3p	1s3s	7438	0.60	8.5
1s3p	1s3d	96000	5.9	19

to occur. Table 16.1 shows some parameters of the relevant transitions. The quantity $\sigma\tau_{SF}$ is the sample column density $\sigma = N_a L$ multiplied by the characteristic superfluorescence time for a cylindrical sample $\tau_{SF} = 8\pi/(3\lambda^2 A_{ki} N_a L)$, where N_a is the number density of excited atoms, L the length of the gain medium, λ the wavelength of the transition, and A_{ki} the spontaneous decay rate (Einstein A coefficient). The quantity $\sigma\tau_{SF}$ is thus a means of comparing the superfluorescence decay rate for different transitions given the same excited state column density. The final column, σ_{th} is the threshold column density defined by equating the superfluorescence decay rate to the wavelength-dependent loss rate due to diffraction (see [16] for details), which is the dominant loss rate for the conditions considered—a cylindrical sample of length 2 mm and diameter 12.6 μm. For yoked superfluorescence to proceed on a particular two-step decay requires [11–14] that (i) the first step transition is that most likely to proceed as superfluorescence, and (ii) the superfluorescence decay rate of the second step is not significantly faster than that of the first step. For the gain medium assumed, the threshold excited state number density required for superfluorescence to proceed on the 1s3d-1s2p transition (the only possible decay route from 1s3d) is around 2×10^{20} m^{-3}. An excitation probability of 0.5 would thus require a ground state number density of 4×10^{20} m^{-3}, well within the range of densities accessible experimentally. The threshold column density for 1s2p-1s1s is significantly lower, and the superfluoresence decay rate is longer, leading to yoked superfluorescence (excitation-induced coherence favours 1s1s over 1s2p). Transitions from the 1s4d and 1s4s states have higher thresholds, but are still well within reach experimentally, and the 1s4d/4s-1s3p-1s1s yoked superfluorescence schemes can also be expected to occur. Preliminary Maxwell–Bloch simulations considering the propagation of the field through the medium suggest that the predicted yoked superfluorescence does indeed occur, and experiments are planned at the SACLA BL1 free-electron laser light source to confirm the predictions. Assuming that the pulse energies, pulse durations, and focus sizes available at a wavelength of 107.4 nm are comparable to those used in single-photon experiments [7], the emissions following two-photon excitation can be expected to be similar in intensity to those following single-photon excitation, since in both cases it is predicted that similar excited state densities can be created, leading to the first observation of superfluoresecence at EUV wavelengths.

Acknowledgements This work was supported in part by the Research Foundation for Opto-Science and Technology, and a JSPS KAKENHI grant (number 15K04707). We are grateful to M. Nagasono, C. Ohae, Y. Miyamoto, K. Nakajima, T. Togashi and N. Sasao for helpful discussions.

References

1. Dicke, R.H.: Coherence in spontaneous radiation processes. Phys. Rev. **93**(1), 99–110 (1954)
2. Gross, M., Haroche, S.: Superradiance: an essay on the theory of collective spontaneous emission. Phys. Rep. **93**(5), 301–396 (1982)
3. Benedict, M.G. (eds.): Super-Radiance: Multiatomic Coherent Emission. CRC Press, Bristol; Philadelphia (1996)
4. Allen, L., Eberly, J.H.: Optical Resonance and Two-Level Atoms. Dover, New York (1987)
5. Cong, K., Zhang, Q., Wang, Y., et al.: Dicke superradiance in solids. J. Opt. Soc. Am. B **33**(7), C80 (2016). arXiv: 1602.04374
6. Röhlsberger, R., Schlage, K., Sahoo, B., et al.: Collective lamb shift in single-photon superradiance. Science **328**(5983), 1248–1251 (2010)
7. Nagasono, M., Harries, J.R., Iwayama, H., et al.: Observation of free-electron-laser-induced collective spontaneous emission (superfluorescence). Phys. Rev. Lett. **107**(19), 193603 (2011)
8. Nakajima, K., Harries, J.R., Iwayama, H., et al.: Simultaneous measurements of super-radiance at multiple wavelengths from helium excited states: I. Experiment. J. Phys. Soc. Jpn. **84**(5), 054301 (2015)
9. Harries, J.R., Iwayama, H., Nagasono, M., et al.: A streak camera study of superfluorescence at multiple wavelengths from helium atoms excited using free electron laser pulses. J. Phys. B: At. Mol. Opt. Phys. **48**(10), 105002 (2015)
10. Ohae, C., Harries, J.R., Iwayama, H., et al.: Simultaneous measurements of superradiance at multiple wavelength from helium excited states: II. Analysis. J. Phys. Soc. Jpn. **85**(3), 034301 (2016)
11. Brownell, J.H., Lu, X., Hartmann, S.R.: Yoked superfluorescence. Phys. Rev. Lett. **75**(18), 3265–3268 (1995)
12. Yi, Z., Jha, P.K., Yuan, L., et al.: Observing the transition from yoked superfluorescence to superradiance. Opt. Commun. **351**, 45–49 (2015)
13. Okada, J., Ikeda, K., Matsuoka, M.: Cooperative cascade emission. Opt. Commun. **26**(2), 189–192 (1978)
14. Lvovsky, A.I., Hartmann, S.R., Moshary, F.: Superfluorescence-stimulated photon echoes. Phys. Rev. Lett. **89**(26), 263602 (2002)
15. Pfeifer, T., Jiang, Y., Düsterer, S., et al.: Partial-coherence method to model experimental free-electron laser pulse statistics. Opt. Lett. **35**(20), 3441–3443 (2010)
16. Harries, J.R., Ohae, C., Kuma, S., et al.: Single-atom response of helium atoms to pulses from an EUV free-electron laser: implications for the subsequent development of superfluorescence. Phys. Rev. A **94**(6), 063416 (2016)
17. Marskar, R., Österberg, U.: Multilevel Maxwell-Bloch simulations in inhomogeneously broadened media. Opt. Express **19**(18), 16784–16796 (2011)

Chapter 17
In Situ Characterization of XFEL Beam Intensity Distribution and Focusability by High-Resolution LiF Crystal Detector

T. A. Pikuz, A. Ya. Faenov, T. Matsuoka, B. Albertazzi, N. Ozaki, N. Hartely, O. Muray Ricardo Arturo, T. Yabuuchi, H. Habara, S. Matsuyama, K. Yamauchi, Y. Inubushi, T. Togashi, H. Yumoto, Y. Tange, K. Tono, Y. Sato, M. Yabashi, M. Nishikino, T. Kawachi, A. Mitrofanov, S. A. Pikuz, D. Bleiner, A. Grum-Grzhimailo, N. N. Rosanov, N. V. Vysotina, M. Harmand, M. Koenig, K. A. Tanaka, T. Ishikawa and R. Kodama

Abstract We present here a new diagnostics based on using LiF crystal detectors that are able to perform measurements an intensity distribution of X-rays beams with diameters ranging from some microns up to some centimetres with high spatial resolution (~1 μm). In situ, 3D visualization of SACLA XFEL focused beam

---

T. A. Pikuz (✉) · B. Albertazzi · N. Ozaki · N. Hartely · O. Muray Ricardo Arturo
H. Habara · S. Matsuyama · K. Yamauchi · Y. Sato · M. Koenig · K. A. Tanaka
R. Kodama
Graduate School of Engineering, Osaka University, Suita, Osaka 565-0871, Japan
e-mail: pikuz.tatiana@gmail.com

T. A. Pikuz · A. Ya. Faenov · A. Mitrofanov · S. A. Pikuz
Joint Institute for High Temperatures, RAS, Moscow 125412, Russia

A. Ya. Faenov · T. Matsuoka · R. Kodama
Open and Transdisciplinary Research Initiatives, Osaka University, Suita, Osaka 565-0871, Japan

N. Ozaki · O. Muray Ricardo Arturo · R. Kodama
Photon Pioneers Center, Osaka University, Suita, Osaka 565-0871, Japan

Y. Inubushi · T. Togashi · H. Yumoto · Y. Tange · K. Tono · M. Yabashi · T. Ishikawa
JASRI/SPring-8, Sayo, Hyogo 679-5198, Japan

T. Yabuuchi · M. Yabashi · T. Ishikawa
RIKEN Harima Institute, Sayo, Hyogo 679-5148, Japan

M. Nishikino · T. Kawachi
Kansai Photon Research Institute, National Institutes for Quantum and Radiological Science and Technology, Kizugawa, Kyoto 619-0215, Japan

D. Bleiner
EMPA Materials Science and Technology, 9014 St. Gallen, Switzerland

T. Kawachi et al. (eds.), *X-Ray Lasers 2016*, Springer Proceedings in Physics 202, https://doi.org/10.1007/978-3-319-73025-7_17

profile along propagation, including propagation inside photoluminescence solid materials, is demonstrated. Also, a high spatial resolution control a quality of targets used in optical laser pump—XFEL probe HEDS experiments is proposed.

17.1 Introduction

At coherent X-ray radiation facilities there is an urgent need for simple and efficient methods for in situ beamline metrology. Additionally, measurements of energy distribution of XFEL beams in the caustic of focusing system are very important both for correct evaluation of X-ray fluence in the different cross sections of such beams and for further improvement of different focusing systems. Here we present a new X-rays diagnostic based on using LiF X-ray detector that is able to perform in situ measurements the intensity distribution of X-rays beams with diameters ranging from some microns up to some centimetres with high spatial resolution ($\sim$1 μm). This diagnostic have been applied for characterization of the SACLA XFEL beam profile and its focusability. Additionally, it was successfully used for single-shot control of quality of targets used in optical laser pump—XFEL probe high energy density science experiments. Obtained experimental results are presented and discussed.

17.2 High-Performance Imaging Using LiF Crystal X-Ray Detector

17.2.1 Main Properties of LiF Crystal X-Ray Detector

LiF is widely used material, in particular, as scintillator detector for different ionizing sources [1]. But as it was shown recently [2–5] the LiF is also a very good

A. Grum-Grzhimailo
Skobeltsyn Institute of Nuclear Physics, Moscow State University, Moscow 119991, Russia

N. N. Rosanov · N. V. Vysotina
Vavilov State Optical Institute, Saint-Petersburg 199053, Russia

M. Harmand
IMPMC, Université Pierre et Marie Curie, CNRS, 75005 Paris, France

B. Albertazzi · M. Koenig
LULI, École Polytechnique, CNRS, CEA, UPMC, 91128 Palaiseau, France

S. A. Pikuz
National Research Nuclear University MEPhI, Moscow 115409, Russia

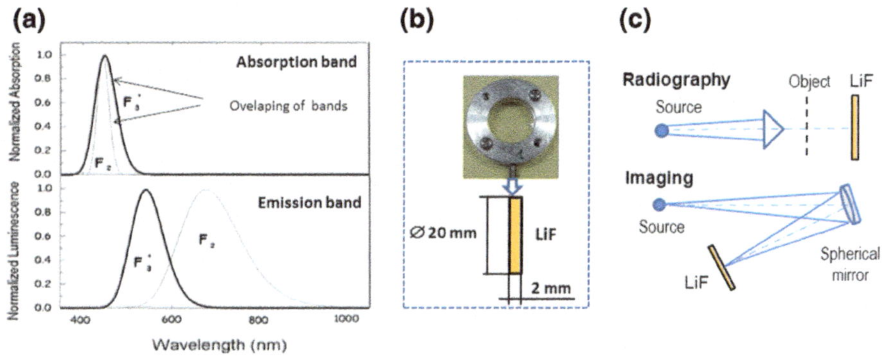

Fig. 17.1 LiF as a high-performance imaging detector for EUV and X-ray spectral range: **a** absorption and emission bands of colour centres generated by X-ray radiation in LiF; **b** view of LiF detector (upper) and dimensions of LiF crystal (below); **c** the use of LiF detector for radiography and imaging (schematic view)

imaging detector for X-ray radiation. Photons with energy greater than of 14 eV generate in LiF so-called F_2- and F_3^+-type colour centres (CCs), which are very stable at room temperature. Those colour centres have absorption and emission bands in visible spectral range (Fig. 17.1a), so photoluminescence (PL) of CCs can be observed by using conventional fluorescent microscope. Our investigations show (see, for example, [2]) that spatial resolution of the PL images measured by LiF detector falls to value of 700 nm, if confocal fluorescent microscope is used for readout, or may be as small as of 75 nm for scanning near-field microscope [3]. LiF detector has large (at least 11 bits) dynamic range and large field of view. It is not sensitive to visible light and does not require any electronic circuit. It is very simple in design and cheap. The view of the LiF detector, which was used in our experiments, is shown in Fig. 17.1b. It is very convenient for high-resolution X-ray radiography and projection imaging (Fig. 17.1c).

17.2.2 3D Visualization of XFEL Beam Focusing Properties

Experiments have been carried out at the hard X-ray beamline BL3 of the SACLA-SPring-8 facility, at the experimental hall EH5. The XFEL beam with photon energy of 10.1 keV and pulse energy at the exit of undulator of 400 mJ was attenuated by a factor of 1500 and focused by a Kirkpatrick-Baez (KB) type HERMES focusing system [5] on the surface of LiF crystal detector. The intensity distribution of the focused XFEL beam was measured at sequences of planes near the focal point (Fig. 17.2). Application of LiF detector consists of the two steps: in the first step, the X-ray radiation produces CCs, which creates a hidden image according with X-ray intensity distribution, in the second step the stored image is

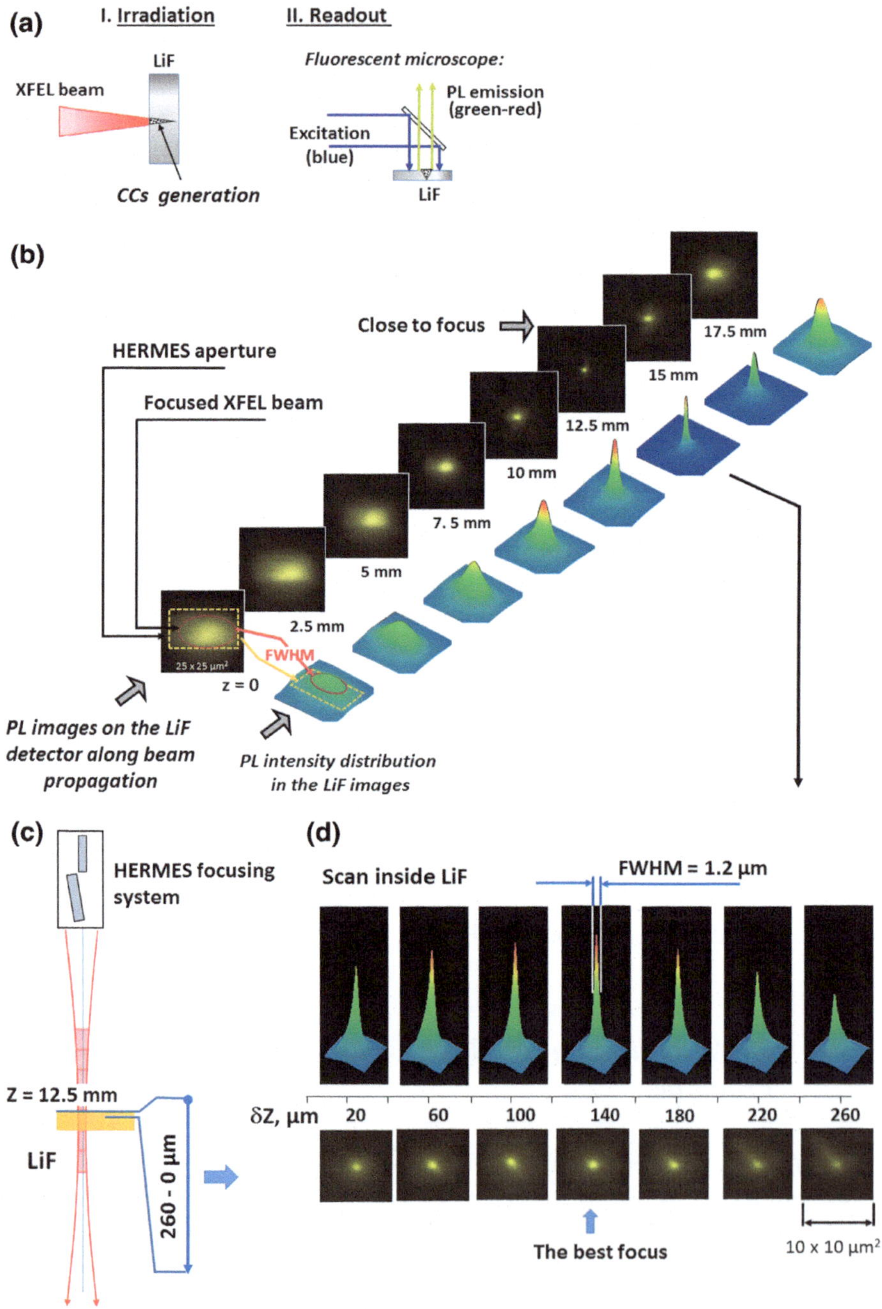

Fig. 17.2 **a** Two-step procedure of application the LiF crystal as imaging detector: I—irradiation by XFEL beam and generation of CCs hidden image; II—visualization of the stored image by means of CCs photoluminescence. **b** Sequences of PL images, obtained on LiF at different planes along XFEL beam (upper diagonal) and plots of PL intensity for correspondent images (below diagonal). **c** Sketch shows the range of measurements inside the crystal. **d** PL images, measured at different distances inside of LiF crystal (below row) and correspondent 3D intensity distribution plots (upper row) [5]

visualized by excitation of CCs photoluminescence (Fig. 17.2a). To readout image, the confocal fluorescent microscope was used. Obtained images are shown in the upper diagonal-sequence in Fig. 17.2b. High sensitivity and large dynamic range of the LiF crystal detector allowed measurements of the intensity distribution at distances far from the best focus as well as near the best focus. Simultaneous observation the image of focusing system aperture and the apparent size of XFEL beam at sequences of planes along the beam propagation allowed us to evaluate divergence of the XFEL beam and the beam quality factor M^2. For our particular experimental conditions, we obtained the vertical divergence $\Theta_v = 3.2 \pm 0.32$ µrad and the horizontal divergence $\Theta_H = 2.8 \pm 0.28$ µrad. The values of the beam quality factor were found $M^2{}_V = 3.8$ in vertical direction and $M^2{}_H = 3.9$ in horizontal direction. The feature of photoluminescence response of the LiF detector to hard X-ray is generation of CCs along the beam propagation inside it. For LiF the attenuation length for photons with energy of 10.1 keV is of ~750 µm. It means that images recorded by LiF actually are 3D images. Figure 17.2c and b show, that scan inside of image, which has smallest size on the surface (position Z = 12.5 mm), allowed to find the best focal plane with high accuracy of ~40 µm. We found the smallest size of the spot of ~1.2 µm. It is necessary to underline, what this value is larger compared to the size of the focal spot of 0.2 × 0.3 µm, measured by the wire-scan method independently. One of the possible reason why the size of the focal spot measured by photoluminescence in LiF is larger, then measured by the wire-scan method is increase of photoluminescent volume due to generation the secondary electron cascade by incoming X-ray photon [6]. Size of photoluminescent area in the best focus also supports such theoretical prediction.

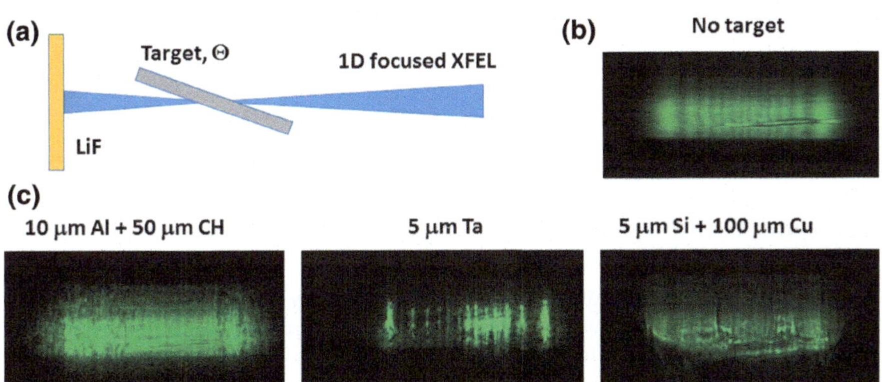

Fig. 17.3 Control of XFEL beam and pump-probe targets imaging control: **a** experimental scheme; **b** image of XFEL beam; **c** transmission images of different targets

17.2.3 Control of Target Quality for Optical Laser Pump—XFEL Probe HEDS Experiments

XFEL beam probe allowed to obtain precise information on material properties and/or on atomic scale processes with an unprecedented time resolution. To reach this goal it is very important to control energy distribution in XFEL focal spot and quality of targets with a high spatial resolution. In most of these experiments, a slit is used to limit the size of the probe beam from ten's to hundred's of microns and it appears to be challenging but necessary to be able to distinguish if the XFEL beam is structured and homogenous. Our experiments at SACLA XFEL facility demonstrated that such high-resolution single-shot in situ control is possible using LiF crystal detectors (see Fig. 17.3). Indeed, distortions of the probe beam due to some damage on KB mirror and diffraction of XFEL beam at stopping aperture (Fig. 17.3b) together with inhomogeneity of used various targets (Fig. 17.3c) is clearly resolved in a single-shot XFEL transmission images.

17.3 Conclusion

In conclusion, we would like to stress that LiF crystal X-ray detector along with unique combination of high spatial resolution and large dynamic range, which provide a high informativeness, is very convenient for in situ radiography or imaging measurements due to its extreme simplicity, absent of any electronic circuits and low cost.

Acknowledgements The XFEL experiments were performed at the BL3 of SACLA with the approval of the Japan Synchrotron Radiation Research Institute (JASRI) (Proposals Nos. 2014A8045, and 2014B8068). This research was partially supported by grants from Grants-in-Aid for Scientific Research (Kakenhi Grant No. 22224012, 25289244, and 15H02153) and the Core-to-Core Program on International Alliance for Material Science in Extreme States with High Power Laser of the Japan Society for the Promotion of Science (JSPS), from the X-ray Free Electron Laser Priority Strategy Program of the MEXT, contract 12005014. This work was carried out within the state assignment of FASO of Russia project #0044-2014-0037 and was partly supported by the Competitiveness Programme of NRNU MEFhI.

References

1. Schulman, J.H., Compton, W.D.: Color Centers in Solids. Pergamon, Oxford (1962)
2. Faenov, A.Ya., et al.: Submicrometer-resolution in situ imaging of the focus pattern of a soft X-ray laser by color center formation in LiF crystal. Opt. Lett. **34**, 941–944 (2009)
3. Ustione, A., et al.: Scanning near-field optical microscopy images of microradiographs stored in lithium fluoride films with an optical resolution of $\lambda/12$. Appl. Phys. Lett. **88**, 14110, 7 (2006)

4. Pikuz, T., et al.: Optical features of a soft X-ray imaging detector based on photoluminescence point defects in LiF crystals irradiated by free electron laser pulses. Opt. Express **20**, 3424–3433 (2012)
5. Pikuz, T.A., et al.: 3D visualization of XFEL beam focusing properties using LiF crystal X-ray detector. Sci. Rep. **5**, 17713 (2015)
6. Caleman, C., et al.: On the feasibility of nanocrystal imaging using intense and ultrashort X-ray pulses. ASC Nano **5**, 139–146 (2011)

Chapter 18
Achieving Laser Wakefield Accelerated Electron Beams of Low Enough Energy Spread for an X-FEL

J. K. Koga, S. V. Bulanov, T. Zh. Esirkepov and M. Kando

Abstract We describe a method to obtain sufficiently low energy spread laser wakefield accelerated electron beams for injection into a conventional undulator and free electron lasing. By using two laser pulses: a moderate power laser to inject and accelerate electrons via its wakefield to moderate energies and a second more powerful laser pulse to generate another wakefield further accelerating them to 1 GeV, and properly matching the acceleration stages an energy spread of 0.2% is possible, leading to the realization of a compact X-ray free electron laser.

18.1 Introduction

A lot of progress in the field of laser wakefield acceleration (LWFA) [1] has been made towards realizing compact electron accelerators (see review [2] and references cited therein). However, LWFA electron beams do not yet have small enough energy spread and transverse emittance for efficient lasing in an X-ray free electron laser (XFEL). There are proposals to achieve such low energy spread beams (see for example [3]). In this paper, we show how to achieve a sufficiently small energy spread for an XFEL via the use of staging [4] and phase matching. The acceleration is done in two stages. The possibility to obtain high-quality beams using such a combination has been previously shown via simulations [5].

18.2 Results

In the first stage, electrons are injected into the wakefield behind a moderate power laser pulse via a steep density gradient [5–8]. The density gradient allows control into which bucket the electrons are injected and further accelerated in a plasma.

J. K. Koga (✉) · S. V. Bulanov · T. Zh. Esirkepov · M. Kando
Kansai Photon Science Institute, National Institutes for Quantum and Radiological Science and Technology, Kizugawa, Kyoto 8 -1 -7, Japan
e-mail: koga.james@qst.go.jp

T. Kawachi et al. (eds.), *X-Ray Lasers 2016*, Springer Proceedings in Physics 202, https://doi.org/10.1007/978-3-319-73025-7_18

Our previous 2D particle-in-cell (PIC) simulations of the injection of electrons into the accelerating phase of a wakefield using the code REMP [9] have shown that an electron bunch with an average electron energy of 24.6 MeV, root mean square (RMS) energy spread of ≈6%, charge of ≈2.45 pC and RMS spatial width of $\approx 1.94\lambda_0$ can be achieved with a moderate power laser pulse [4].

In the second stage a second more powerful laser is used to further accelerate the bunches in a lower density plasma. The key to achieving GeV level energy bunches with small energy spreads near 0.2% is the proper phase matching of the electron bunch into the wakefield. We consider the phase space acceptance working backward from the required parameters for an XFEL using a 1D theoretical model of a laser pulse moving in a plasma with constant phase velocity [4, 10]. The 1D Hamiltonian, $h(X, p)$, is given by [4, 10]:

$$h(X,p) \equiv \sqrt{1 + p^2 + a^2(X)} - \Phi(X) - \beta_w p \tag{18.1}$$

where the momentum in the laser propagation direction, p, is normalized by $m_e c$ and $X = x - \beta_w t$ is normalized by the laser wavelength λ_0 with $\beta_w = \sqrt{\gamma_w^2 - 1}/\gamma_w$ and $\gamma_w = \omega_0/\omega_{pe}\sqrt[4]{1 + a_0^2}$ where ω_0 is the laser frequency and $a_0 \equiv eE_0/m_e\omega_0 c$ with E_0 being the peak field amplitude, $\omega_{pe} = \sqrt{4\pi e^2 n_e/m_e}$ with n_e being the plasma density and e and m_e being the electron charge and mass, respectively, $a(X)$ is the laser amplitude for circular polarization, and $\Phi(X) = e\phi(X)/m_e c^2$ is the normalized electrostatic potential. The equation for $\Phi(X)$ is given by [10]:

$$\begin{aligned}\Phi''(X) = k_p^2\gamma_w^3\beta_w p\{&(1+\Phi)\left[\gamma_w^2(1+\Phi)^2 - 1 - a^2(X)\right]^{-1/2} \\ &-(\mu - \Phi)\left[\gamma_w^2(\mu - \Phi)^2 - \mu^2 - a^2(X)\right]^{-1/2}\}\end{aligned} \tag{18.2}$$

where $k_p = \omega_{pe}/c$ with c being the speed of light, $\mu = m_i/m_e$ with m_i being the ion mass and and the prime refers to differentiation by X. The laser pulse is taken to be Gaussian of the form [10]:

$$a(X) = a_0\{\exp[-4\ln(2)X^2/l_p^2] - 1/16\}\theta(l_p - |X|) \tag{18.3}$$

with l_p being the pulse duration and θ being the Heaviside step function. These equations are solved numerically using the program gauss2 [11, 12]. We choose a laser pulse of duration $l_p/\lambda_0 = 30$ with peak normalized amplitude $a_0 = 1$, mass ratio $\mu = 1836$ and plasma density of $\omega_{pe}/\omega_0 = 0.04$.

Figure 18.1 shows the phase space trajectories of constant h in $p - X$ space where the laser is propagating to the left. We have purposely chosen transient phase space trajectories due to their larger initial spatial acceptance in ΔX versus closed trajectories. The upper inset on the right of the figure shows the sought after final momentum with an energy of ≈1 GeV and spread of ≈0.2% with a spatial extent of $\approx 1.5\lambda_0$. The lower inset on the right of the figure shows the initial injection momentum region

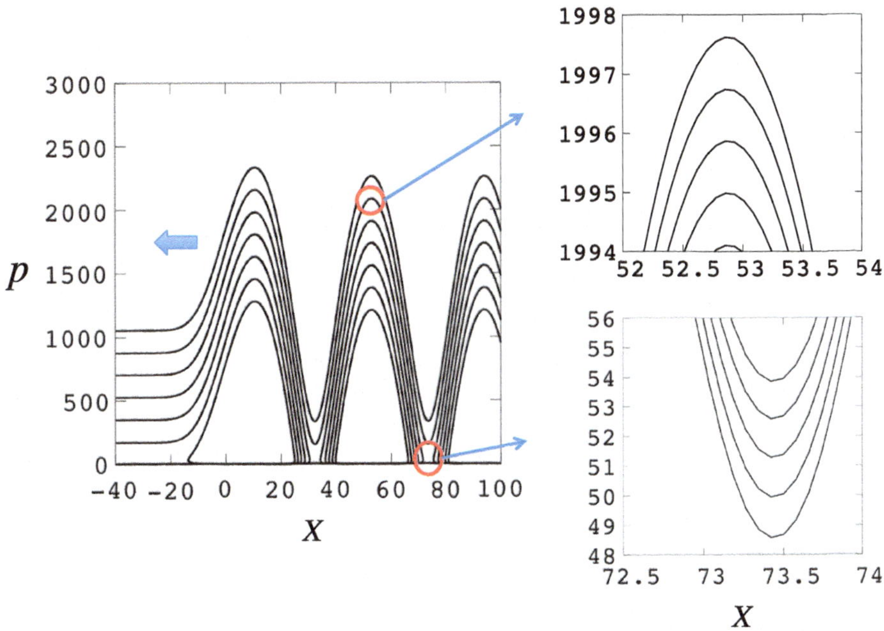

Fig. 18.1 Phase space trajectories in $p-X$ space where the insets show zoom ups of the regions indicated by the circles. The momentum is normalized by $m_e c$ and $X = x - \beta_w t$ normalized by the laser wavelength λ_0 where $\beta_w = \sqrt{\gamma_w^2 - 1}/\gamma_w$ with $\gamma_w = \omega_0/\omega_{pe}\sqrt[4]{1 + a_0^2}$. The arrow indicates the laser propagation direction

corresponding to this final momentum region. It can be seen that the phase space acceptance volume occupies a region with an average energy of ≈26 MeV and a spread of ≈15% with a spatial extent of $\approx\lambda_0$. This phase space acceptance volume is within the parameters of the electron bunch we obtained from our previous PIC simulations with the exception of the spatial extent of the electron bunch (≈2X's larger).

18.3 Conclusions

With further optimization, we expect to satisfy all the required parameters, for example, by choosing an earlier time with larger energy spread and shorter bunch length. With the caveats that we have not included the space charge effects of the electron bunch and that this model is 1D, we anticipate that our acceleration method could provide electrons beams a step closer towards achieving lasing in a compact XFEL.

Acknowledgements This research was funded by the ImPACT Program of the Council for Science, Technology and Innovation (Cabinet Office, Government of Japan).

References

1. Tajima, T., Dawson, J.M.: Laser electron accelerator. Phys. Rev. Lett. **43**, 267–270 (1979)
2. Esarey, E., et al.: Physics of laser-driven plasma-based electron accelerators. Rev. Mod. Phys. **81**, 1229–1285 (2009)
3. Zhang, Z., et al.: Energy spread minimization in a cascaded laser wakefield accelerator via velocity bunching. Phys. Plasmas **23**, 053106-1–053106-7 (2016)
4. Bulanov, S.V., et al.: On some theoretical problems of laser wake-field accelerators. J. Plasma Phys. **82**, 905820308-1–905820308-55 (2016)
5. Tomassini, P., et al.: Production of high-quality electron beams in numerical experiments of laser wakefield acceleration with longitudinal wave breaking. Phys. Rev. ST Accel. Beams **6**, 121301-1–121301-7 (2003)
6. Bulanov, S., et al.: Particle injection into the wave acceleration phase due to nonlinear wake wave breaking. Phys. Rev. E **58**, R5257–R5260 (1998)
7. Brantov, A.V., et al.: Controlled electron injection into the wake wave using plasma density inhomogeneity. Phys. Plasmas **15**, 073111-1–073111-10 (2008)
8. Buck, A., et al.: Shock-front injector for high-quality laser-plasma acceleration. Phys. Rev. Lett. **110**, 185006-1–185006-5 (2013)
9. Esirkepov, T.Zh.: Exact charge conservation scheme for particle-in-cell simulation with an arbitrary form-factor. Comp. Phys. Comm. **135**, 144–153 (2001)
10. Esirkepov, T.Zh., et al.: Electron, positron, and photon wakefield acceleration: trapping, wake overtaking, and ponderomotive acceleration. Phys. Rev. Lett. **96**, 014803-1–014803-4 (2006)
11. González-Pinto, S., et al.: Efficient iterations for Gauss methods on second-order problems. J. Comput. Appl. Math. **189**, 80–97 (2006)
12. Netlib Repository. http://www.netlib.org/ode/ (2015). Accessed 25 May 2015

Chapter 19
Proposal for Experiment Systems Using Laser-Driven Heavy Ions and XFELs to Understand Physical Phenomena Occurring Near the Incident Ion Path

Kengo Moribayashi

Abstract Experiment systems using laser-driven heavy ions and X-ray free electron lasers (XFELs) as pump and probe sources, respectively, are proposed to investigate DNA damage that occurs near the laser-driven heavy ion path. This system is expected to provide high spatial and short time resolution measurements.

19.1 Introduction

In the irradiation of heavy ions on matter, the physical phenomena that occur in the region near the ion path remain largely unknown. These phenomena are expected to play an important role in some scientific field such as radiation biology [1–3]. Study of these phenomena requires short time resolutions because they relax very soon (within few nanoseconds) after heavy ion irradiation.

Aiming to understand these phenomena, we propose experiment systems employing laser-driven heavy ions [4, 5] and X-ray free electron lasers (XFELs) [6] as pump and probe sources, respectively, to provide high spatial and short time resolution measurements. These systems have the following advantages: (i) both sources consist of short pulses, which indicates that short time resolutions can be obtained and (ii) the XFELs attain much higher brightness than synchrotron radiation sources. It is worth mentioning here that 500 TW lasers have been installed in the SACLA XFEL site in Japan. While water was treated as a target in [7], DNA in water [1] was employed in this paper.

K. Moribayashi (✉)
National Institutes for Quantum and Radiological Science and Technology,
8-1-7, Umemidai, 619-0215 Kizugawa-City, Japan
e-mail: moribayashi.kengo@qst.go.jp

T. Kawachi et al. (eds.), *X-Ray Lasers 2016*, Springer Proceedings
in Physics 202, https://doi.org/10.1007/978-3-319-73025-7_19

19.2 Proposal for an Experiment System

Figure 19.1 shows our proposal for an experiment system to measure the damage or double-strand break (DSB) of DNA. As mentioned before, since laser-driven ions and XFELs consist of short pulses, short time resolutions of the order of nanoseconds is expected to be obtained.

We may be able to ignore the effect of radiation damage due to XFELs, which is thought to make the spatial resolution worse [8], because of the following reasons. Water molecules surrounding the DNA can suppress this effect within 50 fs after the XFEL radiation according to a simulation study [8], and the pulse duration of the SACLA XFELs are expected to be controlled to 10 fs.

For 10 keV X-rays, a simulation study [9] estimated that a spatial resolution of a few Å is achieved with an X-ray flux of 5×10^{21}/pulse/mm^2 for a protein of lysozyme. We use this estimation as a standard because the target used in this measurement is expected to have a much simpler structure. Using the maximum X-ray number (5×10^{11}/pulse) of the SACLA XFEL, the X-ray flux mentioned above can be achieved by focusing to a ~10 nm spot-size.

According to the size of DNAs, a spot size of ~100 nm may be sufficient for the measurement of the DSB because a spatial resolution of the order of nanometers is required. Bin et al. [10] measured a number of laser-driven ions of larger than 10^8/cm^2 at $r_l \sim 30$ cm, where r_l is the distance from the focus point of high-intensity lasers. When these ions are focused on a 1 mm^2 area, the average number of ions, which pass through the spot size of XFELs of ~100 nm, is larger than one. Even if this focus spot area of these ions is slightly larger, we may be able to analyze the

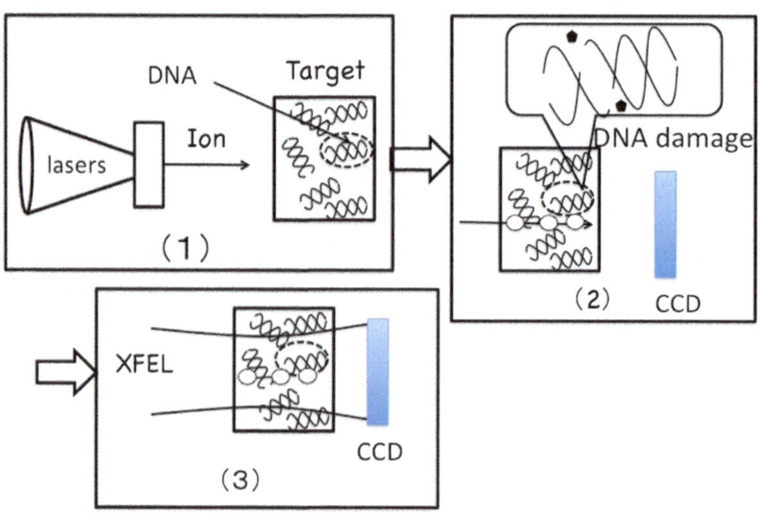

Fig. 19.1 Measurement of the damage or double-strand break (DSB) of DNA. The target is DNA in water

behavior of the ions from the following reason. We can deduce that the number of measured data (N_D) is in proportion to the shot number and the frequency (F_l) of the high-intensity lasers. When $F_l = 1$ Hz, N_D is expected to be about 80,000 in a day. Therefore, even if ions do not appear in 99% of the measured data, we can employ 800 samples for analysis of the behavior of the ions. Further, we can reduce the ion energy to smaller than 1 meV/u for this measurement [1]. This allows us to make the spot size of the high-intensity lasers to larger than that shown in [10]. This increase of the spot size will lead to obtain larger number of ions.

19.3 Summary

We propose experiment systems with high spatial and short time resolutions using laser-driven heavy ions and XFELs. We estimate the required parameters for XFELs and laser-driven ions to study the DSB.

Acknowledgements We wish to thank Dr. N. Shikazono (QST) for his useful supports. This work has been supported by JSPS KAKENHI Grant Number 25390131.

References

1. Akamatsu, K., Shikazono, N.: Localization estimation of ionizing radiation-induced abasic sites in DNA in the solid state using fluorescence resonance energy transfer. Radiat. Res. **183**, 105–113 (2015)
2. Moribayashi, K.: Incorporation of the effect of the composite electric fields of molecular ions as a simulation tool for biological damage due to heavy-ion irradiation. Phys. Rev. A. **84**, 012702/1–012702/7 (2011)
3. Moribayashi, K.: Radial dose calculation due to the irradiation of a heavy ion: role of composite electric field formed from the polarization of molecules and molecular ions. Rad. Phys. Chem. **96**, 211–216 (2014)
4. Nishiuchi, M., et al.: Acceleration of highly charged GeV Fe ions from a low-Z substrate by intense femtosecond laser. Phys. Plasmas **22**, 033107/1–033107/8 (2015)
5. Yogo, A., et al.: Laser ion acceleration via control of the near-critical density target. Phys. Rev. E **77**, 016401/1–016401/6 (2008)
6. Ishikawa, T., et al.: A compact X-ray free-electron laser emitting in the sub-ångström region. Nat. Photonics **6**, 540–544 (2012)
7. Shikazono, N., et al.: Applications of Laser-Driven Particle Acceleration (ALPA). CRC Press (submitted)
8. Hau-Riege, S.P., et al.: Encapsulation and diffraction-pattern-correction methods to reduce the effect of damage in X-ray diffraction imaging of single biological molecules. Phys. Rev. Lett. **98**, 198302/1–198302/4 (2007)
9. Tokuhisa, A., et al.: Classifying and assembling two-dimensional X-ray laser diffraction patterns of a single particle to reconstruct the three-dimensional diffraction intensity function: resolution limit due to the quantum noise. Acta Cryst. A **68**, 366–381 (2012)
10. Bin, J.H., et al.: Ion acceleration using relativistic pulse shaping in near-critical-density plasmas. Phys. Rev. Lett. **115**, 064801 (2015)

Part III
Ultrafast X-Rays and Applications

Chapter 20
Laser-Driven Plasma-Based Incoherent X-Ray Sources at PALS and ELI Beamlines

M. Kozlová, K. Bohaček, V. Horný, K. Ta Phuoc, J. Nejdl, S. Sebban, J. Gautier, M. Krůs and U. Chaulagain

Abstract In this paper, we report on development of incoherent secondary X-ray sources at the PALS Research Center and discuss the plan for the ELI Beamlines project. One of the approaches, how to generate ultrashort pulses of incoherent X-ray radiation, is based on the interaction of femtosecond laser pulses with underdense plasma. This method, known as laser wakefield electron acceleration (LWFA), can produce up to GeV electron beams emitting radiation in the collimated beam with a femtosecond pulse duration. This approach was theoretically and experimentally examined at the PALS Center. The parameters of the PALS Ti: Sapphire laser interaction were studied by extensive particle-in-cell (PIC) simulations with radiation postprocessors in order to evaluate the capabilities of our system in this field. The compressed air, and a mixture of helium and argon were used as accelerating medium. The accelerator was operated in the bubble regime with forced self-injection and resulted in the generation of stable relativistic electron beams with the energy up to 80 meV, hence, the betatron X-ray radiation with critical energy in the keV range was generated. The extensions of this method to the ELI Beamlines facility will enable to generate much higher X-ray energies from 10 keV up to 1 meV with 10 Hz repetition rate. Such source is suitable for various applications like phase contrast imaging or X-ray absorption experiments in a single shot.

M. Kozlová (✉) · K. Bohaček · V. Horný · K. Ta Phuoc · J. Nejdl · S. Sebban
J. Gautier · M. Krůs · U. Chaulagain
IoP ASCR, Na Slovance 2, 182 00 Prague 8, Czech Republic
e-mail: kozlova@fzu.cz

M. Kozlová · V. Horný · J. Nejdl · M. Krůs
IPP ASCR, Za Slovankou 3, 182 00 Prague 8, Czech Republic

K. Bohaček · V. Horný
LOA, ENSTA, Chemin de La Hunière, 91761 Palaiseau, France

K. Ta Phuoc · S. Sebban · J. Gautier
FNSPE of the CTU in Prague, Prague, Czech Republic

T. Kawachi et al. (eds.), *X-Ray Lasers 2016*, Springer Proceedings in Physics 202, https://doi.org/10.1007/978-3-319-73025-7_20

20.1 Introduction

Currently, there are three large-scale laser facilities in the Czech Republic; Prague Asterix Laser System (PALS) Research Centre [1], Extreme Light Infrastructure (ELI) [2], and High average power pulsed LASErs (HiLASE) facility [3]. While the first two-mentioned facilities are focused on the fundamental research, the latest one is focused on the development of high-repetition laser systems, which will find use in industry. Over the past decade, the PALS is running as a user facility that is mostly dedicated to laser plasma studies and development of secondary sources of short-wavelength radiation. Two laser systems, iodine kJ @ 1.3 μ with a pulse length of 350 ps and Ti:Sapphire @ 810 nm with 40 fs pulse length, could be used as drivers for their generation. Moreover, both laser systems could work in synchronized mode. In this paper, the new plasma betatron X-ray source driven by Ti: Sapphire laser will be presented and its future implementation at the new ELI Beamlines facility.

20.2 PALS Beamlines

20.2.1 Ti:Sapphire 20 TW Laser Beamline

The Ti:Sapphire laser system which was installed at the PALS laboratory is a commercial laser system (Coherent Inc.) with peak power of 25 TW [4]. The laser delivers pulses with a duration of 40 fs and the total output energy is up to 1 J with at a repetition rate of 10 Hz. A fundamental wavelength of 812 nm can be converted into second and third harmonics. The laser chain is based on the conventional CPA amplification scheme; it consists of the oscillator, stretcher, regenerative amplifier, pulse picker, multipass (4 passes) amplifier, and compressor. The compressor is designed to use zero order of the first gold reflective grating to produce an auxiliary beam with 50 mJ pulse energy being subsequently compressed by the second compressor. Both beams could be mutually delayed with respect to each other in a range of 0–20 ns. For experiments, all beams could be sent into interaction area, which is equipped with three independent reconfigurable vacuum chambers.

20.2.2 Incoherent Sources at PALS

Besides the generation of coherent sources such as X-ray lasers or HHG [5], the theoretical studies and the simulations shows that with the laser beam described above it is possible to accelerate the electrons up to relativistic energies. Hence, the first experiment was conducted at PALS Center. For the pilot experimental campaign, the compressed air was chosen as accelerating medium. The main beam was focused by off-axis parabolic mirror with f-number 6.5 onto supersonic Laval nozzle gas jet (see in Fig. 20.1). Backing pressure was from 3 to 20 bars.

Two lanxes were used to monitor the pointing stability and the energetic spectrum of the produced electron beam. Compressed air was chosen as an inexpensive accelerating medium. The intensity of the laser pulse was in the order of magnitude of 10^{18} W/cm^2. With such intensity, the oxygen and nitrogen atoms have only upper levels stripped. However, a higher level of ionization can be achieved due to laser self-focusing mechanism, which allows a higher intensity to be reached. These extra electrons can be subsequently trapped and accelerated. Moreover, in order to increase the injected charge, a razor blade was used to create a steep density gradient at the front of the target. To control plasma conditions Mach–Zehnder interferometer and Thomson scattering diagnostics were used. The evaluation of interferometric data shows that stable acceleration conditions with air were reach, when the electron density in the channel was about 5×10^{19} cm^{-3}. Electron acceleration using dry air showed remarkable stability in energy (as can be seen in Fig. 20.2) with a peak electron energy of (17.4 $\pm$ 1.1) MeV, the electron beam was generated in 85% of shots.

Since the production of the stable electron beam was found, the accelerating medium was changed for a mixture of helium and argon. The argon was mixed with helium in a ratio of 1:99 to increase the charge of accelerated electrons using the ionization injection regime. As in the case of air, the backing pressure was set in a

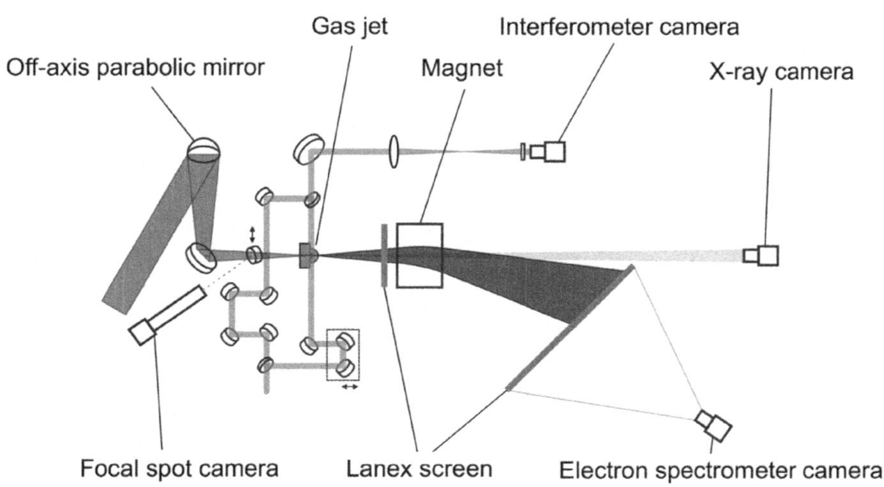

Fig. 20.1 The experimental setup depicts the LWFA and generation of Betatron radiation. The laser pulse is focused onto the gas jet target with OAP mirror (f# = 6.5). Several diagnostics were used to monitor the plasma parameters such as the Mach–Zehnder interferometer. Accelerated electrons are diagnosed by the magnetic spectrometer and the pointing stability is measured before electrons enter the magnetic field of the spectrometer dipole. The electrons were detected by LANEX screens. X-ray beam was detected by X-ray CCD camera

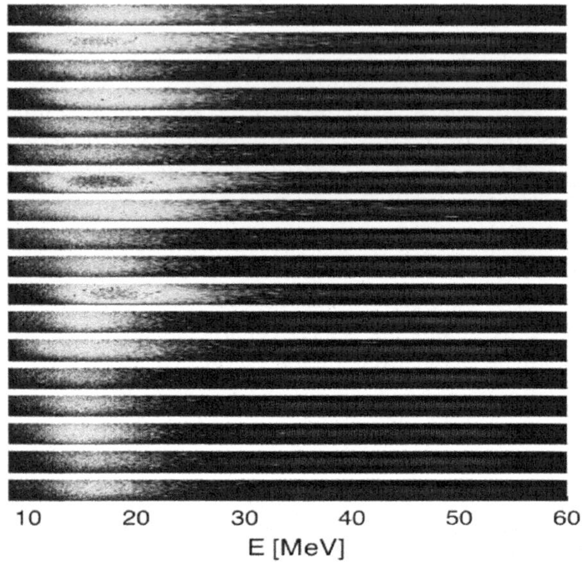

Fig. 20.2 Electron spectra obtained using the dry air target, showing high stability in energy E = (17.4 ± 1.1) MeV

range from 3 to 20 bars. The most stable conditions, for the mixture, of the electron acceleration and betatron X-ray generation were found for electron density of 2.5×10^{19} cm^{-3}. With those plasma conditions, the achieved electron energies were up to 80 meV (see in Fig. 20.2). The spectrum of X-ray radiation generated by betatron motion of the electrons during acceleration was evaluated from records of a cooled 16-bit X-ray CCD camera placed on the laser beam axis working in single photon counting mode [6]. The critical energy of observed radiation has about 1.6 keV (see in Fig. 20.3).

The same laser parameters were used to compare the experimental results and PIC simulations, which were done in the EPOCH 2D code [7] with ADK model for ionization [8]. In case of compressed air, only oxygen and nitrogen were considered as the initially neutral atoms. Hence the simulations showed peak energy of 17 meV for the air target (see in Fig. 20.4), which is in good agreement with measured data (see in Fig. 20.3), and 24 meV for the He–Ar mixture (see in Fig. 20.5). In the latter case, the wake wave starts to form an ion cavity and electrons wiggle in the typical betatron motion with a radius of about 2 μm, indicating a possibility of the generation of X-ray pulses.

20.3 ELI Beamlines Project

Additionally to the well-established PALS facility, and its scientific program, the new ELI Beamlines facility will give an opportunity for the future development and applications of laser-driven ultrashort X-ray sources in the Czech Republic [2].

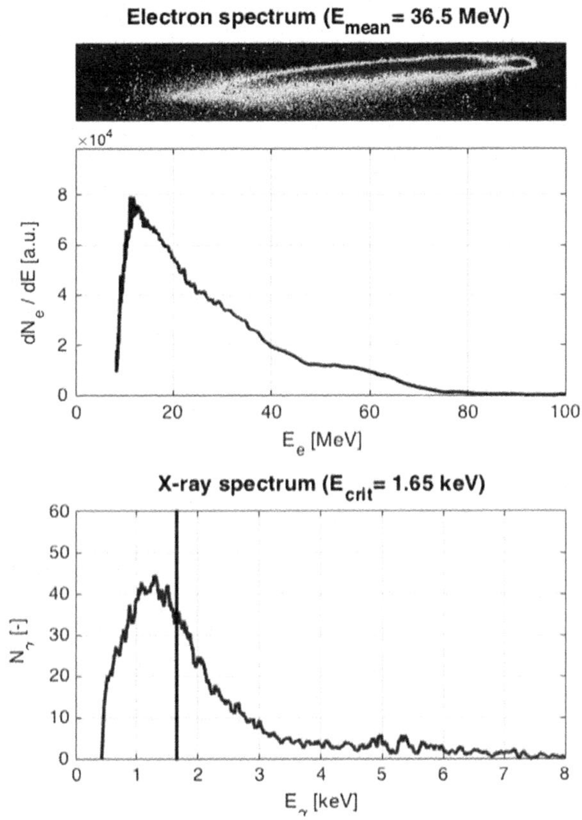

Fig. 20.3 The electron signal (top) and electron spectrum (middle) with the electron energy up to 80 meV and corresponding betatron X-ray spectrum with a critical energy of 1.6 keV

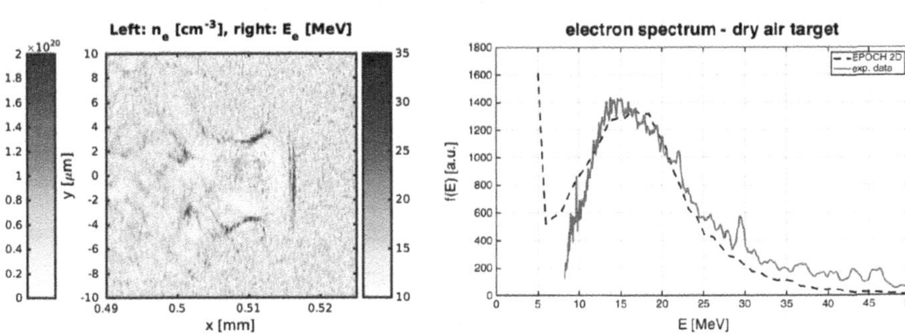

Fig. 20.4 Electron acceleration using dry air. Left: A 2D map of the electron plasma density showing the shape of the nonlinear wake wave. Right: The comparison of the electron spectrum measured experimentally with a simulation. The maximum in the number of electrons in the low energy range obtained from simulations is caused by including the electrons which are momentarily displaced by the laser pulse

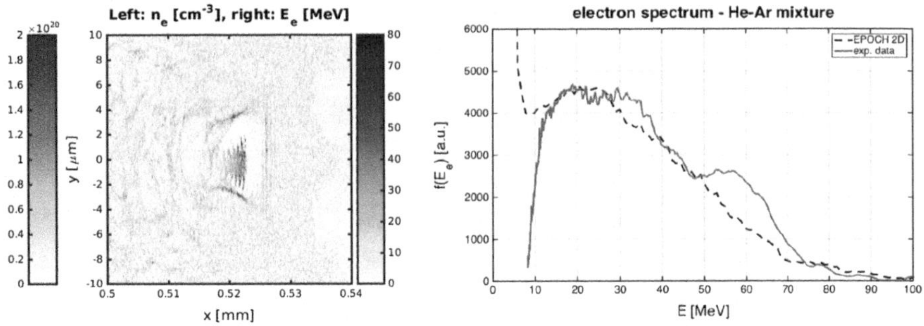

Fig. 20.5 Electron acceleration using the mixture of helium and argon (99:1). Left: A 2D map of the electron plasma density showing the forming ion cavity and electrons undergoing betatron oscillations. Right: The comparison of experimental data with simulations for the electron spectrum. The maximum in the number of electrons in the low-energy range obtained from simulations is caused by including the electrons which are momentarily displaced by the laser pulse

The facility will provides four different laser beams; high-brightness multi-TW ultrashort laser pulses at kHz repetition rate (L1 laser system), PW 10 Hz repetition rate pulses (L2 and L3), and kilojoule 100 fs pulses for generation of 10 PW peak power (L4 laser) [9] (Fig. 20.6).

Plasma-based X-ray sources of coherent or incoherent radiation will be implemented in six experimental halls. More specifically, the plasma X-ray source (PXS) and high harmonics generation (HHG for more details see [5]) driven by the

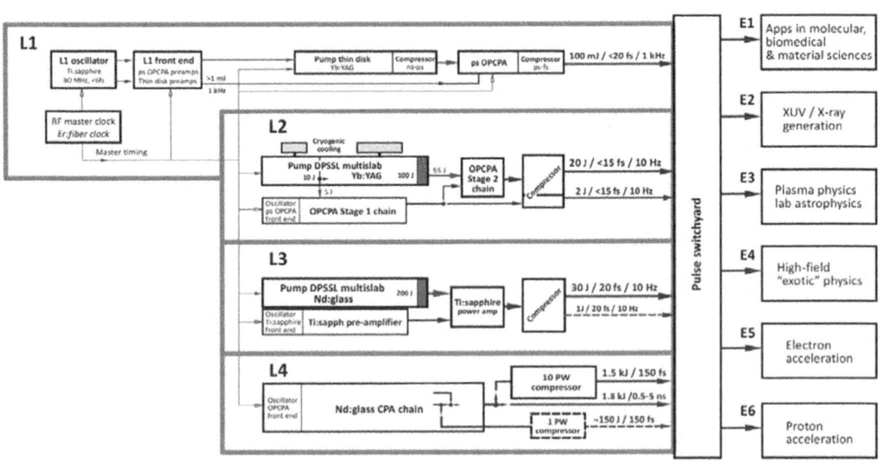

Fig. 20.6 Overview of the scheme of ELI Beamlines laser systems. The facility will consist of four lasers, L1–L4, which will be transported to different experimental hall, E1–E6, where will be used for different experiments. Considered laser upgrades for nearest future are indicated by dotted lines

Fig. 20.7 Schematic overview of experimental hall E2

1-kHz L1 laser system delivering output energy up to 100 mJ in 20 fs pulse duration will be located in experimental hall E1. The PXS source fits onto a standard optical table. The design will allow for ultrafast, hard X-ray Diffraction (XDIFF), X-ray Absorption Spectroscopy (XAS), and X-ray Imaging (XI) experiments. Crystalline, as well as a liquid sample, can be investigated. The PXS beamline is intended to deliver pulses of hard X-rays at selected spectral lines corresponding to 3.3, 11, 24, 77 keV with $>10^9$ photons per pulse (photons/(4π sr line) @ 10 keV) with time envelope shorter than 300 fs. The PXS will have two beamlines that can be used simultaneously or that allow setup of two different experiments. The betatron/Compton beamline in the hall E2 (see in Fig. 20.7) will deliver femtosecond X-ray pulses in a wide spectral range, from tens of keV to tens of MeV with controlled spectral features (broadband or nearly monochromatic tunable spectrum).

The beamline relies on focussing the L3 laser beam into a gas jet or a gas cell. For an appropriate choice of experimental parameters (laser intensity, spot size and duration, electron density in the gas), electrons are accelerated to relativistic energies by plasma wakefield acceleration and wiggled by the plasma itself (Betatron source) or by a second laser pulse (Compton source). This results in the emission of intense femtosecond X-ray or Gamma-ray beams emitted from a micron-sized source. The features of the generated radiation will depend on the need of the end user. The expected performances of Betatron/Compton sources are summarized in Table 20.1.

The Laser-Plasma-Driven Undulator X-ray source (LUX) will be placed in the E5 experimental hall. The driver for the beamline will be PW class L3 laser and this beamline will be focused to generate photons in the “water window” spectral range. X-ray pulses should deliver 10^5–10^6 photons per pulse, using an undulator of less

Table 20.1 The expected performances of either Betatron (left) or Compton (right) source in the E2 hall

Source	Betatron phase A	Betatron phase B	Compto
Energy range	1–10 keV	1–100 keV	100 keV–1 MeV
Photon number	IE+08	1E+08	1E+08
Spectrum	Brodband	Broadband	Broadband/ monochromatic
Pulse duration @ source (fs)	10	10	10
Divergence (mrad)	40	20	20
Source size (μ)	<5	<5	<5

than 1 m long with the undulator period of the order of 5 mm. All the sources will be complimented by auxiliary infrared laser pulses with femtosecond synchronization for pump-probe experiments.

Acknowledgements The authors acknowledge the great support from TiSa technical staff, namely J. Golasowski, J. Hřebíček and T. Medřík. The project is supported by the projects; ELI—Extreme Light Infrastructure—phase 2 (CZ.02.1.01/0.0/0.0/15_008/0000162) from European Regional Development Fund, Czech Science Foundation (project No. 15-03118S), LASERLAB-EUROPE—grant agreement no. 284464, EC's Seventh Framework Programme, OPVK—project no. CZ.1.07/2.3.00/30.0057.

References

1.. Jungwirth, K., et al.: The Prague asterix laser system. Phys. Plasmas (2001)
2. Extreme light infrastructure Whitebook. Science and Technology with ultra-intense lasers, Issue 1, FZU 2010
3. Novak, O., et al.: Status of the high average power diode-pumped solid state laser development at HiLASE. Appl. Sci. **5**, 637–665 (2015). https://doi.org/10.3390/app5040637
4. Hřebíček, J., et al.: 25TW Ti:sapphire laser chain at PALS. In: Proceedings of the SPIE, vol. 8080 (2011)
5. Nejdl, J., et al.: Overview of Laser-Driven Short-Wavelength Sources at PALS and ELI Beamlines (in this proceeding)
6. Fullagar, W., et al.: The use and characterization of a backilluminated charge-coupled device in investigations of pulsed X-ray and radiation sources. Rev. Sci. Instrum. **79**, 103302 (2008)
7. https://ccpforge.cse.rl.ac.uk/gf/project/epoch/
8. Ammosov, M.V., Delone, N.B., Krainov, V.P.: Tunnel ionization of complex atoms and of atomic ions in an alternating electromagnetic field. Sov. Phys. JETP **64**, 1191 (1986)
9. Rus, B., et al.: ELI-Beamlines: development of next generation short-pulse laser systems. In: Proceedings of the SPIE 9515, Research Using Extreme Light: Entering New Frontiers with Petawatt-Class Lasers II, vol. 95150F (2015). https://doi.org/10.1117/12.21849

Chapter 21
Research on Laser Acceleration and Coherent X-Ray Generation Using J-KAREN-P Laser

M. Kando, A. S. Pirozhkov, M. Nishiuchi, H. Kiriyama, A. Kon, H. Sakaki, Y. Fukuda, N. Dover, K. Sekiguchi, K. Nishitani, A. Sagisaka, T. A. Pikuz, A. Ya. Faenov, K. Ogura, Y. Hayashi, H. Kotaki, T. Zh. Esirkepov, K. Huang, N. Nakanii, K. Kondo, J. K. Koga, S. V. Bulanov and K. Kondo

Abstract We present the progress on the upgrade status of the J-KAREN-P, which is a Ti: Sapphire laser aiming at the intensity of 10^{22} W/cm^2 at the repetition rate of 0.1 Hz. The upgrade includes two pilot experiments in order to show the laser

---

M. Kando (✉) · A. S. Pirozhkov · M. Nishiuchi · H. Kiriyama · A. Kon · H. Sakaki
Y. Fukuda · N. Dover · K. Sekiguchi · K. Nishitani · A. Sagisaka · K. Ogura
Y. Hayashi · H. Kotaki · T. Zh. Esirkepov · K. Huang · N. Nakanii · K. Kondo
J. K. Koga · S. V. Bulanov · K. Kondo
Kansai Photon Science Institute, National Institutes for Quantum and Radiological Science and Technology (QST), 8-1-7 Umemidai, Kizugawa, Kyoto 619-0215, Japan
e-mail: kando.masaki@qst.go.jp

A. Kon
Japan Synchrotron Radiation Research Institute (JASRI), Sayo, Hyogo 679-5198, Japan

K. Nishitani
Interdisciplinary Graduate School of Engineering, Kyushu University, 6-1 Kasuga-Koen, Kasuga, Fukuoka 816-8580, Japan

T. A. Pikuz
Graduate School of Engineering, Osaka University, 2-1 Yamadaoka, Suita, Osaka 565-0871, Japan

T. A. Pikuz · A. Ya. Faenov
Joint Institute for High Temperatures of the Russian Academy of Sciences, 13/19 Izhorskaja Street, Moscow 125412, Russia

A. Ya. Faenov
Institutes for Academic Initiatives, Osaka University, Suita, Osaka 565-0871, Japan

S. V. Bulanov
Moscow Institute of Physics and Technology (State University), 9 Institutskii Pereulok, Dolgoprudnyi, Moscow Region 141700, Russia

S. V. Bulanov
A. M. Prokhorov Institute of General Physics of the Russian Academy of Sciences, 38 Vavilov Street, Moscow 119991, Russia

T. Kawachi et al. (eds.), *X-Ray Lasers 2016*, Springer Proceedings in Physics 202, https://doi.org/10.1007/978-3-319-73025-7_21

performance on target. The first experiment is to generate high-energy ions from thin-foil target. The second experiment is the high-order harmonic at a relativistic intensity. Currently, laser acceleration of protons is being tested and we have obtained 32 MeV protons from a 5-μm stainless steel target irradiated by a 14-J, 30-fs laser pulse. In addition, a joint program toward compact X-ray free-electron laser based on laser electron acceleration is presented briefly and the corresponding J-KAREN-P work is presented.

21.1 Introduction

The recent advent of laser technology enables pursuing compact novel beam sources, such as electrons, protons, heavy ions, X-ray, and so on. Especially in high-intensity lasers novel phenomena is expected to happen, e.g., highly efficient ion acceleration, radiation damping effects, vacuum birefringence, vacuum breakdown, etc. In order to boost the field, J-KAREN-P laser is being upgrade since 2013 aiming at ~1 PW, 30 fs, 10^{22} W/cm^2. The contrast ratio of the laser is also an important factor especially for laser-matter interaction because a contained intense preceding component of the laser may ionize material before arriving of the main laser pulse. Thus, the recent development of high-intensity lasers requires not only high peak power but also high quality.

In this paper, first, we present J-KARE-P upgrade status and show the activity report on the pilot experiment on ion acceleration. The planned experiments of electron acceleration and X-ray generation are also presented. The system of J-KAREN-P is almost completed but in trouble-shooting status by employing ion acceleration experiment. Here, we report the results obtained in the first and second experimental campaigns (December 2015 and March 2016).

21.2 J-KAREN-P Upgrade

The J-KAREN laser was a Ti: Sapphire laser system based on the chirped pulse amplification (CPA) with a high-contrast ratio of 10^{12} [1]. The good contrast was due to several novel techniques such as a double CPA system, saturable absorbers, an optical parametric chirped pulse amplifier (OPCPA) instead of a standard regenerative amplifier, and careful adjustment of pump power timing. The system was operated at 10 Hz up to a power amplifier stage (~2 J) and the final booster amplifier (~30 J) was operated at a single shot.

Basic concepts of the upgrade included an increase of a peak power and intensity at an interaction point. Small-diameter optics of the previous system limited the usable laser energy on target and an old pump laser for the final amplifier degraded reproducibility of the output energy. Thus, we replaced the small optics including mirrors, diffraction gratings, vacuum ducts and chambers with larger ones.

Table 21.1 Comparison of typical parameters of the former J-KAREN and the design parameters of the J-KAREN-P upgrade

	Previous J-KAREN	J-KAREN-P upgrade
Peak power on target	200 TW	~1 PW
Pulse duration (fs)	30	30
Repetition rate (Hz)	0.00056	0.1
Contrast ratio	10^{12}	10^{12}
Beam diameter (mm)	150	250
Peak irradiance (W/cm^2)	10^{21}	$\sim 10^{22}$

In addition, the final amplifier was divided into two stages and pumped by six Nd: Glass lasers. Table 21.1 shows the comparison of the previous J-KAREN and the J-KAREN-P upgrade.

The focusability is also very important to explore high-field science; Not only achieving a high intensity with limited power laser but some applications favor that well-controlled qualities. A part of this kind of dependence has been observed in our laboratory using the previous J-KAREN laser.

The strategy to improve the focusability is described here. First, almost all newly installed large flat mirrors and diffraction gratings were characterized using an interferometer (ZYGO GPI) before and after mounting them on optomechanical mounts. Second, the alignment of parallelism of diffraction gratings in a pulse compressor was accomplished within ~100 μrad (final goal: ~10 μrad). Third, a deformable mirror and a wavefront sensor were used to correct the distorted wavefront. In the March 2016 run, the Strehl ratio was ~0.1 on target. Note that, this Strehl ratio was not the expected value of the wavefront sensor but the calculated value from measured far-field and near-field patterns. This poor Strehl ratio was caused by several reasons: an imperfect algorithm of closed loop in the deformable mirror and the wavefront sensor, imperfect beam expander mirrors (reveals higher order spherical aberration), and several lenses used in expanders (chromatic aberration). After the run, we improved the above problems and obtained the Strehl ratio of ~0.5.

21.3 Laser Acceleration of Particles

Laser acceleration of charged particles is one of the applications using J-KAREN-P laser [2, 3]. In this section, we introduce our ion and electron activities. Laser ion acceleration uses mainly target normal sheath acceleration (TNSA), where energetic ions are accelerated along the normal to target, while energetic electrons tend to the laser direction. In our configuration, the laser pulse irradiates targets (either tapes or foils) at the incidence angle of 45° with respect to the target normal in order to avoid possible back reflection from the target.

21.3.1 Laser-Driven Ion Acceleration

After the completion of the laser part upgrade, we started shooting targets since December 2015 as a pilot experiment to test the system performance. The laser energy was limited to a moderate energy of 14 J on target. The pulse duration was 36 fs (effective width), the contrast ratio was 10^{10} at −500 ps, and the focal spot was 1.48 ± 0.06 μm × 1.57 ± 0.09 μm full width at half maximum (FWHM) containing 10.11 ± 0.82% of the total energy. The peak irradiance was 2×10^{21} W/cm^2. The target was a 5-μm-thickness tape of stainless steel. An experimental setup inside the target chamber is shown in Fig. 21.1.

Detectors used were an electron spectrometer (ESM), a time-of-flight ion spectrometer (TOF), a scintillator-based ion footprint monitor, plasma monitors (shadowgraphy and interferometry), reflection and transmission light diagnostics (spatial distribution and spectrum) and an insertable stack of radiochromic films (RCF). All detectors except the RCF were capable of an online readout after each shot. In addition, on-shot input laser diagnostics were prepared, such as near and far-field pattern imagers, a prepulse monitor with fast photodiodes, and an optical spectrometer. The ESM was composed of a permanent magnet, a phosphor scintillating screen and an imaging camera that covers electron energies up to 100 MeV. The TOF measured proton energies from a few MeV to ~50 MeV, consisting of a vacuum tube and a plastic scintillator with a photomultiplier read with an oscilloscope.

Fig. 21.1 Short-F chamber inside. A tape target is irradiated by an alignment red laser

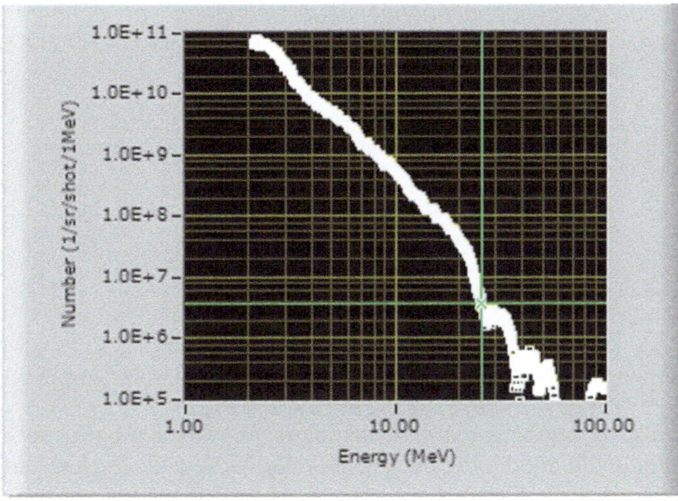

Fig. 21.2 A typical time-of-flight proton energy spectrum

The TOF was used to optimize the target position and the maximum proton energy of 32 MeV was obtained as shown in Fig. 21.2. The proton energy was also confirmed with the RCF detector. The direction of the protons was almost normal to the target showing the main acceleration process was target normal sheath acceleration.

21.3.2 Laser-Driven Electron Acceleration

We are planning electron acceleration experiments using J-KAREN-P at the Long-F target area. The main purpose of the experiments is to establish stable electron acceleration conditions using 10 J class laser for booster acceleration in multistage laser wakefield acceleration (MSLWFA) scheme [4].

The MSLWFA concept is adopted in the ImPACT program led by Y. Sano having collaboration with Osaka University, High Energy Accelerator Research Organization (KEK), RIKEN-Harima, Nagaoka University of Technology, and QST. The concept of MSLWFA is depicted in Fig. 21.3. An injector, a phase rotator, and a booster will be employed like conventional linear accelerators. The resonant frequencies used in the stages differ by changing the laser pulse duration. In the project, we will construct a Ti:Sapphire laser system that can deliver three different pulse duration laser pulses. Here, the duration is adjusted by adding the group velocity dispersion. The three pulses are used to optimize the electron beam parameters.

The mission of the planned experiment using J-KAREN-P is to simulate booster acceleration with J-KAREN-P by chirping the pulse up to ~100 fs (The Fourier transform limit duration is ~30 fs). The focal length of an off-axis parabola is 2.6 m

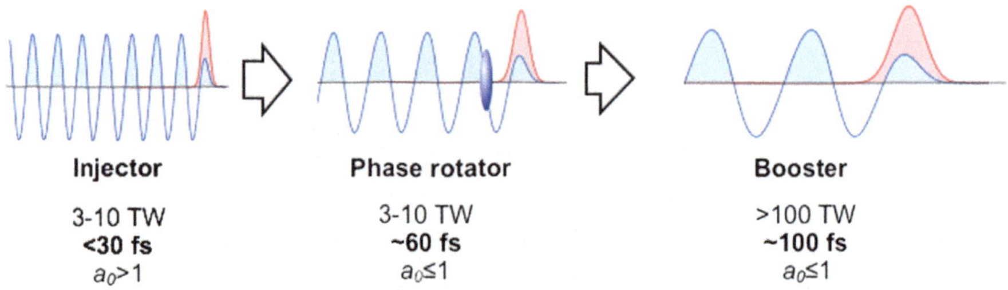

Fig. 21.3 A schematic of multistage laser wakefield acceleration concept

(f/10). In parametric particle-in-cell simulations, electron acceleration up to 1 GeV is possible when a 1-cm long waveguide is used.

We are also developing real-time monitors for wakefield structures, betatron X-rays, and pulse duration of electron beams. The wakefield structures are observed by using a sub 10 fs shadowgraphy and phase-sensitive imaging and by a frequency domain holography.

Betatron X-rays are generated by wiggling motion of energetic electrons in the wakefield structure. If an electron beam is injected offset to a next stage, the amplitude of transverse oscillation becomes significant and thus strong X-ray emission is expected (betatron radiation). We will monitor such X-rays using a Laue-type spectrograph. A prototype spectrograph is under construction in our institute.

The pulse duration of the electron beam is a crucial parameter for obtaining high brightness. Note that the energy spread is also related to the injected electron beam duration. We have chosen two methods; one is using electro-optic sampling and another is based on the coherent transition radiation detection.

21.4 Coherent X-Ray Generation

Laser-based coherent X-ray sources are highly demanded and several mechanisms based on the intense, short laser pulses are known. Here, we introduce the two cases; relativistic high-order harmonics and laser-accelerated electron-based free-electron laser that are underdevelopment using J-KAREN-P.

21.4.1 Relativistic High-Order Harmonics from Underdense Plasma

High-order harmonics in the extreme ultraviolet (XUV) can be generated in the interaction of an intense laser pulse with a gas target [5, 6]. These phenomena were found in our previous experiment using J-KAREN. A simple scaling law suggests that keV coherent emission is possible if more intense laser pulse is used. Such fully coherent X-rays are useful for coherent imaging of single molecules and nonlinear X-ray optics.

Currently, the experimental setup is constructed using another target chamber of the J-KAREN-P, so-called Long-F chamber, together with the electron acceleration experiment.

21.4.2 Laser-Driven Electron Acceleration-Based X-Ray Free-Electron Laser

Laser-driven acceleration toward X-ray free-electron laser has been started in ImPACT program as partly introduced in the Sect. 21.3.2 [7]. The goal of the project is to generate 1 keV photons from a short-period undulator [8] using 1 GeV electron beams accelerated with a multistage laser wakefield accelerator. Preliminary parameters are given in Table 21.2. The challenging part is to obtain monochromatic electron beam with high pointing and energy stability.

21.5 Concluding Remarks and Future Prospects

We introduced our J-KAREN-P laser facility at Kansai Photon Science Institute, National Institutes for Quantum and Radiological Science and Technology (QST). The upgrade is almost completed and we have started test experiments on ion acceleration. We observed energetic protons up to 32 MeV. We are further improving the quality of J-KAREN-P and expect higher proton energy. Heavier ion acceleration such as carbon ion is also planned.

Electron acceleration experiments and relativistic high-order harmonics experiments are also being in progress.

Table 21.2 Example XFEL parameters for 1-keV using standard undulator used in SACLA and a short-period undulator being developed in ImPACT program

	Standard undulator	Short-period undulator
Undulator period (cm)	1.8	0.4
K-parameter	2.1	0.15
Electron energy (GeV)	2.5	0.65
Pierce parameter, ρ	2.5×10^{-3}	5.9×10^{-4}
Required emittance, ε_n (mm mrad)	<0.48	<0.13

The plan toward laser-driven X-ray free-electron laser is briefly introduced in the scope of the ImPACT program led by Dr. Sano. Several test experiments and developments are underway at KPSI, QST.

Acknowledgements We thank J-KAREN-P operators for the help with construction and for the operation during the experimental campaigns. The electron acceleration and XFEL development are supported by ImPACT Program of Council for Science, Technology, and Innovation (Cabinet Office, Government of Japan). M. K. acknowledges Prof. Kodama, Prof. Hosokai, Prof. Yamamoto, Dr. Yabashi and Prof. Sasaki for collaborative work in the development of laser-driven XFEL sources and Program Manager Dr. Sano for the encouragement and support. A part of the work is supported by Japan Society for the Promotion of Science (JSPS) KAKENHI JP 26707031 and Ministry of Education, Culture, Sports, Science, and Technology (MEXT).

References

1. Kiriyama, H., Shimomura, T., Sasao, H., et al.: Temporal contrast enhancement of petawatt-class laser pulses. Opt. Lett. **37**, 3363–3365 (2012)
2. Ogura, K., Nishiuchi, M., Pirozhkov, A.S., et al.: Proton acceleration to 40 MeV using a high intensity, high contrast optical parametric chirped-pulse amplification/Ti:sapphire hybrid laser system. Opt. Lett. **37**, 2868–2870 (2012)
3. Nishiuchi, M., Sakaki, H., Esirkepov, T.Z., et al.: Acceleration of highly charged GeV Fe ions from a low-Z substrate by intense femtosecond laser. Phys. Plasmas **22**, 033107 (2015)
4. Hosokai, T., Masuda, S., Zhidkov, A., et al.: Staging laser wakefield acceleration research at practical accelerators. In: Proceedings of the 13th Annual Meeting of Annual Meeting of Particle Accelerator Society of Japan, MOOLP01, pp. 1–3 (2016)
5. Pirozhkov, A.S., Kando, M., Esirkepov, T.Z., et al.: Soft-X-ray harmonic comb from relativistic electron spikes. Phys. Rev. Lett. **108**, 135004–135005 (2012)
6. Pirozhkov, A.S., Kando, M., Esirkepov, T.Z., et al.: High order harmonics from relativistic electron spikes. New J. Phys. **16**, 093003–093030 (2014)
7. http://www.jst.go.jp/impact/en/program/03.html
8. Yamamoto, S.: Undulator development towards very short period lengths. Synch. Rad. News **28**, 19–22 (2015)

Chapter 22
X-Rays Driven by Single-Cycle, Petawatt Lasers: A Path to Exawatt Pulses

J. Wheeler, G. Cojocaru, R. Ungureanu, R. Banici, R. Gonin, S. Mironov, M. Serbanescu, R. Dabu and G. Mourou

Abstract The post-compression of high-energy Petawatt (PW) scale laser pulses offers the promise of maximizing the peak intensity deliverable by a given laser system for the amplified energy produced within the pulses. One recent proposal made by Mourou et al. relied upon the flat-top property of modern high-energy laser systems to suggest a solution utilizing self-phase modulation within a thin plastic film material to generate the bandwidth necessary to recompress to a shorter pulse duration through compensation with negative dispersion mirrors. The so-called thin film compressor (TFC) promised to offer an efficient and affordable method to boost the peak pulse achievable with existing facilities. Recent measurements including experiments done at the CETAL PW laser based at the National Institute for Lasers, Plasma, and Radiation Physics (INFLPR) in Magurele, Romania demonstrate the feasibility of the TFC configuration on a small-scale and encourage continued pursuit at increasing laser levels. The applications of such high energy, ultrashort pulses are already being considered in regard to particle acceleration and X-ray generation.

J. Wheeler (✉) · G. Mourou
DER-IZEST, École Polytechnique, Palaiseau Cedex, France
e-mail: jonathan.wheeler@polytechnique.edu

J. Wheeler · R. Dabu
National Institute for Nuclear Physics-HH, ELI-NP, Măgurele, Romania

G. Cojocaru · R. Ungureanu · R. Banici · M. Serbanescu
CETAL, INFLPR, Măgurele, Romania

R. Gonin
Laserix, Université Paris-Sud, Orsay Cedex, France

S. Mironov
Institute of Applied Physics RAS, Nizhny Novgorod, Russia

T. Kawachi et al. (eds.), *X-Ray Lasers 2016*, Springer Proceedings in Physics 202, https://doi.org/10.1007/978-3-319-73025-7_22

22.1 Introduction

Petawatt (PW = 10^{15} W) and multi-PW laser facilities are encouraging scientists to explore novel applications to a broad range of exciting new science covering fundamental physics in vacuum structure, nuclear physics, and astrophysics as well as applications in these areas, such as radioactive waste management, material science, and life sciences. It is within the realm of exploring novel uses of extreme laser pulses that this current work expands upon to explore a pulse compression technique that promises to efficiently transform the PW multi-cycle pulses to single-cycle pulses. This offers access to even more opportunities including not only unprecedented pulse intensities but also the possibility to drive ultra-relativistic, plasma processes.

The proposed Thin Film Compression technique (TFC) [1] is a dramatic gain in the compressible pulse energies as compared with the gas-filled hollow capillary methods and single-cycle TW (10^{12} W) laser systems regularly demonstrated today. The TFC relies upon the nonlinear intensity-dependence of the index of refraction due to both the spatial and temporal beam profiles: $n(\boldsymbol{x}, t) = n_0 + n_2 I(\boldsymbol{x}, t)$. Specifically, the flat-top nature of PW pulses allows for uniform intensity-dependent spectral broadening across the beam profile due to self-phase modulation (SPM) induced by the pulse while avoiding complications such as self-focusing and filamentation. The resultant pulse is then recompressed to shorter pulse duration through phase compensation with negative dispersion mirrors.

The proposed compression scheme proposes a thin, submillimeter thick, thermoplastic film (i.e., PMMA, Cellulose Acetate, and PET, among others) to act as the nonlinear medium and is attractive due to the accommodation of a clear aperture

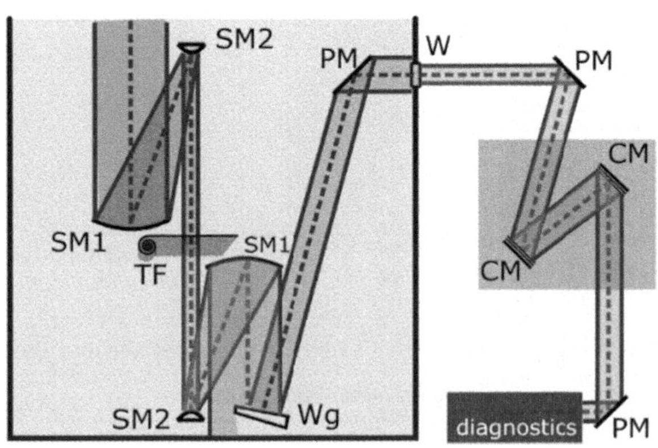

Fig. 22.1 The beam from the frontend of the CETAL PW laser is reduced to an energy density typical of PW transport using spherical mirrors (SM1,2) before interaction with a thin film (TF) followed by expansion (SM2,1) and attenuation by uncoated wedge (Wg). The resulting pulse is recompressed by negative dispersion mirrors (CM) outside of a vacuum. A WIZZLER was used to measure the pulse duration

for a 20–50 cm diameter beam while being affordably replaceable. The nonlinear response is often greater than that of glass with initial studies on PET [2] suggesting to have a n_2 approximately double that of fused silica: $n_2 = 3 \cdot 10^{-7}$ cm^2/GW. This range suggests adequate SPM will occur at the propagation intensity near 1 TW/cm^2 for PW pulses through the spectral phase had yet to be measured to confirm the ability to recompress the resultant pulse.

22.2 Methods and Results

The 1PW CETAL laser system in Bucharest-Magurele, Romania is now available and offers the opportunity to begin experimental work on new PW class pulses. Here, the peak power is specified at ≥ 1 PW, with a pulse duration <25 fs, repetition rate of 0.1 Hz and a picosecond pre-pulse contrast $>10^{10}$ at 100 ps. There is also the possibility to run the system without the final amplifier for a TW output specified with a peak power ≥ 45 TW, and a repetition rate of 10 Hz. In the work, the maximum energy delivered on target was 200 mJ and the beam diameter of the system was reduced to a beam area of 1 cm^2 allowing for the use of standard 25 mm optics while maintaining expected PW transport fluence at 100 mJ/cm^2 (Fig. 22.1). The initial temporal duration for these studies was measured to be 45 fs and the intensities of interaction are therefore up to 3.0 TW/cm^2. The variation of the pulse spectrum after passing through 500 μm of cellulose acetate with increasing laser intensity is shown in Fig. 22.2. The transmission of the films is greater than 90% at normal incidence. A series of chirp mirrors were installed to find an ideal re-compression of a sampled portion of the pulse outside of the vacuum chamber. Figure 22.3 shows the best-measured post-compression with a single pair of chirped mirrors (−250 fs^2/bounce) from approximately 45 fs down to 29 fs. This is under conditions of non-optimized initial pulse compression

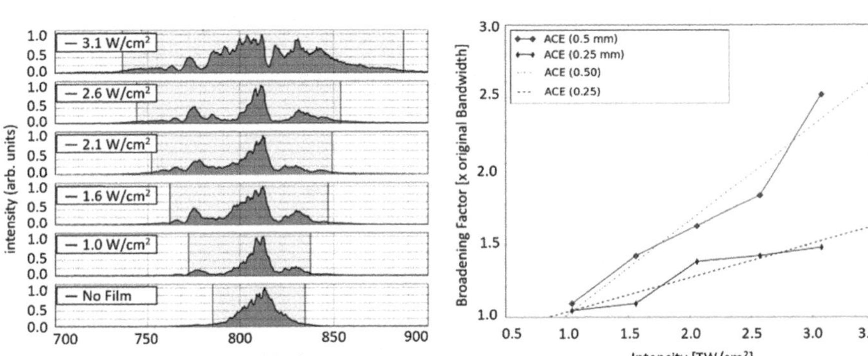

Fig. 22.2 Spectral response in cellulose acetate (0.5 mm thick) with increasing intensity

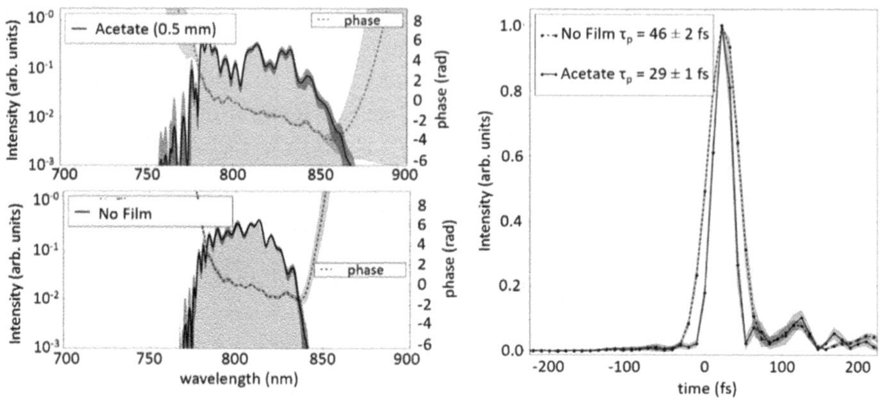

Fig. 22.3 Spectral amplitude and phase with cellulose acetate compared with the original pulse characteristics (No Film) [3]

(expected to be 25 fs) and additional limitations of the WIZZLER diagnostic which seemed unable to measure pulses under 20-fs duration within our conditions.

22.3 Perspectives

It is expected that such efficient compression will allow for several stages to reach single-cycle pulse duration and with appropriate focusing can reach its minimum volume as defined by the laser wavelength (λ) and enters the so-called λ^3 regime [4]. Such a compressed, high-energy pulse offers access to novel laser intensities—above 10^{19} W/cm^2—becoming the gateway to study relativistic laser–plasma interactions and promises to open up an entirely new realm of fundamental and applied physics. One potential application will be the up-conversion of NIR laser pulses through a plasma-based relativistic oscillating mirror [5] (ROM) to coherent, hard X-ray pulses of keV to MeV energy photons with pulse durations on the order of attoseconds to zeptoseconds (10^{-18} to 10^{-21} s) with estimates of over 10^{10} photons/pulse. At these pulse durations, even modest conversion efficiency leads to exawatt (10^{18} W) level pulses.

References

1. Mourou, G., Mironov, S., Khazanov, E., Sergeev, A.: Single cycle thin film compressor opening the door to Zeptosecond-Exawatt physics. Eur. Phys. J. Spec. Top. **223**, 1181 (2014)
2. Mironov, S.Y., Ginzburg, V.N., Gacheva, E.I., Silin, D.E., Kochetkov, A.A., Mamaev, Y.A., Shaykin, A.A., Khazanov, E.A., Mourou, G.A.: Use of polyethylene terephthalate for temporal recompression of intense femtosecond laser pulses. Laser Phys. Lett. **12**, 25301 (2015)

3. Mironov, S.Y., Wheeler, J., Gonin, R., Cojocaru, G., Ungureanu, R., Banici, R., Serbanesu, M., Dabu, R., Mourou, G., khazanov, E.A.: 100J-level pulse compression for peak power enchancement. Quantum Electron. **47**, 173 (2017)
4. Naumova, N.M., Nees, J.A., Sokolov, I.V., Hou, B., Mourou, G.A.: Relativistic generation of isolated attosecond pulses in a λ^3 focal volume. Phys. Rev. Lett. **92**, 3 (2004)
5. Bulanov, S.V., Naumova, N.M., Pegoraro, F.: Interaction of an ultrashort, relativistically strong laser pulse with an overdense plasma. Phys. Plasmas **1**, 745 (1994)

Chapter 23
Ultra-intense X-Ray Radiation Photopumping of Exotic States of Matter by Relativistic Laser–Plasma in the Radiation-Dominated Kinetic Regime (RDKR)

A. Ya. Faenov, J. Colgan, S. A. Pikuz, A. Zhidkov, T. A. Pikuz, J. Abdallah Jr., E. Tubman, N. M. H. Butler, R. J. Dance, I. Yu. Skobelev, M. Z. Alkhimova, N. Booth, J. Green, C. Gregory, A. Andreev, M. Nishiuchi, H. Sakaki, A. Sagisaka, A. S. Pirozhkov, K. Ogura, Y. Fukuda, M. Kanasaki, N. Hasegawa, M. Nishikino, M. Kando, T. Kawachi, K. Kondo, P. McKenna, G. J. Tallents, N. Woolsey and R. Kodama

Abstract In the present report, we discussed our recent results concerning the investigations of high energy density matter. By using of high-resolution X-ray spectroscopic measurements and kinetic simulations, we demonstrate that the energy of femtosecond laser pulses is efficiently converted to X-ray radiation and produces exotic states in solid density plasma regions. It is shown that exotic hollow multicharged ions states can be obtained and studied not only by

A. Ya. Faenov (✉) · A. Zhidkov · R. Kodama
Open and Transdisciplinary Research Initiatives, Osaka University, Suita, Osaka, Japan
e-mail: anatolyf@hotmail.com

A. Ya. Faenov · S. A. Pikuz · T. A. Pikuz · I. Yu. Skobelev · M. Z. Alkhimova
Joint Institute for High Temperatures, Russian Academy of Sciences, Moscow, Russia

J. Colgan · J. Abdallah Jr.
Theoretical Division, Los Alamos National Laboratory, Los Alamos, USA

T. A. Pikuz · R. Kodama
PPC and Graduate School of Engineering, Osaka University, Yamadaoka, Suita, Osaka, Japan

E. Tubman · G. J. Tallents · N. Woolsey
Department of Physics, York Plasma Institute, University of York, York, UK

N. M. H. Butler · R. J. Dance · P. McKenna
Department of Physics, SUPA, University of Strathclyde, Glasgow, UK

T. Kawachi et al. (eds.), *X-Ray Lasers 2016*, Springer Proceedings in Physics 202, https://doi.org/10.1007/978-3-319-73025-7_23

ultra-intense X-ray radiation of XFEL lasers, but also upon using optical laser technology. We have concluded that observable features of the hollow-ion spectra are sensitive to such plasma parameters as density, temperature, hot-electron fraction, and intensity of the X-ray pumping radiation and could be used for effective diagnostics of warm dense matter parameters.

23.1 Introduction

The radiation properties of high energy density plasma [1–4] and, especially, relativistic laser-produced plasma [5–7] are under increasing interests in recent years due to their importance to our understanding of stellar interiors, the cores of giant planets, and the properties of hot plasma in inertial confinement fusion devices [8–15]. Recently, ultra-intense X-ray free-electron laser sources (XFELs) are being used (See Fig. 23.1) to create directly such high energy density matter [3, 16] using X- rays at intensities greater than 10^{17} W/cm^2. The X-ray radiation intensity dominates standard collisional atomic processes creating exotic states of matter with empty K or L-shells and composed of hollow atoms [3, 4, 17–19], which are diagnosed through the observation of unique spectral lines [20–27].

We have concluded in [19, 23–26] that conventional modern optical lasers (See Figs. 23.1 and 23.2) with short pulse duration could generate very bright X-ray radiation with intensities exceeding 10^{17} W/cm^2. In this case, exotic states of matter are efficiently produced (See Fig. 23.3a). In the present report, we give overview of obtained results. The very important conclusion means that exotic hollow ions states can be obtained and studied not only by ultra-intense X-ray radiation of XFEL lasers, but also upon using optical laser technology. It should be mentioned

N. Booth · J. Green · C. Gregory
CLF, STFC Rutherford Appleton Laboratory, Didcot, Oxfordshire, UK

A. Andreev
Max Born Institute, Berlin 12489 Max-Born Str. 2a, Berlin, Germany

A. Andreev
ELI-ALPS, Szeged, Hungary

M. Nishiuchi · H. Sakaki · A. Sagisaka · A. S. Pirozhkov · K. Ogura
Y. Fukuda · M. Kanasaki · N. Hasegawa · M. Nishikino · M. Kando
T. Kawachi · K. Kondo
Kansai Photon Science Institute, National Institutes for Quantum and Radiological Science and Technology, Kyoto, Japan

S. A. Pikuz · I. Yu. Skobelev · M. Z. Alkhimova
National Research Nuclear University MEPhI, Moscow 115409, Russia

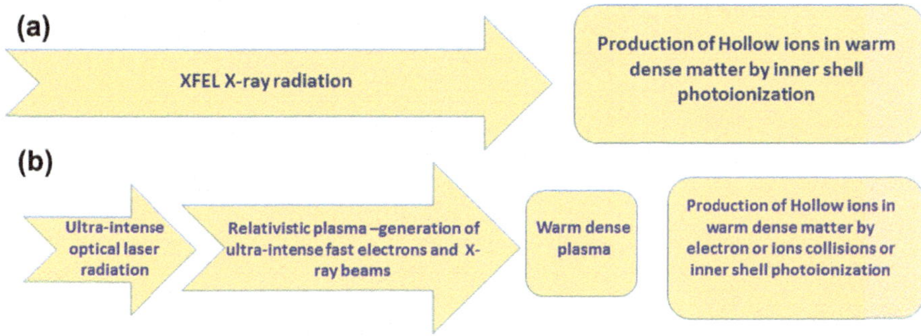

Fig. 23.1 Ultra-intense X-ray radiation of XFEL laser beams or plasma under relativistic laser irradiation could efficiently produce warm dense matter with states of hollow ions, which are exotic for atomic physics. **a** Direct excitation by ultra-intense X-ray laser or XFEL pulses; **b** Indirect excitation by visible ultra-intense laser pulses interacting with thin foils and causing generation of ultra-intense X-ray radiation

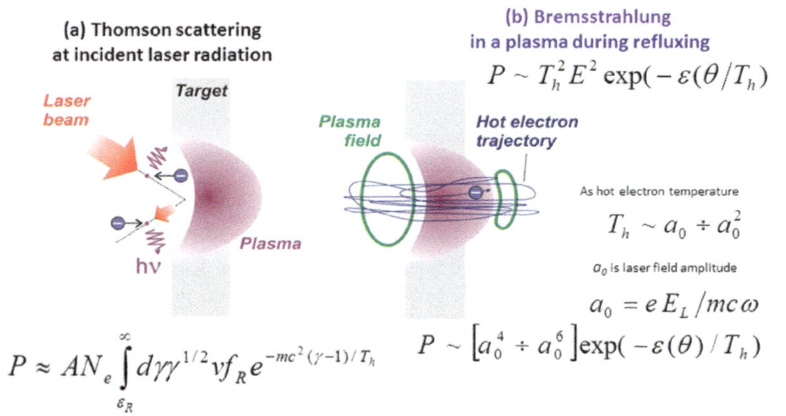

Fig. 23.2 Generation of ultra-intense X-ray source with intensities P ~ 10^{17-18} W/cm^2 in plasma irradiated by relativistic laser pulses with intensities above 10^{20} W/cm^2. **a** X-ray generation by Thomson scattering at incident laser radiation; **b** X-ray generation by bremsstrahlung in a plasma during electron refluxing

also that radiation dominated kinetic regime (RDKR) for multicharged ions [19, 23–26] could be reached with the help of optical lasers with ultra-high flux densities.

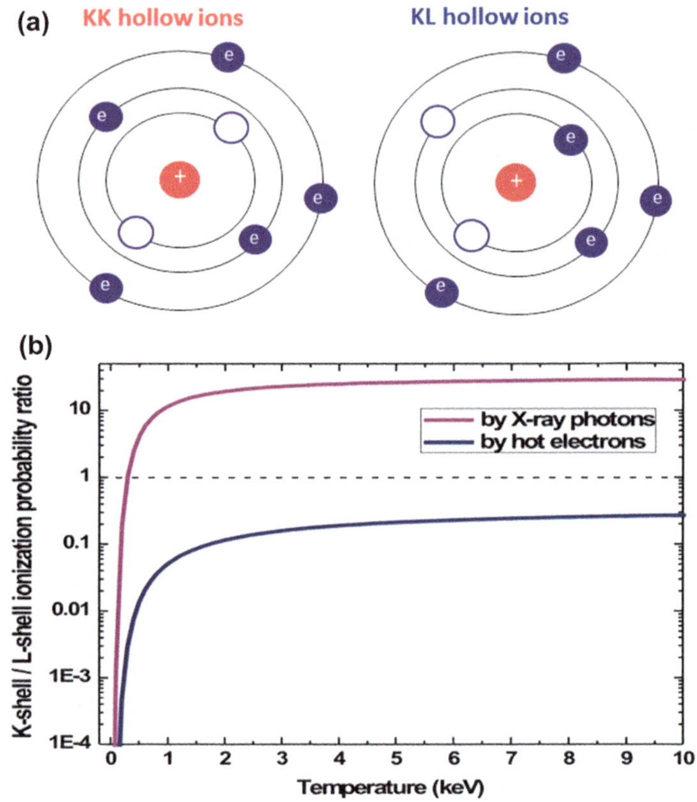

Fig. 23.3 Hollow ions spectroscopy principles: the role of external X-ray photo pumping and hot electrons collisional excitation on the KK and KL hollow ions production. **a** Exotic hollow atoms states, in which K- or L-shell internal electrons are removed, but external electrons are still existed; **b** Ionization K-shells/L-shells probability ratio for ionization by hot electrons collisions and by hard X-ray photons photopumping. Such ratio clear demonstrated that under hot electrons collisions the generation of KK hollow ions is very weak, but KL hollow ions produced rather efficiently. Meanwhile, KK hollow ions efficiently generated by hard X-ray photopumping

23.2 Driving Mechanism of Ultra-intense X-Ray Source Generation in Relativistic Laser–Plasma Interaction and its Photopumping of Hollow Ions States

The basic idea of ultra-intense X-ray source generation under interaction of relativistic laser pulses with thin foils is as follows. Laser light of ultra-high intensity is focused onto the target, which is a thin metal foil. Due to field ionization [28], the valence electrons are ionized and then quickly accelerated to high (MeV) energy if the laser beam intensity is high enough. During interaction of an intense laser pulse with the foil, some of these hot electrons will oscillate when many times passing through the foil (this effect known as refluxing) [29], and to radiate the X-ray photons of very different hardness, until γ-radiation (see, for example, [29–34]). Possible mechanisms of radiation are Thomson scattering of the incident beam or scattered laser radiation and bremsstrahlung in a strong field at the boundaries of the plasma foil (See Fig. 23.2). According to [19, 25], if the intensity of laser light exceeds 10^{20} W/cm^2, these processes can lead to the formation of the flow of X-ray photons with energies of a few keV and an X-ray intensity of higher than 10^{17} W/cm^2. For petawatt laser (PWL) intensities, which could be above 10^{22} W/cm^2, the radiation process begins to dominate over the electron dynamics reaching so-called radiation damping regime (RDR). In the case of lower laser intensities (10^{20}–10^{22} W/cm^2) radiation intensity does not yet dominate over the electron dynamics, but already dominates over radiation from standard collisional atomic processes. Such RDKR plasma conditions

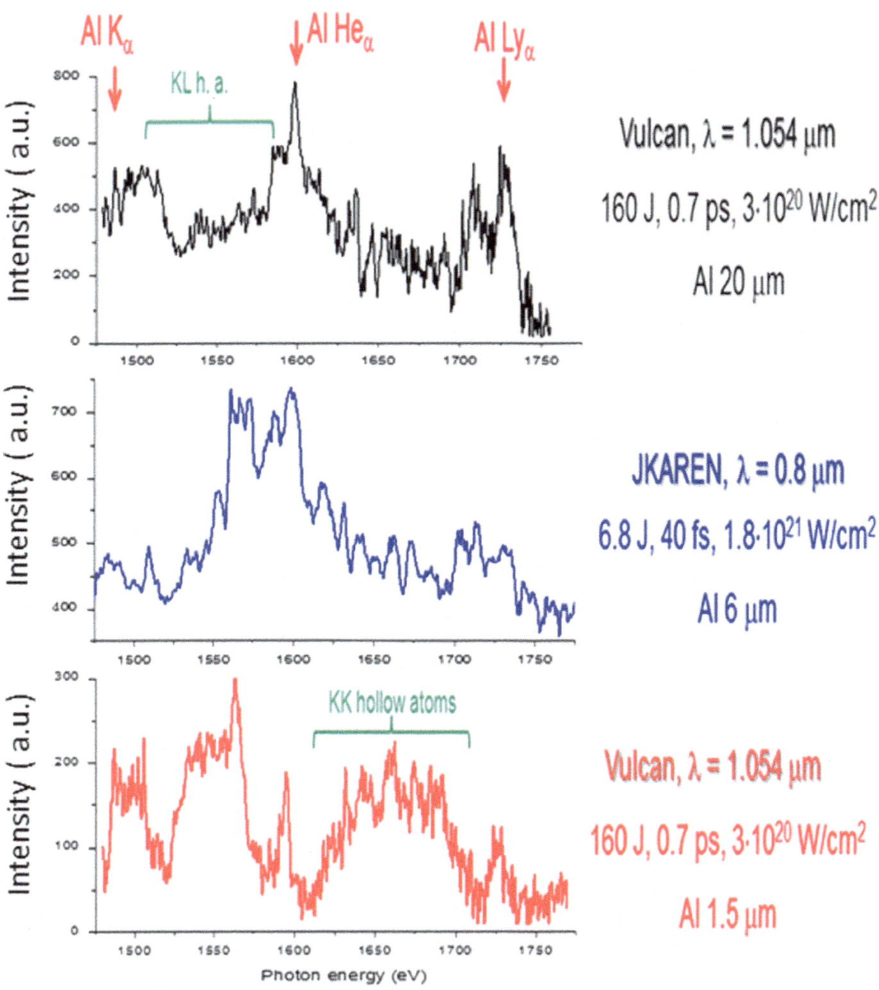

Fig. 23.4 Observation of X-ray spectra of KK Hollow ions in plasmas irradiated by ultra-intense relativistic optical laser pulses [19, 23–25]

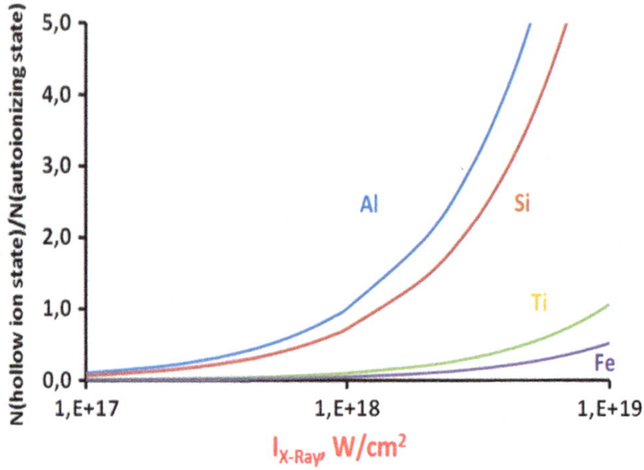

Fig. 23.5 Efficiency of hollow ions photoionization by external X-ray source for different ionic charges. The fact of observation of hollow ions of different ions allowed to estimate the value of X-ray source intensity

begin to be responsible for creating in the plasma periphery exotic states of matter composed of hollow ions with multiple inner-shell excitations (See Fig. 23.3b), radiation of which is observed as KK hollow ions spectra. Typical spectra of KK hollow ions, observed in different relativistic laser-produced plasma experiments are presented in Fig. 23.4. It is clearly seen that the KK hollow ions of Al could be observed when optical laser intensity exceeds 3×10^{20} W/cm^2. It is necessary to underline that efficiency of hollow ions photoionization by external X-ray sources dramatically falls down with increasing nuclear charge of target Z. Indeed, the ratio of production of usual autoionizing states versus hollow ions ionization states is $N(2l^n)/N(1s2l^n) \sim Z^{-4}$ (See also Fig. 23.5). It means that, if for generation of hollow ions states of Al the X-ray source intensity should be in the order of 10^{17} W/cm^2 or higher, than for production of the same states in Fe the X-ray intensity should be at least of 10^{19} W/cm^2. Additionally, the energy of photons, which photo pump mater should be higher than potential of K-shell ionization.

23.3 Observation of Al and Si Hollow Ions Spectra in Plasma Irradiated by Ultra-intense Femto- and Sub-picosecond Relativistic Laser Pulses

Recently, it was shown [19, 23–25] (See also Fig. 23.4) that conventional high-power optical lasers with short pulse duration could efficiently produce so-called hollow multicharged ions. Even more exotic states, namely Rydberg hollow ions have been observed in the X-ray emission of Si-tagets, heated by

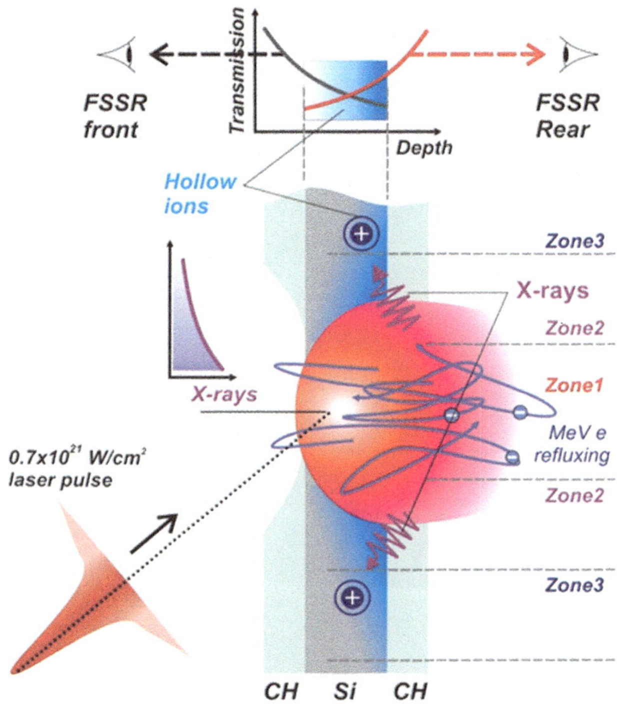

Fig. 23.6 A schematic diagram of hollow ion formation by the ultra-intense optical laser pulse and its detection from both sides of the target in experiments [26]

sub-picosecond laser pulses with intensity of 0.3×10^{21} W/cm^2 at Vulcan laser facility [26]. In Fig. 23.6 a schematic diagram of such experiment is presented. It should be noted that, in this case, we observed spectra radiated from front and rare side of target. The comparison of experimental data [26] with theoretical results produced by ATOMIC code are presented in Fig. 23.7. It was found that the main spectral features can be reasonably explained, if we consider 3 plasma zones, as shown in Fig. 23.6. The first, central zone contains bulk plasma at high electron temperatures $T_e \sim 400$ eV (for the rear side of the target) or 550 eV (front side) and an electron density $N_e \sim 10^{22}$ cm^{-3}. Just in this region, the most part of the laser energy is directly absorbed. The second plasma zone surrounds the central zone. It has lower T_e of 180 eV and N_e of 3×10^{23} cm^{-3} (close to solid density conditions). This zone feels a radiation field at a temperature of 2 keV, which arises from the radiation, emitted by high-energy refluxing electrons formed in the focus of the laser spot. The third plasma zone is placed beyond the second plasma zone. It has a cooler temperature of 10 eV and solid density. Plasma in this zone is also excited by external X-ray radiation.

It should be stressed that emission around 5.0 Å, as it is prominent from calculations, arise from what was termed as *Rydberg hollow ions*, that is, transitions

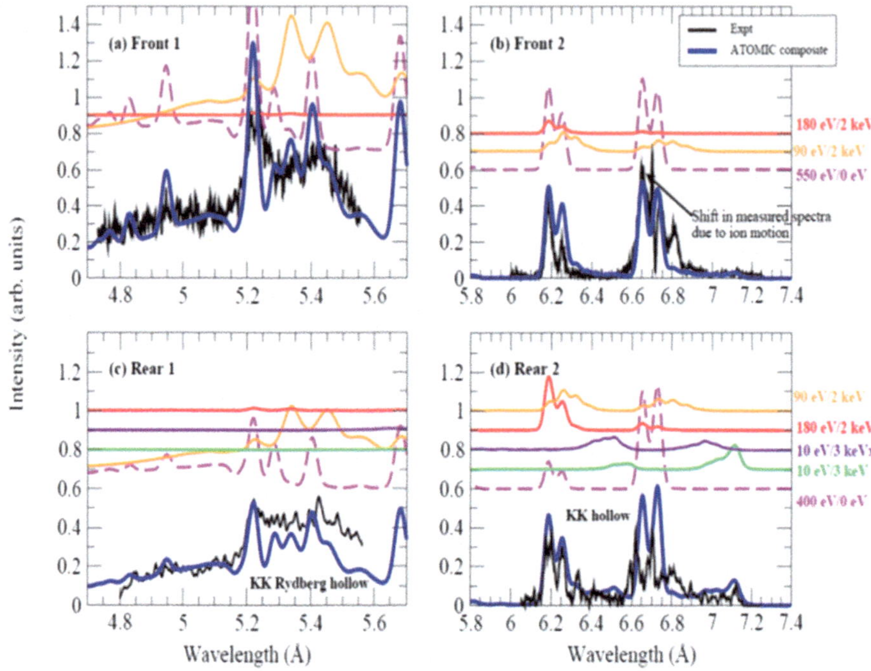

Fig. 23.7 Comparison of measured spectra from the four spectrometers and composite ATOMIC calculations [26], assuming emission from various plasma zones as it mentioned in Fig. 23.6

from the $3p$ subshell to the empty $1s$ (K) shell of Li-like (and others) Si ions. It follows from our calculations that for the Si XIII ion the X-ray radiation will produce a sufficient number of hollow ions when $I_{X\text{-ray}}$ is of about 10^{18} W/cm^2 or greater (See also Fig. 23.5).

Another important results were obtained at the J-KAREN laser facility [25] when Ti:Sa laser irradiated Al foils with different thickness (Fig. 23.8). It is clearly seen that with increasing of laser intensity up to 10^{21} W/cm^2 intensity of observed X-ray spectra increases proportional to $\sim E^{4-5}$ of laser power (See Fig. 23.7a, b). It means that the RDKR is reached only at laser intensities around 10^{21} W/cm^2.

We have concluded that observable features of the hollow-ion spectra are sensitive to such plasma parameters as density, temperature, hot-electron fraction, and intensity of the X-ray pumping radiation and could be used for effective diagnostics of warm dense matter parameters. Our results strongly support theoretical predictions [29–34] that with increasing laser intensities up to 10^{22}–10^{24} W/cm^2 a conversion efficiency of visible light both to soft and to hard X-rays may achieve of 30–50%, which opens a path towards many applications of recently developing petawatt-class lasers, particular with a high repetition rate and femtosecond pulse duration.

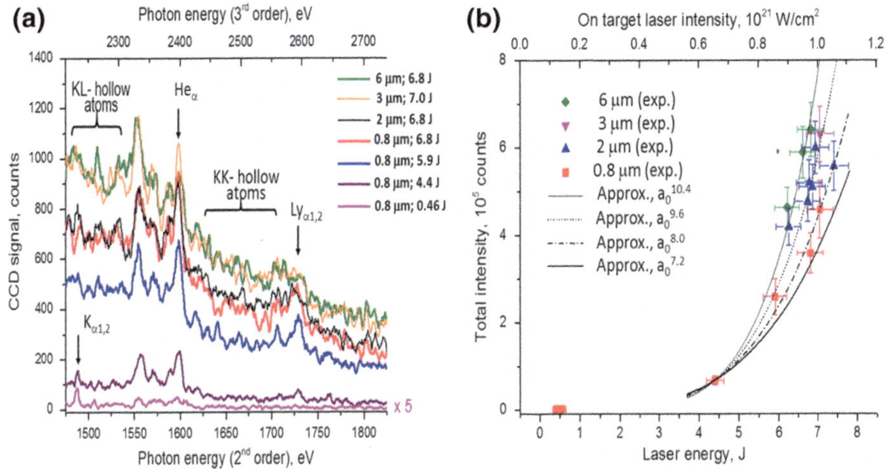

Fig. 23.8 **a** Single shot K-shell spectra of Al ions [25] emitted from foil targets with different thickness and laser energies. **b** Intensity of X-ray emission observed at X-ray CCD integrated over the energy range 1450–1850 eV and 2210–2730 eV (raw data) versus the laser intensity for Al foils with different thickness

Acknowledgements The research leading to these results has received funding from the Science and Technology Facilities Council, and the Engineering and Physical Science Research Council (Grant No. EP/J003832/1) of the UK. The Los Alamos National Laboratory is operated by Los Alamos National Security, LLC for the NNSA of the U.S. DOE under Contract No. DEAC52-06NA25396. This work was supported by JSPS KAKENHI (Grant No. 24740280 and Grant No. 25390135), by funds of Institute for Academic Initiatives, Osaka University, and partly is supported by RAS Presidium Program for Basic Research Program#13 and Competitiveness Program of NRNU MEPhI.

References

1. Drake, R.P.: High-Energy-Density Physics: Fundamentals. Springer, Berlin, Heidelberg, New York (2006)
2. Fortov, V.E.: Extreme States of Matter: on Earth and in the Cosmos. Springer, Berlin, Heidelberg, New York (2011)
3. Vinko, S.M., et al.: Creation and diagnosis of a solid-density plasma with an X-ray free-electron laser. Nature (London) **482**, 59–63 (2012)
4. Skobelev, I.Yu., Faenov, A.Ya., Pikuz, T.A., Fortov, V.E.: Spectra of hollow ions in an ultra-dense laser plasma. Phys. Usp. **55**, 47 (2012)
5. Mourou, G.A., Tajima, T., Bulanov, S.V.: Optics in the relativistic regime. Rev. Mod. Phys. **78**, 309–371 (2006)
6. Pukhov, A.: Strong field interaction of laser radiation. Rep. Prog. Phys. **66**, 47–101 (2003)
7. McKenna, P., Neely, D., Bingham, R., Jaroszynski, D. (eds): Laser-Plasma Interactions and Applications. Springer (2013)
8. Di Matteo, T., Springler, V., Hernquist, L.: Energy input from quasars regulates the growth and activity of black holes and their host galaxies. Nature **433**, 604–606 (2005)

9. Beiersdorfer, P.: Laboratory X-ray astrophysics. Ann. Rev. Astron. Astrophys. **41**, 343–390 (2003)
10. Chabrier, G.: Plasma physics and planetary astrophysics. Plasma Phys. Control. Fusion **51**, 123014 (2009)
11. Bouquet, S., et al.: From lasers to the universe: scaling laws in laboratory astrophysics. High Energy Density Phys. **6**, 368–380 (2010)
12. Bingham, R., et al.: Laboratory astrophysics: Investigation of planetary and astrophysical maser emission. Space Sci. Rev. **178**, 695–713 (2013)
13. Giardi, A., et al.: Astrophysics of magnetically collimated jets generated from laser-produced plasmas. Phys. Rev. Lett. **110**, 025002 (2013)
14. Savin, D.W., et al.: The impact of recent advances in laboratory astrophysics on our understanding of the cosmos. Rep. Prog. Phys. **75**, 036901 (2012)
15. Albertazzi, B., et al.: Laboratory formation of a scaled protostellar jet by coaligned poloidal magnetic field. Science **346**, 325–328 (2014)
16. Yoneda, H., et al.: Saturable absorption of hard X-rays iron. Nat. Commun. **5**, 5080 (2014)
17. Faenov, A.Ya., et al.: Diagnostics of warm dense matter by high-resolution X-ray spectroscopy of hollow ions. Laser Part. Beams **33**, 27–39 (2015)
18. Pikuz, S.A (Jr.), Faenov, A.Ya., Skobelev, I.Yu., Fortov, V.E.: Production of exotic states of matter with the use of X-rays generated by focusing a petawatt laser pulse onto a solid target. Phys.—Uspekhi **57**, 702–707 (2014)
19. Colgan, J., et al.: Exotic dense-matter states pumped by a relativistic laser plasma in the radiation-dominated regime. Phys. Rev. Lett. **110**, 125001 (2013)
20. Faenov, A.Ya., et al.: High-resolution X-ray spectroscopy of hollow atoms created in plasma heated by subpicosecond laser radiation. Proc. SPIE **3157**, 10–21 (1997)
21. Faenov, A.Ya., et al.: High-resolved X-ray spectra of hollow atoms in a femtosecond laser-produced solid plasma. Phys. Scr. **T80**, 536–538 (1999)
22. Faenov, A.Ya., et al.: Hollow ion spectra in warm dense laser-produced plasma: observation and modeling. J. Phys.: Conf. Ser. **163**, 012016 (2009)
23. Pikuz, S.A., et al.: Measurement and simulations of hollow atom X-ray spectra of solid-density relativistic plasma created by high-contrast PW optical laser pulses. High Energy Density Phys. **9**, 560–567 (2013)
24. Hansen, S.B., et al.: Detailed analysis of hollow ions spectra from dense matter pumped by X-ray emission of relativistic laser plasma. Phys. Plasmas **21**, 031213 (2014)
25. Faenov, A., et al.: Nonlinear increase of X-ray intensities from thin foils irradiated with a 200 TW femtosecond laser. Sci. Rep. **5**, 13436 (2015)
26. Colgan, J., et al.: Evidence of high-n hollow-ion emission from Si ions pumped by ultraintense X-rays from relativistic laser plasma. Eur. Phys. Lett. **114**, 35001 (2016)
27. Cricosta, O., et al.: Detailed model for hot-dense aluminum plasmas generated by an X-ray free electron laser. Phys. Plasmas **23**, 022707 (2016)
28. Zhidkov, A., Sasaki, A.: Effect of field ionization on interaction of an intense subpicosecond laser pulse with foils. Phys. Plasmas **7**, 1341–1344 (2000)
29. Zhidkov, A., et al.: Radiation damping effects on the interaction of ultraintense laser pulses with an overdense plasma. Phys. Rev. Lett. **88**, 185002 (2002)
30. Ridgers, C.P., et al.: Dense electron-positron plasmas and ultraintense γ-rays from laser-irradiated solids. Phys. Rev. Lett. **108**, 165006 (2012)
31. Nakamura, T., et al.: High-power γ-rays flash generation in ultraintense laser-plasma interactions. Phys. Rev. Lett. **108**, 195001 (2012)
32. Ji., L.L., et al.: Radiation-reaction trapping of electrons in extreme laser fields. Phys. Rev. Lett. **112**, 145003 (2014)
33. Sentoku, Y., et al.: Kinetic effects and nonlinear heating in intense X-ray laser-produced carbon plasmas. Phys. Rev. E **90**, 051102 (2014)
34. Capdessus, R., d'Humières, E., Tikhonchuk, V.T.: Influence of ion mass on laser-energy absorption and synchrotron radiation at ultrahigh laser intensities. Phys. Rev. Lett. **110**, 215003 (2013)

Chapter 24
Enhanced Coherent Thomson Scattering in the Few-Cycle Regime

Ke Hu and Hui-Chun Wu

Abstract We study an enhancement to the coherent Thomson scattering by scattering off a few-cycle incident laser pulse from a relativistic electron layer. Here the existing theory for long incident lasers is extended, which allows us to calculate coherent Thomson scattering in the few-cycle regime. From the analytic theory and PIC simulations, we find two new nonadiabatic effects: amplitude enhancement and frequency downshift. The scattering efficiency can be increased by two orders of magnitude due two these two effects when the thickness of the layer is larger than the wavelength of the generated X-ray pulse.

24.1 Introduction

The advent of ultrashort laser sources, such as high harmonics from gas targets, relativistic laser–plasma interactions, and free electron lasers are opening a new area of attosecond optics and ultrafast science. Among them, coherent Thomson scattering (CTS) from relativistic electron layers has the potential to produce isolated X-ray pulses as powerful as 10 GW and as short as 10 as. The electron layer has a relativistic velocity β_0 vertical to the layer surface. When an oppositely directed incident laser impinges on this layer, its frequency ω_i is boosted due to the relativistic Doppler effect.

CTS of long-pulse incident lasers has been discussed in previous studies [1, 2]. An analytic theory is built, from which we know that the scattering efficiency has nothing to do with the laser duration. In this paper, we extend the previous theory to the few-cycle regime. It is striking that the conversion efficiency can rise considerably

K. Hu (✉) · H.-C. Wu
Department of Physics, Institute for Fusion Theory and Simulation (IFTS), Zhejiang University, Hangzhou 310027, China
e-mail: ke_hu@zju.edu.cn

H.-C. Wu
IFSA Collaborative Innovation Center, Shanghai Jiao Tong University, Shanghai 200240, China

T. Kawachi et al. (eds.), *X-Ray Lasers 2016*, Springer Proceedings in Physics 202, https://doi.org/10.1007/978-3-319-73025-7_24

when the incident laser becomes ultrashort. We use both PIC simulations and analytic calculations to obtain the scattering amplitude and frequency, as well as the enhancement condition.

24.2 Amplitude Enhancement

In the analytic model, suppose the electron layer has a Gaussian density distribution $n_e(x) = n_0 \exp(-\pi x^2/l_0^2)$, where n_0 and l_0 represent the maximum density and characteristic thickness, respectively. The layer flies along the x-axis with β_0 and the corresponding relativistic factor is $\gamma_0 = (1 - \beta_0^2)^{-1/2}$. Such a layer can be produced from the laser–plasma interaction in an ultra-thin (nm-scale) solid foil [2, 3]. Its energy spread can be as small as ~0.1%, thus we assume it to be monoenergetic. The circularly polarized incident laser has a normalized vector potential $\mathbf{a}_i(\tau) = e\mathbf{A}_i/mc = a_i f(\tau) \times [\hat{y}\cos(\omega_i\tau) + \hat{z}\sin(\omega_i\tau)]$, where m is the electron mass, c is the speed of light, $a_i = eA_i/mc, f(\tau) = \exp[-(\tau/T_i)^2]$, $\omega_i = 2\pi/\tau_i$, $\tau = t + x/c$, T_i represents the pulse duration, and $\tau_i = \lambda_i/c$ is the light wave cycle. Under this convention, the wavelength of the generated X-ray can be expressed by $\lambda_s = \lambda_i/D_{nl}$, where $D_{nl} = 4\gamma_0^2/(1 + a_i^2)$ is the relativistic Doppler factor.

In the flat-top laser pulse ($f(\tau) = 1$), every electron has almost the same dynamics in the longitudinal direction. However, the situation becomes quite different if a few-cycle incident laser is adopted, since the effect of its envelope $f(\tau)$ must be considered. And the electron layer undergoes a deformation during the scattering process [4]. After numerically integrating the current density of the layer, we can obtain peak amplitude of the scattering X-ray.

Figure 24.1a displays how the scattering amplitude a_s changes with the laser duration T_i for the cases of $a_i = 5$ and 3.5. The initial parameters are $\lambda_i = 800$ nm, $l_0 = 0.6$ nm, $n_0 = 10n_c$, and $\gamma_0 = 70$, where $n_c = \varepsilon_0 m\omega_i^2/e^2$ is the critical density. We see that the PIC results agree well with those obtained from our analytic theory. For $a_i = 5$, as T_i goes down, the X-ray strength remains almost unchanged at first,

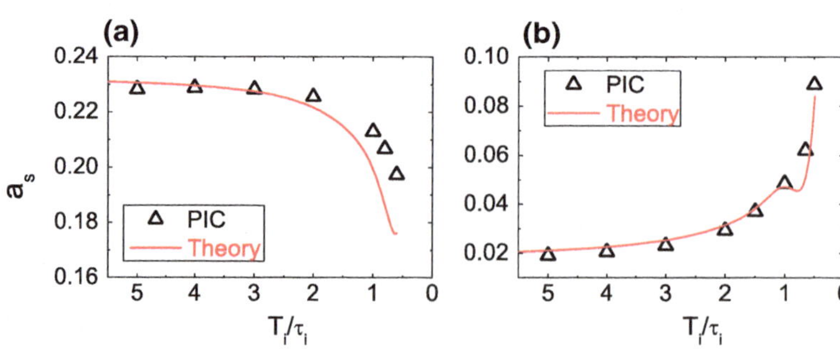

Fig. 24.1 Peak scattering amplitude a_s versus the duration of the incident laser T_i for $a_i = 5$ (**a**) and $a_i = 3.5$ (**b**)

then slightly decreases when $T_i < 3\tau_i$. For $a_i = 3.5$, however, the peak scattering amplitude increases dramatically.

Physically, the enhancement of a_s in Fig. 24.1 is due to the suppression of the destructive interference. We can calculate from the analytic theory that the enhancement emerges at

$$l_0 > 0.225\frac{\lambda_i(1+a_i^2)}{\gamma_0^2} \simeq 0.9\lambda_s. \tag{24.1}$$

24.3 Frequency Downshift

For long incident lasers, the central frequency of the generated X-ray signal $\omega_s = \omega_i D_{nl}$ is independent of the laser duration. Nevertheless, in the few-cycle regime, the central frequency is related to T_i. It is obvious in Fig. 24.2a that $\omega_s/(\omega_i D)$ decreases with T_i for both $a_i = 5$ and 3.5. We call this effect frequency downshift.

From the theory, central scattering frequency can be quantitatively calculated:

$$\omega_s = \omega_i D_{nl}\frac{\pi(T_i/\tau_i)^2}{\pi(T_i/\tau_i)^2 + (l_0/\lambda_s)^2}. \tag{24.2}$$

Similar to (24.1) presented above, (24.2) demonstrates that the frequency downshift shows up only when $l_0 \sim \lambda_s$. And in Fig. 24.2a, (24.2) agrees well with the results obtained by PIC simulations.

Figure 24.2b illustrates the X-ray amplitude for $a_i = 3.5$. One can see that the X-ray wavelength increases as the laser duration shrinks from $T_i = 5\tau_i$ to $T_i = 0.8\tau_i$. At $T_i = 0.8\tau_i$, the generated pulse is shorter than 8 as long and its intensity is about 1.3×10^{16} W/cm^2. If the emission spot has a radius of 5 μm, the X-ray's peak power can be as large as 10 GW.

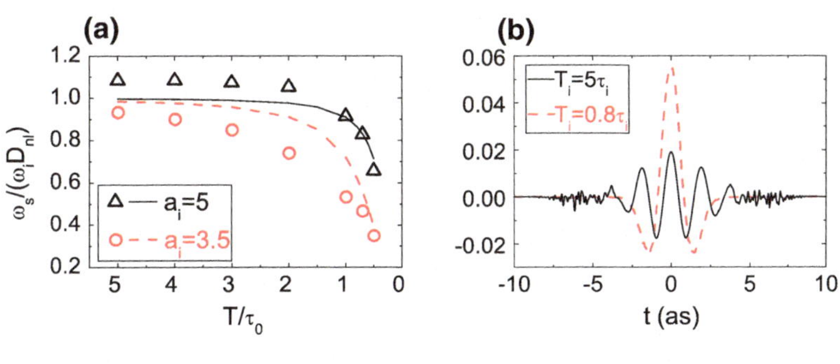

Fig. 24.2 **a** Central frequency ω_s of the generated pules versus the duration of the incident laser T_i for $a_i = 5$ and 3.5. **b** Electric field $a_{s,y}$ of the generated X-rays for $a_i = 3.5$

At last, we analyze the efficiency of energy conversion. For $a_i = 5$, the efficiency increases by about 46% when $T_i \rightarrow 0.5\tau_i$, which is mainly due to the frequency downshift as shown in Fig. 24.2a. For $a_i = 3.5$, both amplitude enhancement and frequency downshift play important roles in the increase. The efficiency rise is over 100 times, from 6.8×10^{-9} at $T_i = 5\tau_i$ to 7.9×10^{-7} at $T_i = 0.5\tau_i$.

24.4 Summary

In summary, we have discussed new nonadiabatic effects of nonlinear CTS in the few-cycle regime and found that the efficiency increases more than one order of magnitude within a region of laser intensities. The dramatic scattering enhancement is due to two facts. First, peak scattering amplitude rises due to the weakening of destructive interference. Second, as the laser duration shrinks, the lower frequency components are reflected more efficiently, which causes the scattering frequency to become lower and the duration to become larger. Both scattering amplitude and frequency are predicted by a revised analytic model, and agree well with PIC simulations. Moreover, the enhancement condition is also obtained: the thickness of the electron layer should be larger than the wavelength of the generated X-rays.

References

1. Wu, H.C., Meyer-ter-Vehn, J.: Eur. Phys. J. D **55**, 443 (2009)
2. Wu, H.C., Meyer-ter-Vehn, J., Fernández, J., Hegelich, B.M.: Phys. Rev. Lett. **104**, 234801 (2010)
3. Wu, H.C., Meyer-ter-Vehn, J., Hegelich, B.M., Fernández, J.: Phys. Rev. ST Accel. Beams **14**, 070702 (2011)
4. Hu, K., Wu, H.C.: Opt. Lett. **41**, 4586 (2016)

Chapter 25
Plasma Channel Undulator for Narrow-Bandwidth X-Ray Generation

S. G. Rykovanov, J. W. Wang, V. Yu. Kharin, B. Lei, C. B. Schroeder, C. G. R. Geddes, E. Esarey and W. P. Leemans

Abstract A new plasma channel undulator concept based on the wakefields generated by short intense laser pulse undergoing centroid oscillations inside parabolic plasma channel is presented. The period of such an undulator is proportional to the Rayleigh length of the laser pulse and can be in the submillimeter range, while its strength can reach unity. Two-dimensional particle-in-cell simulations of the laser pulse propagation and wakefields are presented. Spontaneous radiation produced by the electron beam inside the plasma undulator is calculated.

25.1 Introduction

Nowadays, it is hard to imagine scientific progress without UV and X-ray sources for studying the properties of materials, and dynamics of atoms and molecules [1]. The brightest X-ray pulses are generated using free-electron lasers using magnetic undulators to force electron beam oscillations and microbunching. Typically, the period of state-of-the-art magnetic undulators that can reach undulator strength a_u on the order of unity, is on the order of several centimetres. Undulators with period on the order of a millimetre capable of reaching undulator strength of unity can help to significantly shrink the size and the cost of FEL facilities. Several interesting microundulator concepts have been proposed [2–6].

Recently, a novel method to produce a plasma undulator based on the laser pulse centroid oscillations inside the parabolic plasma channel was proposed [7, 8]. Wakefields in the underdense plasma created by short intense laser pulse injected off-axis or under some angle with respect to the parabolic plasma channel serve as undulator fields and force electrons to wiggle and emit radiation. Here we present the results of the 2D3V particle-in-cell (PIC) simulations using the code LAPINE [9], demonstrating the feasibility of the proposed concept.

S. G. Rykovanov (✉) · J. W. Wang · V. Y. Kharin · B. Lei
Helmholtz Institute Jena, Fröbelstieg 3, 07743 Jena, Germany
e-mail: s.rykovanov@gsi.de

C. B. Schroeder · C. G. R. Geddes · E. Esarey · W. P. Leemans
Lawrence Berkeley National Laboratory, Berkeley, CA 94720, USA

T. Kawachi et al. (eds.), *X-Ray Lasers 2016*, Springer Proceedings in Physics 202, https://doi.org/10.1007/978-3-319-73025-7_25

25.2 Plasma Wakefield, Electron Trajectories and Radiation

Laser pulse guiding for distances on the order of several centimetres is necessary for achieving the GeV level electron beam energies [10]. In the case of a matched laser pulse injected off-channel axis or under some angle into the parabolic plasma channel, its centroid will exhibit harmonic oscillations with the characteristic period $\lambda_u = 2\pi Z_R$, while its spot size will remain constant. For the laser pulse centroid, assuming for simplicity 2D geometry with z the channel longitudinal axis and x the transverse coordinate, one can write $x_c = x_{ci} \cos\left(z/Z_R + \varphi\right)$, where x_{ci} is initial laser pulse centroid displacement. Propagation of the moderately intense laser pulse, with dimensionless amplitude $a_0 = eA_0/mc^2 = 0.3$, where A_0 is the peak of the laser pulse vector potential, c—speed of light in vacuum and e and m electron absolute charge and mass, inside the parabolic channel obtained from the 2D PIC simulations is presented in Fig. 25.1 (top), where the instantaneous intensity in arbitrary units is presented as a function of z and x at different moments of time. Oscillatory behaviour is clearly observed. The plasma density profile was chosen to be $n = n_0\left(1 + x^2/R_{ch}^2\right)$, where $R_{ch}^2 = \pi e^2 n_0 w_0^4/mc^2$ is the matched channel radius, and n_0 is the on-axis plasma density.

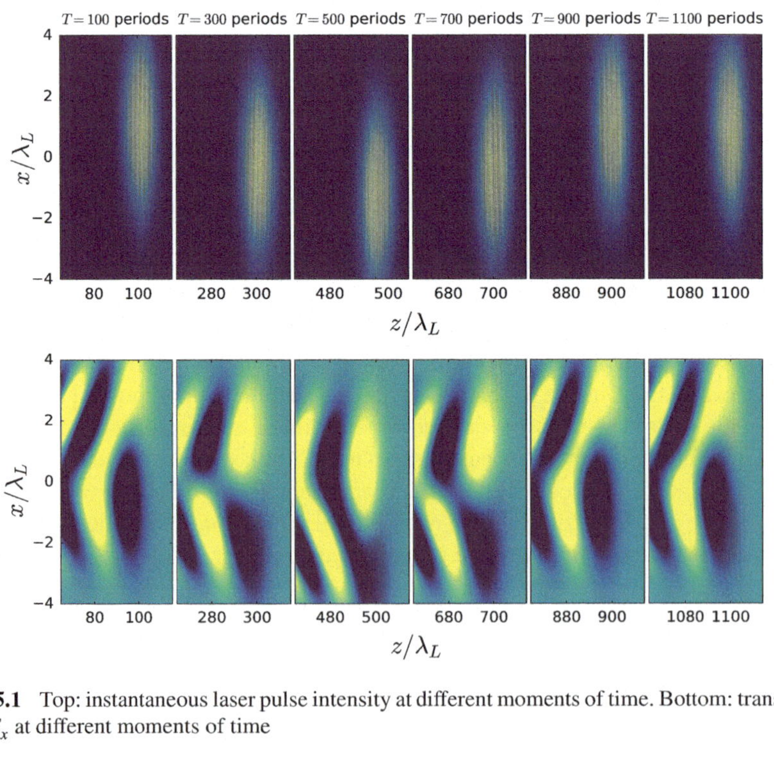

Fig. 25.1 Top: instantaneous laser pulse intensity at different moments of time. Bottom: transverse field E_x at different moments of time

Using linear plasma theory, for the case of the optimal laser pulse duration $k_p\tau_L = 2$ with k_p—plasma wavelength, one obtains the following expressions for the focusing field E_x

$$\frac{eE_x}{mc^2k_p} = -\frac{4a_0^2C}{k_p^2w_0^2}\left(x - x_c(z)\right) e^{\left(-2\frac{(x-x_c(z))^2}{w_0^2}\right)} \sin\left[k_p(z - ct)\right], \tag{25.1}$$

where $C = \sqrt{e^{-1}\pi/8} \approx 0.38$ in the case of the linearly polarized laser pulse. Results of the simulations showing the transverse field E_x as a function of z and x at different moments of time is presented in Fig. 25.1 (bottom). The focusing or defocusing field E_x exhibits oscillatory structure, which can be used for wiggling the electrons and generating the radiation. For the electron correctly injected [8] in the phase where the accelerating field E_z is zero and transverse field E_x is focusing, and assuming that both electron wiggling amplitude and laser pulse centroid amplitude are much smaller than the laser spot size, the trajectory is given by

$$p_x/mc = a_u \sin\left(k_u z + \varphi\right), \tag{25.2}$$

where a_u is the undulator strength and the amplitude of the normalized transverse momentum of an electron and $k_u = 1/Z_R$. The strength of the undulator is equal to

$$a_u = \frac{\gamma_0 k_u k_\beta^2 x_{ci}}{k_u^2 - k_\beta^2} \approx \frac{4\pi a_0^2 C x_{ci}}{\lambda_L}, \tag{25.3}$$

with λ_L the laser wavelength, and can reach values on the order of unity or higher. Here, $k_\beta = \left(\frac{4a_0^2C}{\gamma_0 w_0^2}\right)^{1/2}$ is the betatron wavenumber and γ_0 is the Lorentz factor of injected electron.

On-axis spontaneous radiation spectrum generated by an electron beam with central Lorentz factor $\gamma_0 = 1000$ (with 1% energy spread and normalized emittance $\varepsilon_n = 0.1$ µm), injected into the plasma undulator structure of Fig. 25.1, created by

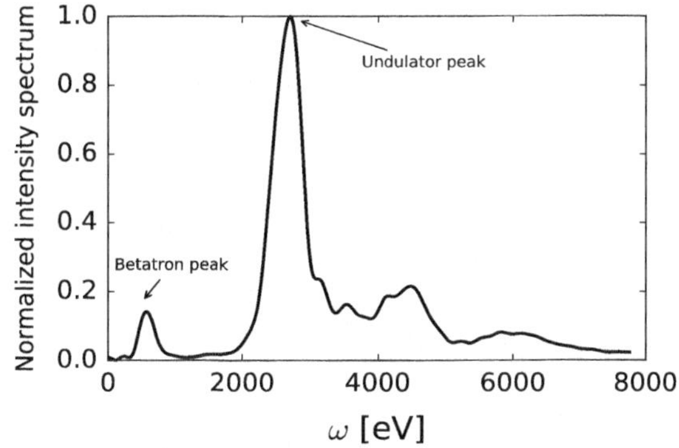

Fig. 25.2 Normalized on-axis radiation spectrum generated by the electron beam in the plasma undulator

the linearly polarized laser pulse with $a_0 = 0.3$, $w_0 = 7\,\mu\text{m}$ and initial centroid displacement of 2.5 μm, is presented in Fig. 25.2. The undulator strength is $a_u \approx 1$ and its period $\lambda_u \approx 1$ mm.

25.3 Conclusions

In conclusion, the concept of the plasma undulator based on the wakefields generated by the laser pulse undergoing centroid oscillations in the plasma channel was presented and supported by the 2D PIC simulations. Results of the PIC simulations fit with results of the analytical calculations (trajectories, betatron and undulator peak positions fit (24), (25), (27) of [8], respectively). Detailed analytical and numerical calculations can be found in [8].

References

1. Attwood, D.: Soft X-rays and Extreme Ultraviolet Radiation. Cambridge University Press (1999)
2. Tatchyn, R.: Rev. Mod. Phys. **60**, 2571 (1989)
3. Tantawi, S., et al.: Phys. Rev. Lett. **112**, 164802 (2014)
4. Joshi, C., et al.: IEEE J. Quantum Electron. **QE-23**, 1571 (1987)
5. Andriyash, I.A., et al.: Nat. Commun. **5**, 4736 (2014)
6. Balcou, P.: Eur. Phys. J. D **59**, 525 (2010)
7. Rykovanov, S.G., et al.: Phys. Rev. Lett. **114**, 145003 (2015)
8. Rykovanov, S.G., et al.: Phys. Rev. Accel. Beams **19**, 090703 (2016)
9. Xu, H., et al.: Chin. J. Comput. Phys. **19**, 305 (2002) (in Chinese)
10. Leemans, W.P., et al.: Nat. Phys. **2**, 696 (2006)

Chapter 26
Three-Dimensional Supersonic Sapphire Micronozzles for Laser-Plasma Wakefield Accelerators

V. Tomkus, L. Rimgaila, V. Stankevič and G. Račiukaitis

Abstract Analysis of three-dimensional supersonic micronozzles laser-machined in a bulk sapphire with femtosecond laser pulses for tailored electron beam injection into Laser-Plasma Wakefield Accelerators is presented. The designed sub-millimetre 3D structures were inscribed inside the transparent sapphire plate and etched in hydrofluoric acid to remove the modified material. Micronozzles are used for the investigation of the helium gas critical flow parameters like pressure, temperature and longitudinal profile formation for density-triggered injection of electrons.

26.1 Introduction

The laser-plasma wakefield particle accelerator provides a significant reduction of the accelerator length, compared to conventional RF accelerators, however due to the lack of proper control over the injection of electrons into the wakefield, it is difficult to produce monoenergetic bunches of the accelerated electron [1, 2]. Micro-sized components with the feature size in the order of tens to few hundred micrometres are required to control the plasma density gradient. The behaviour of fluid flow in supersonic micronozzles of sub-millimetre-scale significantly differs from classical nozzles due to the relative importance of the viscous forces.

V. Tomkus (✉) · L. Rimgaila · V. Stankevič · G. Račiukaitis
Center for Physical Sciences and Technology, Savanoriu Ave. 231, LT-02300 Vilnius, Lithuania
e-mail: vidmantas.tomkus@ftmc.lt

T. Kawachi et al. (eds.), *X-Ray Lasers 2016*, Springer Proceedings in Physics 202, https://doi.org/10.1007/978-3-319-73025-7_26

26.2 Methods

In this report, several configurations of single and dual supersonic converging–diverging micronozzles in the range of the throat diameter between 25 and 500 μm laser-machined in a bulk sapphire for the controlled electron beam injection into a laser-plasma wakefield accelerators were analysed. Dual helium gas jets induce shocks with sharper density transition triggering electron injection into the plasma bubble. The micronozzles were integrated with the capillary channels into the single lab-on-a-chip device. Implementation of gases with different ionisation potential allows a tailored control of electron injection and the resulting acceleration in the guided hydrogen plasma channel (Fig. 26.1).

Based on numerical simulation, the critical flow parameters like pressure, temperature, velocity and divergence were optimised for the helium density profile formation of the gas target (Fig. 26.2). The He gas concentration of the single nozzle with the inlet pressure of P_{in} 50 bar and hydrogen pressure in the acceleration channel of 50 mbar along the axis of laser beam propagation 300 μm below the outlet drops from $2 \cdot 10^{19}$ to $5 \cdot 10^{18}$ cm^{-3} within the distance of 150 μm for the nozzle with the throat diameter d_{thr} of 80 μm and within the distance of 80 μm for the nozzle with the throat diameter d_{thr} of 25 μm. The dual nozzle design of the nozzles with the throat diameter d_{thr} of 25 and 80 μm separated by 100 and 300 μm allows implementation of the longitudinal density profile with the He gas concentration drop within the distance of 25–80 μm depending on the inlet pressure and velocity difference.

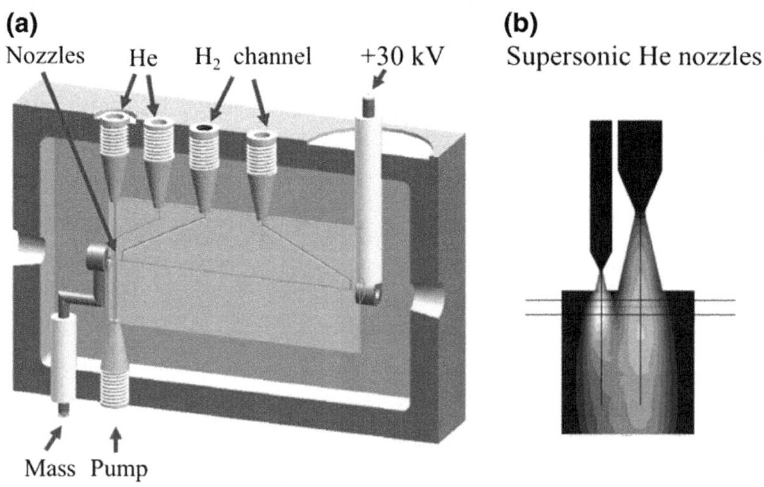

Fig. 26.1 The principle of operation of the lab-on-chip device for tailored electron acceleration (**a**), dual micronozzle for the electron injection into the capillary (**b**)

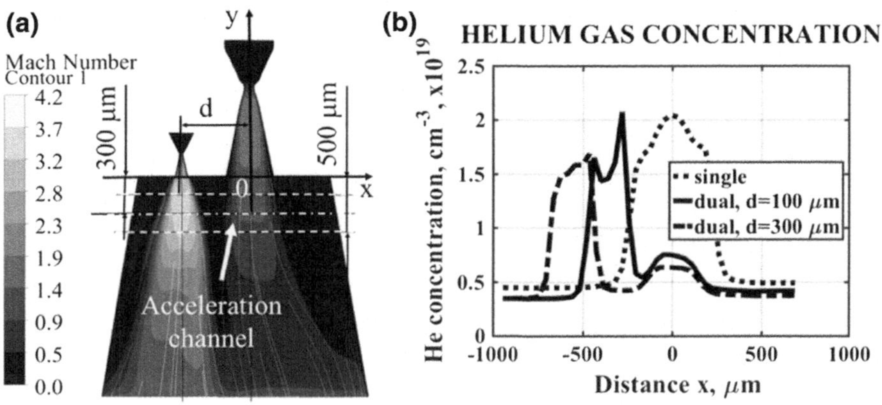

Fig. 26.2 Numerical simulation of helium gas velocity profile (**a**), and concentration distribution along the axis of laser beam propagation 300 μm below the outlet of nozzles with d_{thr} = 25 μm and d_{thr} = 80 μm separated by of 100 and 300 μm (**b**)

26.3 Fabrication of Sapphire Micronozzles

The microfluidic devices were formed in a commercially available sapphire block with 40 × 40 × 2 mm^3 dimension. The designed sub-millimeter 3D structures were inscribed inside the transparent sapphire material using a direct laser writing technique [3, 4]. The femtosecond Yb:KGW laser (Pharos, Light Conversion) beam with 300 fs pulse duration, 500 kHz repetition rate and 50–200 nJ pulse energy was used for the formation of 3D structure in the sapphire block. The beam was focused by the 50x (NA = 0.55) objective at a depth of ~550 μm below the sample surface. The sample was moved perpendicularly to the beam propagation direction by a precision translation stage at a speed of 0.5–6 mm/s. The linear polarisation was set perpendicular to the x-axis direction. The optimal translation speed and laser pulse energy were as follows: 3 mm/s and 75 nJ (Fig. 26.3).

Modification of sapphire under laser irradiation facilitates in different etching rates between the laser-modified and the unmodified material of more than 10,000:1 allowing the formation of structures with the high length to diameter aspect ratio more than 5000 [5].

For the first etching approach, the modified sample was submerged in the diluted 15% concentration HF acid for 24 h. Improvement in the etching time was achieved by immersing a jar with the acid in an ultrasonic water bath at 40 °C. The image of the irradiated microstructure is presented in the Fig. 26.4a. The image of the 3D micronozzle with the length of 200 μm and the throat diameter of 50 μm etched for 24 h is provided in the Fig. 26.4b. The etched surface roughness was optimised to get roughness of 100–200 nm in the plane perpendicular to laser beam radiation and less than 100 nm in the plane collinear to the laser beam radiation.

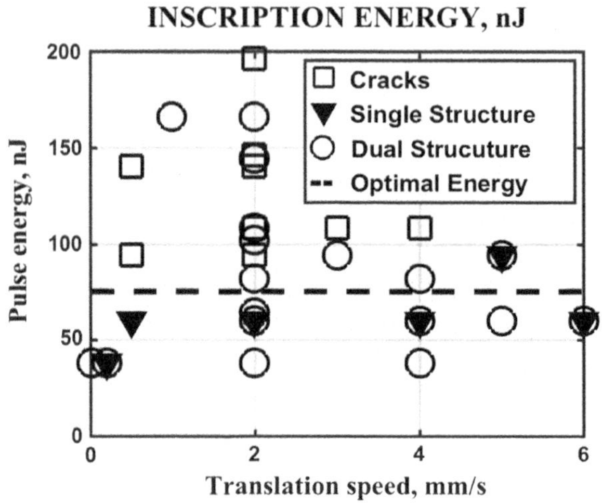

Fig. 26.3 The optimal ratio of the laser pulse energy and translation speed

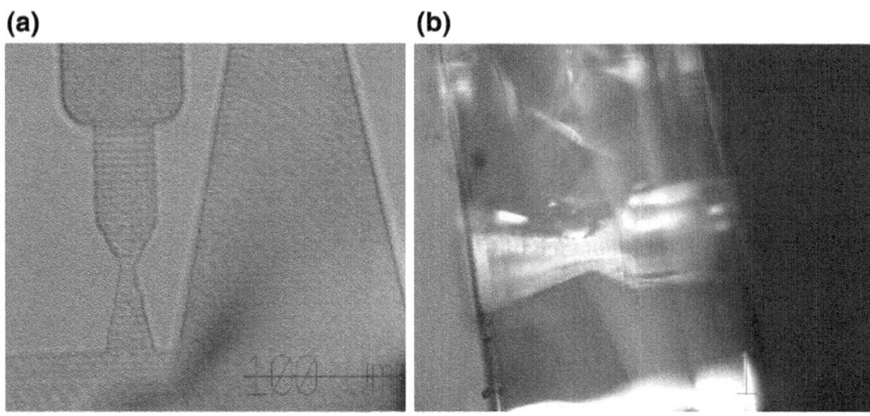

Fig. 26.4 Images of the irradiated (**a**) and etched (**b**) 3D sapphire nozzle

26.4 Conclusions

The critical flow parameters of helium gas like pressure, temperature, velocity and divergence for the micronozzles with the throat diameter between 25 and 500 μm were analysed. The dual nozzle with the throat diameter of 25 and 80 μm separated by 100 μm allows implementation of the longitudinal profile with sharp He concentration drop along the axis of laser beam propagation from $2 \cdot 10^{19}$ to $5 \cdot 10^{18}$ cm^{-3} within the distance of 25–80 μm depending on the inlet pressure and velocity difference and density-triggered electron injection into laser-plasma wakefield accelerator. Supersonic three-dimensional micronozzles were inscribed in a sapphire block with ultrashort femtosecond laser pulses and etched in the diluted 15% concentration hydrofluoric acid for 24 h. The etched surface roughness was optimised to get roughness of 100–200 nm.

References

1. Matlis, N., et al.: Transient behaviour of a supersonic three-dimensional micronozzle with an intersecting capillary. J. Appl. Phys. **119**, 074501 (2016)
2. Schmid, K., et al.: Supersonic gas jets for laser-plasma experiments. Rev. Sci. Instrum. **83**, 053304 (2012)
3. Stankevic, V., et al.: Internal to external microfluidic device for ellipsometric biosensor application. J. Laser Micro/Nanoeng. **11**, 53 (2016)
4. Wortmann, D., et al.: Micro- and nanostructures inside sapphire by fs-laser irradiation and selective etching. Opt. Express **16**(3), 1517 (2008)
5. Juodkazis, S., et al.: Control over the crystalline state of the sapphire. Adv. Mater. **18**, 1361 (2006)

Chapter 27
Generating Ultrahigh Brilliance Quasi-monochromatic MeV γ-Rays with High-Quality LWFA Electron Beams

C. H. Yu, R. Qi, W. T. Wang, J. S. Liu, W. T. Li, C. Wang, Z. J. Zhang, J. Q. Liu, Z. Y. Qin, M. Fang, Y. Xu, Y. X. Leng, F. L. Wei, Z. H. Song, R. X. Li and Z. Z. Xu

Abstract By designing a cascaded laser wakefield accelerator to generate high-quality electron beams, which were bound to head-on collide with the intense driving laser pulse via reflection of a thin foil, by using the self-synchronized all-optical Compton scattering scheme, we have produced the ultrahigh brilliance MeV γ-rays. This compact γ-ray source may provide applications in X-ray radiology and nuclear resonance fluorescence, a pump-probe study in ultrafast science.

27.1 Introduction

Laser wakefield accelerators (LWFA) have achieved significant progress recently owing to the sophisticated injections, cascade and guiding technologies, and they can produce monoenergetic, GeV-class femtosecond e-beams with tens of pC charge over a distance of centimeter-scale [1, 2]. Holding the potential of becoming compact alternatives to conventional accelerators, high-quality LWFA electron beams (*e*-beams) may lead to the realization of the compact, near-monochromatic, ultra-short and high peak brilliant X- and γ-ray sources up to a few MeV.

C. H. Yu · R. Qi · W. T. Wang · J. S. Liu (✉) · W. T. Li · C. Wang · Z. J. Zhang
Z. Y. Qin · M. Fang · Y. Xu · Y. X. Leng · R. X. Li · Z. Z. Xu
State Key Laboratory of High Field Laser Physics, Shanghai Institute of Optics and Fine Mechanics, CAS, Shanghai 201800, China
e-mail: michaeljs_liu@mail.siom.ac.cn

J. S. Liu · R. X. Li · Z. Z. Xu
IFSA Collaborative Innovation Centre, Shanghai Jiao Tong University, Shanghai 200240, China

F. L. Wei · Z. H. Song
State Key Laboratory of Intense Pulsed Radiation Simulation and Effect, Northwest Institute of Nuclear Technology, Xi'an 710024, China

T. Kawachi et al. (eds.), *X-Ray Lasers 2016*, Springer Proceedings in Physics 202, https://doi.org/10.1007/978-3-319-73025-7_27

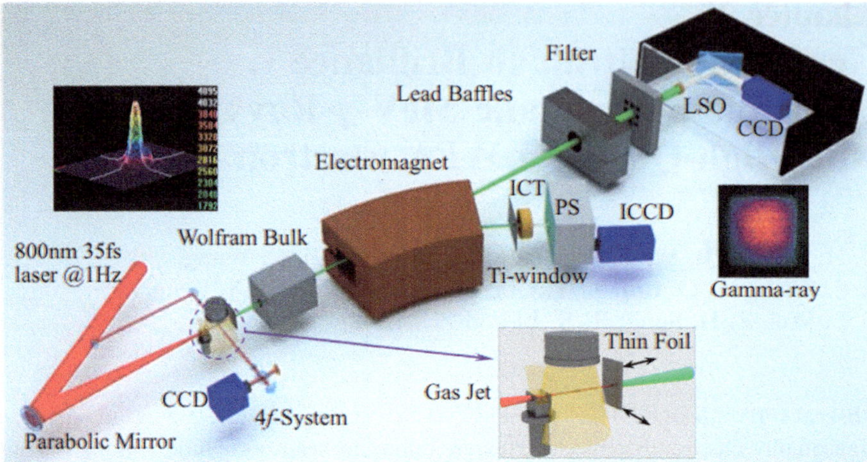

Fig. 27.1 Experimental setup. The pulses were focused into the two-segment (0.8 + 3 mm) pure helium gas jets. About 5% of the pump laser was split off as a probe pulse into a Michelson-type interferometer system to measure the plasma density profile. The *e*-beams are deflected by a 90-cm-long dipole electromagnet with a maximum magnetic field of 1.1 T, and measured by a Lanex phosphor screen (PS) imaged onto a ICCD camera. A 10-cm-thick wolfram collimator with a 2-mm hole is placed to block the irrelevant radiation. The filter and Lu_2SiO_4(LSO)-crystal scintillator with a diameter of 50 mm were placed 5.2 m downstream to measure the γ-rays

The experiments were carried out at the laser system based on the chirped-pulse amplification with a pulse duration of roughly 30 *fs*, a peak power of 200 TW @1 Hz repetition rate [3]. The experimental setup was shown in Fig. 27.1. A cascaded laser accelerator [4] was redesigned to generate high-quality *e*-beams. In order to reduce the difficulty of laser-electron temporal and spatial synchronization, we employed a self-synchronized all-optical Compton scattering scheme [5–7], in which the reproducible *e*-beams would collide with the intense driving laser pulse via the reflection of a thin foil (20 μm Ti) in the naturally overlapped region.

27.2 High-Quality LWFA *e*-Beams Generation

By designing a structured dual-stage LWFA to manipulate electron seeding and energy chirp reversal for compressing the energy spread, we first produced the high-brightness high-energy *e*-beams [7, 8], in which the maximum six-dimensional brightness $B_{6D,n}$ is very close to that of the typical e-beams from the state-of-the-art linac drivers. And, the *e*-beam peak energy could be manipulated by adjusting the horizontal span between the two gas jets to control the acceleration length (1.5–3 mm). These *e*-beams possessed rms energy spread of ~1%, average

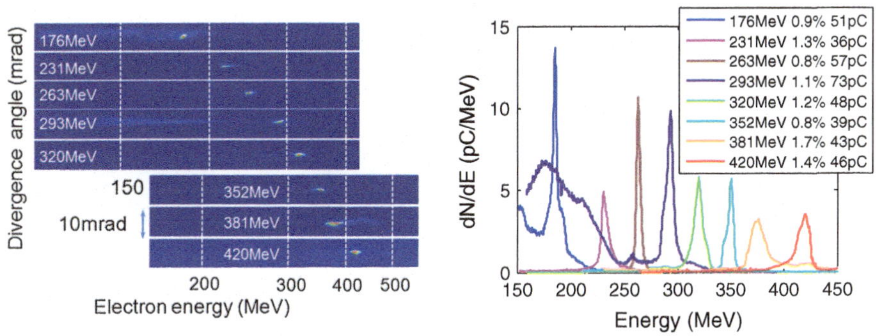

Fig. 27.2 Angle-integrated energy spectra of tunable high-quality *e*-beams

integrated charge of ~50 pC and rms divergence of <0.5 mrad as shown in Fig. 27.2.

Owing to the shot-to-shot fluctuation in laser power within 3% and small jitter in gas density, the *e*-beam peak energy had a fluctuation of ±5% and the uncertainty of the measured *e*-beam charge was estimated within 12%. However, for the inherently laser-electron overlap in the self-synchronized Compton scattering scheme, the scattered laser pulse and the high-quality *e*-beams interacted almost in the focus region (several μm^2) to achieve the high brightness of the γ-ray source.

27.3 Ultrahigh Brilliance Quasi-monochromatic MeV γ-Rays

The γ-ray beam pattern was recorded with and without inserting the filter in front of the LSO-crystal scintillator as shown in Fig. 27.3a, b, which had a near-Gaussian profile with the FWHM divergence angles of 3.8 and 4.1 mrad in horizontal and vertical directions, respectively. The γ-ray detection system was encased by the lead baffles to minimize the background noise and calibrated using a ^{60}Co (1.25 MeV) radiation source of known activity. The total photon number could be measured precisely within the error range of 3%. As shown in Fig. 27.3d, the Compton scattering γ-ray signal intensity was ~40× higher than the background signal, indicating that a high signal-to-noise ratio Compton source could be produced. Bremsstrahlung radiation could also be minimized and less than 3.8×10^5 photons (<3 MeV) within 8 mrad were detected, corresponding to 1.1×10^{-3} photons for each electron in the Monte Carlo simulations. By further optimizing the collision point, the photon yield was improved up to $(4.7 \pm 0.3) \times 10^7$ per shot, in a good agreement with the predicted one $\sim 8 \times 10^7$ estimated by [9]:

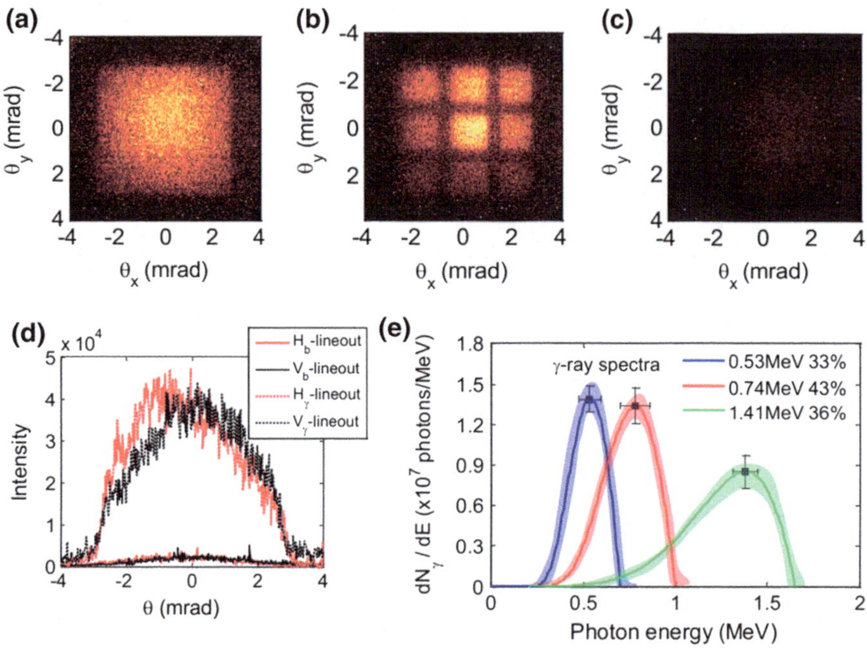

Fig. 27.3 **a** Typical background-subtracted Gaussian-profile γ-ray beam; **b** Imaged γ-ray beam profile transmitted through a 3 × 3 squared lead-grid filter; **c** Background profile when the foil was placed at z = 10 mm; **d** Horizontal and vertical lineouts of the γ-ray profile and the background profile; **e** Reconstructed γ-ray spectra under different *e*-beam energies, and the error bars and shaded bands represent the spectra uncertainties

$$N_p = \frac{\pi}{3}\alpha N_e N_l a_r^2 \frac{\left(1+a_r^2/2\right)\left(1+\beta_e\right)}{\langle n\rangle\left(1+\beta_z\right)}. \tag{27.1}$$

The interaction angle between the scattered laser and e-beams was about 178° to prevent the damage to the laser system by backpropagating light. It need to be noted here that the properties of the scattering laser light cannot be controlled independently and thus it was not flexible in some ways. And the radiated photon energy is given by $\hbar\omega_{\gamma\text{-ray}} \approx \hbar\omega_{\text{laser}} \times 4\gamma^2/(1 + a_r^2/2 + \gamma^2\theta_o^2)$ eV and at its harmonics, with a typical divergence $\theta \cong a_r/\gamma$ around the electron velocity vector. Therefore, the laser intensity and off-axis observation in the experiments can induce a red-shift, resulting in lower photon energies.

The spectra could be deduced from indirect filter-pack measurements [5, 10, 11]. The γ-ray spectra with the peak energy from 0.3 to 2 MeV and a narrow bandwidth of ~33% (FWHM) were measured by tuning the *e*-beam energy from 160 to 420 MeV. According to the 3D PIC simulations, it was assumed that the *e*-beam had the duration of 8 *fs* and the diameter of 3 μm, indicating our γ-ray source had the brilliance of ~3.1 × 10^{22} photons s^{-1} mm^{-2} $mrad^{-2}$ 0.1% BW.

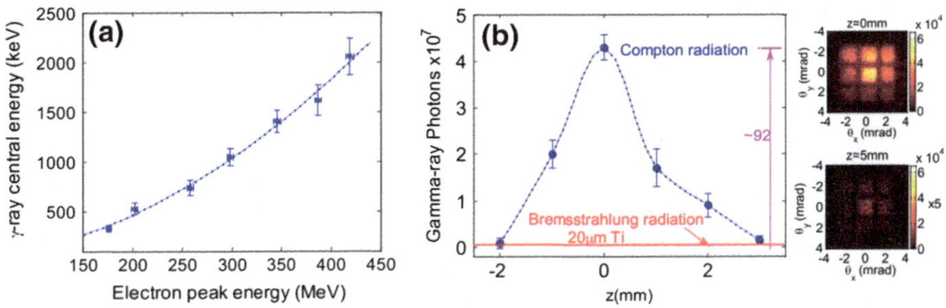

Fig. 27.4 **a** Measured γ-ray peak energy plotted versus with the tunable *e*-beam peak energy, is fitted with the blue dash line; **b** Detected γ-ray photons as a function of the foil position. Each data point corresponds to an average over five consecutive shots at different z

Furthermore, the foil position (z) played an important role in the γ-ray flux as shown in Fig. 27.4b. Move the foil inside the gas jet ($z < 0$), the γ-ray photon yield diminished. When the foil was moved away from the exit of the gas jet ($z > 0$), the self-focusing laser pulse would diverge to have a smaller laser intensity at the collision position, leading to the obvious decrease of Compton scattering radiation. Both the experiments and relevant simulations have confirmed that γ-ray emission arose substantially from electrons wiggling in the laser field.

27.4 Conclusion

Based on the self-synchronized Compton scattering scheme, quasi-monochromatic and ultrahigh brilliant γ-rays have been generated with high-quality *e*-beams from the specially designed LWFA. It is believed that this efficient and compact MeV γ-ray source, based on the perfect combination of a laser-plasma accelerator and a plasma mirror, may be of interest to provide some practical applications in the future.

References

1. Liu, J.S.: All-optical cascaded laser wakefield accelerator using ionization-induced injection. Phys. Rev. Lett. **107**, 035001 (2011)
2. Leemans, W.P.: Multi-GeV electron beams from capillary-discharge-guided subpetawatt laser pulses in the self-trapping regime. Phys. Rev. Lett. **113**, 245002 (2014)
3. Xu, Y.: A stable 200TW/1 Hz Ti:sapphire laser for driving full coherent XFEL. Opt. Laser Technol. **79**, 141 (2016)
4. Wang, W.T.: Control of seeding phase for a cascaded laser wakefield accelerator with gradient injection. Appl. Phys. Lett. **103**, 243501 (2013)
5. Ta Phuoc, K.: All-optical Compton gamma-ray sources. Nat. Photonics **6**, 308–311 (2012)

6. Tsai, H.D.: Compact tunable Compton X-ray source from laser-plasma accelerator and plasma mirror. Phys. Plasmas **22**, 023106 (2015)
7. Yu, C.H.: Ultrahigh brilliance quasi-monochromatic MeV γ-ray based on self-synchronized all-optical Compton scattering. Sci. Rep. **6**, 29518 (2016)
8. Wang, W.T.: High-brightness high-energy electron beam from a laser wakefield accelerator via energy chirp control. Phys. Rev. Lett. **117**, 124801 (2016)
9. Ride, S.K.: Thomson scattering of intense lasers from electron beams at arbitrary interactions angles. Phys. Rev. E. **52**, 5425–5442 (1995)
10. Chen, S.: MeV-energy X rays from inverse Compton scattering with laser-wakefield accelerated electrons. Phys. Rev. Lett. **110**, 155003 (2013)
11. Powers, N.D.: Quasi-monoenergetic and tunable X-rays from a laser-driven Compton light source. Nat. Photonics **8**, 28–31 (2013)

Chapter 28
Features of Induced X-Ray Radiation and Possible Tandem FEL Realization on Channeling Particles

V. I. Vysotskii and M. V. Vysotskyy

Abstract The features of spontaneous and induced radiation of moving charged particles in focusing fields in mediums with frequency dispersion and the possibilities of creation of multiphoton inversionless short wave FEL are considered. These processes are based on modified Doppler effect in 1D and 2D focusing fields.

28.1 Introduction

It is well known that the main method of control and optimization of radiative processes for accelerated charged particles in the material environment is connected with the Doppler effect. The "traditional" expression

$$\omega_{rad,\,abs} = \Omega_0/\gamma(1-\beta n(\omega)\,\cos\theta),\ \gamma = 1/\sqrt{1-\beta^2},\ \beta = v/c \tag{28.1}$$

for this effect can be obtained from the kinematic relations and is the result of the invariance of the 4-scalar phase at relativistic transformations of coordinates and time (4-component vector $\{\vec{r},\, ct\}$) in the special theory of relativity. This formula is incorrect in the most interesting Cherenkov's area $\beta n(\omega)\,\cos\theta \approx 1$. In our works [1, 2], the features of full Doppler effect including area $|\beta n(\omega)\,\cos\theta| \to 1$ were investigated. The general formulas for the full Doppler effect at radiation and absorption are the following:

$$\omega_{rad} = \frac{\gamma mc^2}{\hbar(n^2-1)}\left[\pm\sqrt{(1-\beta n\,\cos\theta)^2 + 2\frac{\varepsilon_2^* - \varepsilon_1^*}{\gamma^2 mc^2}(n^2-1)} - (1-\beta n\,\cos\theta)\right], \tag{28.2}$$

V. I. Vysotskii (✉) · M. V. Vysotskyy
Kiev National Shevchenko University, Kiev, Ukraine
e-mail: vivysotskii@gmail.com

T. Kawachi et al. (eds.), *X-Ray Lasers 2016*, Springer Proceedings in Physics 202, https://doi.org/10.1007/978-3-319-73025-7_28

$$\omega_{abs} = \frac{\gamma mc^2}{\hbar(n^2-1)}\left[\pm\sqrt{(1-\beta n\cos\theta)^2 - 2\frac{\varepsilon_2^* - \varepsilon_1^*}{\gamma^2 mc^2}(n^2-1)} + (1-\beta n\cos\theta)\right]. \quad (28.3)$$

In these formulas $\varepsilon_{1,2}^* = \varepsilon_{1,2}(1+\varepsilon_{1,2}/2mc^2)$, $n \equiv n(\omega)$, $\varepsilon_{1,2}$—energy levels of moving particle (e.g., levels of relativistic charged particle in crystal channel).

These results were obtained using the law of momentum conservation $\Delta\vec{p} \equiv \vec{p}_1 - \vec{p}_2 = n(\omega)\hbar\vec{k}$ at radiation and absorption. This law is valid when the particle (source or absorber) moves in a medium with dielectric permittivity dispersion $n(\omega)$, if in this system there are no force fields acting on a moving particle and the total momentum of the system "particle + photon" is an integral of motion. For such objects, the radiation or absorption corresponds to the internal electromagnetic transitions in the moving particle and environment affects only the formation and parameters of the emitted (absorbed) photons.

Another type of system is connected with the processes in which the external force field plays an important role in the movement of the source. It should be noted that the experimental realization of the extreme condition $\beta n(\omega)\cos\theta \approx 1$ can be achieved, as a rule, only for a motion of light relativistic particles, like electrons and positrons, in which there are no internal transitions. For these particles, radiation and absorption are connected with their interaction with the environment in the form of different force fields that influence their state. A typical example of such effect is a quasi-characteristic radiation during channeling. It is obvious that at the presence of such fields the total momentum of the system "particle + photon" is not an integral of motion, the law of its conservation in the form of equality $\Delta\vec{p} \equiv \vec{p}_1 - \vec{p}_2 = n(\omega)\hbar\vec{k}$ is not fulfilled. This problem is considered below.

28.2 The Feature of Full Doppler Effect for Moving Particles and Sources in Transverse-Focusing Fields and During Channeling

Let us consider the features of the Doppler effect at the motion of the emitting (absorbing) particles in the transverse-focusing field, in particular, on transverse—nonuniform potential (including channeling in crystals) at the critical (Cherenkov) condition $\beta n(\omega)\cos\theta \approx 1$. The fundamental difference from the case of a homogeneous medium is connected with the need to use, together with the laws of conservation of energy and momentum, only those components that are integrals of motion in the field. In particular, for the motion of a fast particle in a two-dimensional (2D) potential field $V(x, y) \equiv V(\theta, \varphi)$ the gradient of which is directed perpendicular to the direction of motion of the particle along the crystal axis oz (axial channeling), particle integrals of motion are energy and the longitudinal component of the particle momentum p_z. Accordingly, at the motion along the same axis oz in a 1D inhomogeneous potential field of the crystal plane $V(x)$ (planar channeling), particle integrals of motion are energy and two components (p_y and p_z) of the total momentum of the system. In these cases, the law of conservation of the transverse component of momentum $\Delta\vec{p}_\perp$ is fulfilled only for the whole system (in the channeling case: for moving electrons or positrons and for the system of atoms which form the crystal planes or axis and which are lying within the area of the longitudinal coherence, where all atoms interact coherently with the channeled particle). With these remarks, the process of photon emission with frequency ω_{rad} and momentum $\hbar\vec{k}$ by a particle with initial momentum $p = m_2c(\gamma^2 - 1)^{1/2}$, $\vec{p} = \vec{e}_z p$, mass $m_2 = m + \varepsilon_2/c^2$ and two energy levels of internal state (initial upper level ε_2 and final lower level ε_1 for the case of quasi-characteristic radiation at channeling) is described by the law of energy conservation

$$\sqrt{p_z^2c^2 + (m_2c^2)^2} = \sqrt{(\vec{p}_z - \Delta\vec{p})^2c^2 + (m_1c^2)^2} + \hbar\omega_{rad} \tag{28.4}$$

and the laws of conservation of corresponding components of moving particle momentum for motion in 2D and 1D potential field (axial and planar motion):

$$\Delta\vec{p}_{2D} = \vec{e}_z\Delta p_z = \vec{e}_z\hbar k(\omega)\cos\theta;\ k(\omega) = n\omega/c,\ n \equiv n(\omega),\ 0 \le \theta \le \pi; \tag{28.5a}$$

$$\Delta\vec{p}_{1D} = \vec{e}_y\hbar k(\omega)\sin\theta\sin\varphi + \vec{e}_z\hbar k(\omega)\cos\theta,\ 0 \le \varphi \le 2\pi. \tag{28.5b}$$

Here, φ is the angle between the projection of the wave vector $\vec{k}$ of emitted quanta on the plane xy and axis oy, perpendicular to the momentum $\vec{p} = \vec{e}_z p$ of free motion of the particle. At $\varphi \equiv 0$ relation (28.5b) turns to (28.5a).

Eliminating the value $\Delta\vec{p}$ from (28.4) by substitutions (28.5b), we obtain the equation for determining the frequency of radiation in a 1D nonuniform potential field (e.g., at the planar channeling) that has the solution

$$\omega_{rad}^{(1d)} = \frac{\gamma mc^2(\beta n\cos\theta - 1)\left\{1 \pm \sqrt{1 + 2\frac{(\varepsilon_2^* - \varepsilon_1^*)}{\gamma^2mc^2}\frac{[n^2(1 - \sin^2\theta\cos^2\varphi) - 1]}{(\beta n\cos\theta - 1)^2}}\right\}}{\hbar[n^2(1 - \sin^2\theta\cos^2\varphi) - 1]}. \tag{28.6}$$

The frequency of radiation in a 2D nonuniform field (case of axial channeling) is obtained from (28.6) with the formal replacement $\varphi \equiv 0$:

$$\omega_{rad}^{(2d)} = \frac{\gamma mc^2(\beta n\cos\theta - 1)}{\hbar[n^2\cos^2\theta - 1]}\left\{1 \pm \sqrt{1 + 2\frac{(\varepsilon_2^* - \varepsilon_1^*)}{\gamma^2mc^2}\frac{[n^2\cos^2\theta - 1]}{(\beta n\cos\theta - 1)^2}}\right\}. \tag{28.7}$$

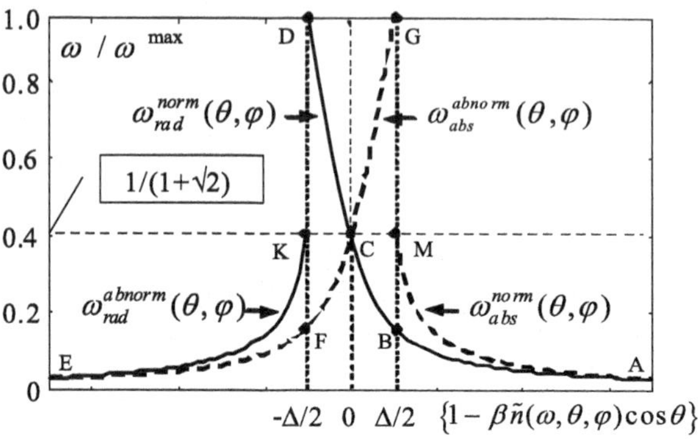

Fig. 28.1 Frequencies of absorption (AM and GE branches) and radiation (AD, KE) for normal (AM, AD) and abnormal (GE, KE) Doppler effect in transverse 1D ($\tilde{n}(\omega, \theta, \varphi) \to n(\omega)\sqrt{1 - \sin^2\theta\cos^2\varphi}$) or 2D ($\tilde{n}(\omega, \theta, \varphi) \to n(\omega)\, cos\theta$) field

In the same way, we can find expressions for the absorption frequency in a 1D nonuniform field (planar channeling) and 2D field (axial channeling)

$$\omega_{abs}^{(1d)} = \frac{\gamma mc^2(1-\beta n\cos\theta)\left\{1 \pm \sqrt{1 - 2\frac{(\varepsilon_2^* - \varepsilon_1^*)}{\gamma^2 mc^2}\frac{[n^2(1-\sin^2\theta\cos^2\varphi)-1]}{(1-\beta n\cos\theta)^2}}\right\}}{\hbar[n^2(1-\sin^2\theta\cos^2\varphi)-1]}; \tag{28.8}$$

$$\omega_{abs}^{(2d)} = \frac{\gamma mc^2(1-\beta n\cos\theta)}{\hbar[n^2\cos^2\theta - 1]}\left\{1 \pm \sqrt{1 - 2\frac{(\varepsilon_2^* - \varepsilon_1^*)}{\gamma^2 mc^2}\frac{[n^2\cos^2\theta - 1]}{(1-\beta n\cos\theta)^2}}\right\}. \tag{28.9}$$

Signs in formulas (28.6)–(28.9) are selected according to the requirement $\omega_{rad,\, abs} \geq 0$.

The principal difference between formulas (28.6)–(28.9) for full Doppler effect in transverse fields from formulas (28.2)–(28.3) for free motion is that frequencies of radiation and absorption depend not only on the magnitude and the sign of the expression $(1 - \beta n\cos\theta)$, but also on the angles θ and φ. The structure of dependencies $\omega_{rad,\, abs}^{(1d,\, 2d)}(\theta, \varphi)$ (28.6)–(28.9) is presented in Fig. 28.1, where

$$\Delta = \sqrt{8|\varepsilon_2^* - \varepsilon_1^*|(\tilde{n}^2 - 1)/\gamma^2 mc^2}, \tag{28.10}$$

$$\hbar\omega_{\max} = (2 + \sqrt{2})\sqrt{|\varepsilon_2^* - \varepsilon_1^*|mc^2/(\tilde{n}^2 - 1)} \tag{28.11}$$

28.3 The Possibility of Creation of Tandem Self-supported Laser Inversionless System on the Basis of Channeling Particles

The analysis of the Doppler effect in transverse fields is easy to carry out taking into account that after the appropriate formal replacements $n(\omega)(1-\sin^2\theta\cos^2\varphi)^{1/2}\rightarrow \tilde{n}$; $n(\omega)\cos\theta\rightarrow\tilde{n}$ the (28.6)–(28.9) of full Doppler effect in transverse fields are transformed into appropriate equations for isotropic medium (28.2), (28.3). The scheme of possible radiative transitions at different directions of emission and absorption of quanta is shown in Fig. 28.2 and it is based on the results shown in Fig. 28.1. From these graphs follows that the use of alternating normal and abnormal Doppler effects leads to the possibility of existence of closed radiation cycle. It consists of two consecutive processes: the photon emission in space direction $-\Delta/2<\{1-\beta\tilde{n}(\omega,\theta,\varphi)\cos\theta\}<0$ (see Fig. 28.2b) during transition from upper level to down one (normal Doppler Effect) and photon emission in another direction $\{1-\beta\tilde{n}(\omega,\theta,\varphi)\cos\theta\}<-\Delta/2$ (see Fig. 28.2a) at reverse transition from the down level to the same upper one (anomalous Doppler Effect). The scheme of the repetitive cycle is shown in Fig. 28.2 (a ring with an arrow).

These results suggest the possibility of realization of tandem inversionless FEL with multiple radiative transitions between same two levels (upper and lower) with an emission of a great number of X-ray photons in two directions. The frequencies of photons in these directions may be same or different. The source of energy for these repetitive cycles of quanta radiation is the kinetic energy of the particle. At typical transition energy $\varepsilon_2^*-\varepsilon_1^*\approx 1-10\,\text{eV}$, the maximum frequency of radiation in such tandem system corresponds to points K and C (Fig. 28.1) and equals $\omega_{rad(K,C)}=\{2(\varepsilon_2^*-\varepsilon_1^*)mc^2/(\tilde{n}^2-1)\}^{1/2}/\hbar\geq 1\,\text{MeV}$. Unfortunately, such value of ω_{rad} is not realistic, because in such frequency range it is impossible to satisfy the condition $n(\omega_{rad})>1$. The easiest way for the realization of this condition may be achieved in the low-frequency range $\lambda_{rad}\approx 0.5-2\,\text{A}$ at dynamical Bragg

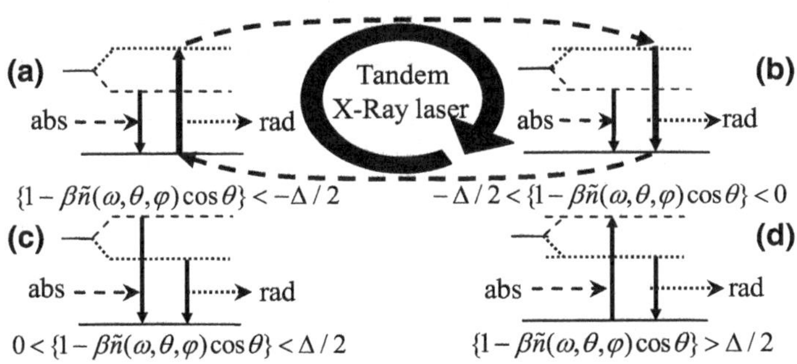

Fig. 28.2 The change of the directions of radiative transitions depending on modified Cherenkov parameter $\{1-\beta\tilde{n}(\omega,\theta,\varphi)\cos\theta\}$ and sequence of repeated emission of photons in tandem X-ray FEL

diffraction of generated X-radiation in the same crystal, in this case $n - 1 \approx 10^{-6} - 10^{-4}$. One more interesting problem is connected with radiation at channeling when the particle moves in periodically modulated channels. In this regime, there is a synchronization of its own wave functions of transverse motion and formation of coherent correlated states [3–6] that may substantially affect the emission processes.

References

1. Vysotskyy, M.V., Vysotskii V.I.: The Doppler effect in a medium for radiation sources in motion at extreme conditions. Nucl. Instrum. Methods Phys. Res. B **252**, 75–80 (2006)
2. Vysotskii, V.I., Vysotskyy, M.V.: Peculiarities of induced radiation and controlled dechanneling of relativistic particles in crystals under the Cherenkov-Vavilov extreme condition. J. Surf. Invest. **4**, 162–169 (2010)
3. Vysotskii, V.I., Adamenko, S.V., Vysotskyy, M.V.: Subbarrier interaction of channeling particles under the self-similar excitation of correlated states in a periodically deformed a crystal. Ibid **6**, 369–374 (2012)
4. Vysotskii, V.I., Adamenko, S.V., Vysotskyy, M.V.: Acceleration of low energy nuclear reactions by formation of correlated states of interacting particles in dynamical systems. Ann. Nucl. Energy **62**, 618–625 (2013)
5. Vysotskii, V.I., Vysotskyy, M.V.: Coherent correlated states and low-energy nuclear reactions in non stationary systems. Eur. Phys. J. A **49**, 99 (2013)
6. Vysotskii, V.I., Vysotskyy, M.V.: Correlated states and transparency of a barrier for low-energy particles at monotonic deformation of a potential well with dissipation and a stochastic force. J. Exp. Theor. Phys. **118**, 534–549 (2014)

Chapter 29
Features of Resonant Absorption and Short-Wavelength Laser Amplification in Realistic Media

V. I. Vysotskii and M. V. Vysotskyy

Abstract In the work, the influence of space distribution of electrons, atoms, and molecules in any realistic material target on effective electromagnetic characteristics (susceptibility and permittivity) is discussed and investigated. It was shown that these characteristics depend not only on the averaged values of the local susceptibility and permeability, but also on their spatial derivatives. It was shown that these effects are important for Cherenkov radiation and Short-wavelength laser amplification .

29.1 Introduction

The problem of creation of optimal sources of coherent and directed X-ray and gamma-ray radiation [1] is one of the most important in modern physics. This problem can be solved by using different physical systems and effects, including Cherenkov radiation and realization of short-wavelength lasers.

The main problem at the creation of X-ray Cherenkov generator is connected with the satisfaction of threshold conditions (e.g. [2]) $\beta n(\omega)\cos\theta = 1,\ \beta = v/c$.

According to the standard point of view, the effective susceptibility $\chi_{eff}(\omega)$ and permittivity $\varepsilon_{eff}(\omega)$ in any realistic material environment is determined by averaging $\varepsilon_{eff}(\omega) = \langle\varepsilon(\vec{r},\omega)\rangle \equiv 1 + \langle\chi(\vec{r},\omega)\rangle$, $\chi_{eff}(\omega) \equiv \langle\chi(\vec{r},\omega)\rangle$, the local characteristics $\chi(\vec{r},\omega)$, $\varepsilon(\vec{r},\omega)$ within the spatial interval, which exceeds the average size of atomic (molecular) inhomogeneities. It is supposed that this method corresponds to the transition to macroscopic electrodynamics. In this approach, the wave equation for X-ray waves is postulated in the form of the "standard" wave equation

$$\Delta\langle\vec{E}(\vec{r},\omega)\rangle + (\omega/c)^2\varepsilon_{eff}(\omega)\langle\vec{E}(\vec{r},\omega)\rangle = 0, \tag{29.1}$$

V. I. Vysotskii (✉) · M. V. Vysotskyy
Kiev National Shevchenko University, Kiev, Ukraine
e-mail: vivysotskii@gmail.com

T. Kawachi et al. (eds.), *X-Ray Lasers 2016*, Springer Proceedings in Physics 202, https://doi.org/10.1007/978-3-319-73025-7_29

determining the evolution of the locally averaged field $\langle \vec{E}(\vec{r}, \omega) \rangle$ in a medium with the locally averaged characteristics $\chi_{eff}(\omega)$ and $\varepsilon_{eff}(\omega)$.

The simple analysis has shown that this "traditional" method of $\chi_{eff}(\omega)$ and $\varepsilon_{eff}(\omega)$ introduction is incorrect, because it is based on the incorrect assumption $\langle \varepsilon_{eff}(\vec{r}, \omega)\vec{E}(\vec{r}, \omega) \rangle = \langle \varepsilon_{eff}(\vec{r}, \omega) \rangle \langle \vec{E}(\vec{r}, \omega) \rangle$ in (29.1). It is shown below that the heuristic assumption $\chi_{eff}(\omega) \equiv \langle \chi(\vec{r}, \omega) \rangle$, which is usually used in the analysis of wave propagation, is incorrect for shortwave and the actual value of $\chi_{eff}(\omega)$ depends not only on $\chi(\vec{r}, \omega)$ but also on $\nabla\chi(\vec{r}, \omega)$.

29.2 Wave Equation for X-Ray Radiation in Realistic (Actual) Media

From the system of Maxwell equations for nonmagnetic material media

$$\nabla \times \vec{H}(\vec{r}, \omega) = -i(\omega/c)\vec{D}(\vec{r}, \omega),\ \nabla \times \vec{E}(\vec{r}, \omega) = i(\omega/c)\vec{H}(\vec{r}, \omega),\ \nabla\vec{D}(\vec{r}, \omega) = 0 \quad (29.2)$$

in the case of harmonic electromagnetic fields

$$\vec{E} = \vec{E}(\vec{r}, \omega)e^{-i\omega t},\ \vec{H} = \vec{H}(\vec{r}, \omega)e^{-i\omega t},\ \vec{D} = \varepsilon(\vec{r}, \omega)\vec{E}(\vec{r}, \omega)e^{-i\omega t} \quad (29.3)$$

follows the equation for the electric field

$$\nabla \times \nabla \times \vec{E}(\vec{r}, \omega) \equiv \nabla\nabla\vec{E}(\vec{r}, \omega) - \Delta\vec{E}(\vec{r}, \omega) = (\omega/c)^2\varepsilon(\vec{r}, \omega)\vec{E}(\vec{r}, \omega). \quad (29.4)$$

This equation can be transformed using the relation

$$\nabla\vec{E}(\vec{r}, \omega) = -\{\nabla\varepsilon(\vec{r}, \omega)/\varepsilon(\vec{r}, \omega)\}\vec{E}(\vec{r}, \omega) \quad (29.5)$$

that can be received from the "material" Maxwell (29.2) in the absence of uncompensated electric charges. After substituting (29.5) to (29.4), we obtain the modified wave equation in realistic medium with the nonuniform distribution of electrons, atoms, and molecules in space [2]

$$\Delta\vec{E}(\vec{r}, \omega) + \tilde{k}^2(\vec{r}, \omega)\vec{E}(\vec{r}, \omega) = 0,\ \tilde{k}^2(\vec{r}, \omega) = (\omega/c)^2\tilde{\varepsilon}(\vec{r}, \omega), \quad (29.6)$$

$$\tilde{\varepsilon}(\vec{r}, \omega) = \varepsilon(\vec{r}, \omega) + (c/\omega)^2\left\{\Delta\varepsilon(\vec{r}, \omega)/\varepsilon(\vec{r}, \omega) - 2(\nabla\varepsilon(\vec{r}, \omega)/\varepsilon(\vec{r}, \omega))^2\right\}. \quad (29.7)$$

Here, $\tilde{\varepsilon}(\vec{r}, \omega)$ is the effective permittivity of the nonuniform realistic medium.

The most simple solution of (29.6) can be obtained by the method of slowly varying amplitudes (WKB method). In the one-dimensional case, we have

$$E(x) = E_0\sqrt{\tilde{k}(0)/\tilde{k}(x)}\,\exp\{\pm i\,\langle\tilde{k}(x)\rangle x\} \approx E_0\exp\{\pm i\,\langle\,\tilde{k}(x)\,\rangle x\},$$
$$\tilde{k}(x) = \frac{\omega}{c}\left[\varepsilon(x,\omega) + \frac{c^2}{\omega^2}\frac{1}{\varepsilon(x,\omega)}\frac{d^2\varepsilon(x,\omega)}{dx^2} - 2\frac{c^2}{\omega^2}\left(\frac{1}{\varepsilon(x,\omega)}\frac{d\varepsilon(x,\omega)}{dx}\right)^2\right]^{1/2}. \quad (29.8)$$

In X-ray and gamma-ray ranges $\tilde{\varepsilon}(x,\omega) \approx 1 + \tilde{\chi}(x,\omega)$, $|\tilde{\chi}(x,\omega)| \ll 1$ (the value 4π is inserted in $\tilde{\chi}(x,\omega)$). Solution (29.8) can be simplified taking into account that in a periodic medium the following conditions are met

$$\chi(x,\omega) = \chi(x \pm L,\omega),\ \frac{d\chi(x,\omega)}{dx}\Big|_x = \frac{d\chi(x,\omega)}{dx}\Big|_{x\pm L},\ \langle\frac{d^2\chi(x,\omega)}{dx^2}\rangle = 0. \quad (29.9)$$

Taking into account these relations, from (29.8) follows the final expressions for electric field and effective dielectric susceptibility inside any 1D crystal

$$E(x) = E_0\,\exp\left\{\pm i\,\frac{\omega}{c}\left(1 + \frac{\chi_{eff}(\omega)}{2}\right)x\right\};\ \chi_{eff}(\omega) = \bar{\chi}(\omega) - \Delta\chi_{eff}(\omega),$$
$$\Delta\chi_{eff}(\omega) \equiv 2(c/\omega)^2\,\langle(d\chi(x,\omega)/dx)^2\rangle\,;\ \bar{\chi}(\omega) \equiv \langle\chi(x,\omega)\rangle. \quad (29.10)$$

In the typical model of the one-dimensional periodical susceptibility

$$\chi(x,\omega) = \sum_n\left\{\bar{\chi}(\omega)L/\sqrt{\pi u^2}\right\}\exp\left[-(x-nL)^2/u^2\right] \quad (29.11)$$

in a medium with the period $L \gg u$, we have the final result for $\chi_{eff}(\omega)$

$$\chi_{eff}(\omega) = \bar{\chi}(\omega) - \tfrac{c^2}{\omega^2}(\bar{\chi}(\omega))^2\tfrac{1}{L}\int_{-\infty}^{\infty}\tfrac{4x^2L^2}{\pi u^6}e^{-2x^2/u^2}dx = \bar{\chi} - \Delta\chi_{eff}(\omega),$$
$$\Delta\chi_{eff} = (\bar{\chi}(\omega))^2K,\ K \approx 0.01(\lambda^2L/u^3)\,. \quad (29.12)$$

Here, u is the *rms* amplitude of thermal oscillations of atoms.

In the case of actual crystal parameters ($L \approx 2A, u \approx 0.2A$) and for two different kinds of shortwave radiation (X-ray with $\lambda \approx 1\,A$, $|\bar{\chi}| \approx 10^{-5}\ldots10^{-6}$ and radiation of Mossbauer Fe^{57} isotope with $E_\gamma = 14.4\,keV$, $\lambda \approx 0.7\,A$ and $|\bar{\chi}| \approx 10^{-4}$) the WKB conditions for the considered solution (29.8) are the following:

$$\left|\frac{d}{dx}\frac{1}{\tilde{k}(x)}\right|_{XRay} \approx \frac{\lambda|\bar{\chi}|L}{2\pi^{3/2}u^2} \approx 10^{-4}\ldots\ 10^{-5} \ll 1; \left|\frac{d}{dx}\frac{1}{\tilde{k}(x)}\right|_{Fe^{57}} \approx 10^{-3} \ll 1. \tag{29.13a}$$

These conditions confirm the correctness of the use of both the WKB method and the obtained results (29.8–29.12). In the opposite case of longer wave radiation ($\lambda \geq 10^{-6}$ cm and $|\bar{\chi}| \geq 10^{-1}\ldots 10^{-3}$), this criterion is not fulfilled and

$$\left|\frac{d}{dx}\frac{1}{\tilde{k}(x)}\right| \approx 10^1 - 10^3 \gg 1. \tag{29.13b}$$

This result shows that solutions (29.8–29.12) are inapplicable in "ordinary" optics in the case of propagation in crystals of radiation with wavelengths that are longer than X-ray radiation and do not change previously known optical laws!

The results are similar in three-dimensional crystals.

29.3 Features of Resonant Absorption and Shortwave Laser Amplification in Realistic Media

The above discussed features of the electrodynamics of material media are based on the assumption that the sensitivity $\chi(\vec{r}, \omega)$ of material environment is the real (without imaginary part) function of coordinates and frequency. The presence of absorption leads to significant changes of these features. According to (29.8–29.12), the effective susceptibility $\chi_{eff}(\omega)$ is a nonlinear function of local dielectric characteristics of the material environment. In the result, the presence of absorption (at $\bar{\chi}''(\omega) \neq 0$) leads to essential modification of the expression for real and imaginary part of the susceptibility $\chi_{eff}(\omega)$

$$\begin{gathered}\chi_{eff}(\omega) = \bar{\chi} - \bar{\chi}^2 K = \chi'_{eff}(\omega) + i\chi''_{eff}(\omega);\\ \chi'_{eff}(\omega) = \bar{\chi}' - (\bar{\chi}')^2 K + (\bar{\chi}'')^2 K, \chi''_{eff}(\omega) = \bar{\chi}''(\omega)(1 - 2\bar{\chi}'(\omega)K).\end{gathered} \tag{29.14}$$

From these formulas follows that the absorption and dispersion of realistic material medium have a strong mutual influence and significantly change the character of the different radiation processes.

Three important results follow from these expressions.

(a) The presence of absorption in a spatially inhomogeneous media increases the real part of the effective susceptibility χ'_{eff} (29.14) by the value $(\bar{\chi}'')^2 K$, which assists in the generation of Cherenkov radiation;
(b) The presence of dispersion in a spatially inhomogeneous medium changes resonant absorption by *decreasing* $|\chi''_{eff}|$ in the range of frequencies $\omega < \omega_0$

with $\bar{\chi}' > 0$ and by *increasing* it for $\bar{\chi}' < 0$ (at $\omega > \omega_0$). In the result, the presence of the spatial structure of realistic environment leads to the breaking of the symmetry of absorption resonance.

(c) The presence of dispersion of $\bar{\chi}'(\omega)$ essentially changes the condition of X-ray (gamma-ray) [1] laser amplification in spatially inhomogeneous media

$$G_{inhomog} = (\omega/c)\chi''_{eff} \equiv (\omega/c)\bar{\chi}''(1 - 2\bar{\chi}'K)$$
$$\approx \frac{\langle \Delta n_n(r) \rangle \sigma_0 \gamma^2}{(\omega_0 - \omega)^2 + \gamma^2} \left\{ 1 + K \left(\frac{\langle \Delta n_n \rangle c\sigma_0 [4(\omega - \omega_0)/\gamma]}{\omega_0 \{1 + [2(\omega_0 - \omega)/\gamma]^2\}} + \frac{8\pi \langle n_e(\vec{r}) \rangle e^2}{m\omega^2} \right) \right\}. \tag{29.15}$$

Here $\langle \Delta n_n \rangle = \langle n_n^{(e)} \rangle - \langle n_n^{(g)} \rangle$—inverse population on resonant transition.

References

1. Vysotskii, V.I., Kuz'min, R.N.: Gamma-Ray Lasers. MSU Publications, Moscow (1989)
2. Vysotskii, V.I., Vysotskyy, M.V.: Conditions for generating X-ray Cherenkov radiation during the charge motion in actual mediums. J. Surf. Invest. **7**, 51–55 (2013)

Part IV
Alternative X-Rays and Applications

Chapter 30
Observation and Investigation of Intensive Directional Quasi-coherent X-Ray Generated at Interaction of Cavitating Liquid Jet with a Target

V. I. Vysotskii and A. A. Kornilova

Abstract In the chapter, the results of a detailed study of the generation of directed controlled quasi-coherent X-ray radiation with energy 1–5 keV stimulated by the cavitation of a liquid jet are presented and discussed. This radiation is characterized by a significant transverse coherence and a small angular divergence.

30.1 Introduction

It is well known that at the creation of controlled X-ray sources using both charged particle beams and fast-excitation atom systems, it is necessary to solve very complex technological problems. A much simpler alternative to such systems is the shock wave excited by cavitation of a liquid jet [1]. The main elements that generate such waves are cavitation bubbles. Each bubble is a dynamic self-controlled object that is capable of exerting a very strong influence on the surrounding liquid in the collapse mode. Similar processes are possible at Coulomb collapse of the target [2–4]. During our studies of the cavitation processes in machine oil and water, we discovered previously unknown phenomena [5–7] associated with the generation of directed X-rays outside the cavitation chamber. These phenomena are discussed below.

V. I. Vysotskii (✉)
Kiev National Shevchenko University, Kiev, Ukraine
e-mail: vivysotskii@gmail.com

A. A. Kornilova
Moscow State University, Moscow, Russia

T. Kawachi et al. (eds.), *X-Ray Lasers 2016*, Springer Proceedings in Physics 202, https://doi.org/10.1007/978-3-319-73025-7_30

30.2 Cavitation-Induced X-Ray in Spindle Oil in Closed Chamber

The first part of our research was carried out on the basis of an experimental setup in which a jet of spindle oil came under pressure 1…100 atm from a narrow (~1 mm) long (2 cm) capillary into the volume of cavitation chamber (sizes: D = 8 cm and L = 15 cm) made of Plexiglas (Fig. 30.1).

At a low pressure of the oil jet ($P < 20$ atm), luminescence and X-ray radiation were absent. X-ray emission occurred when pressure $P \approx 20$–60 atm was reached. At such pressure, the chamber volume was filled with a finely dispersed fraction of liquid that corresponded to the beginning of the cavitation process [6]. The X-ray radiation was detected by amplitude *Amptek X-123 CdTe X-ray and gamma-ray spectrometer*, the entrance window of which was placed outside the cavitation chamber near the thick wall with thickness about 3 cm. As the pressure of the liquid increases, the frequency of the radiation increases, and its intensity decreases (see Fig. 30.1).

Detailed studies have shown that the mechanism of X-ray emission is associated with several successive phenomena: (a) the formation of shock waves in a liquid during cavitation, (b) the interaction of these waves with the inner surface of the chamber, (c) the excitation of secondary shock waves in the walls of the chamber, (d) the reflection of secondary shock waves from the outer wall of the chamber and the excitation of atoms on this external surface, (e) synchronized spontaneous emission of these atoms with large transverse coherence [6, 7]. These shock waves

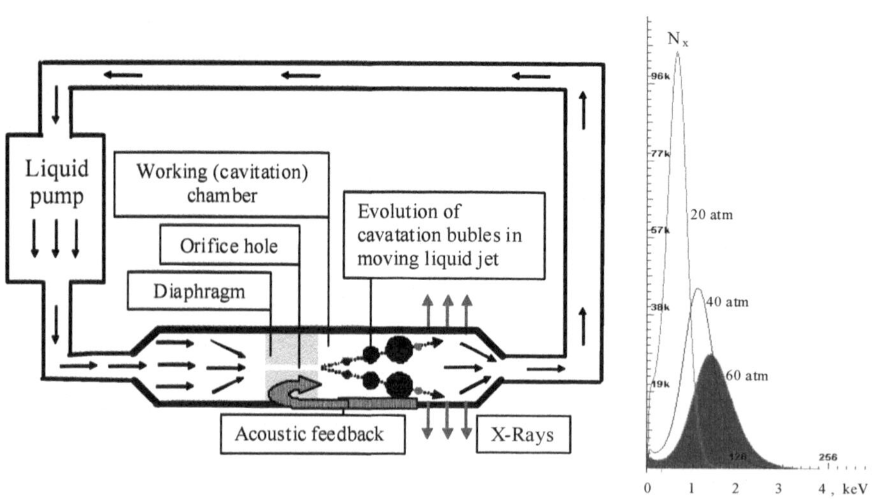

Fig. 30.1 The scheme of the experimental setup for study of cavitation phenomena in spindle oil jet (left) and spectrum of X-ray detected outside the cavitation chamber at different oil pressure (right)

on the surface of the chamber were also investigated using an acoustic piezoelectric detector.

It is very important to note that the process of excitation of shock waves in different cavitation bubbles and the subsequent excitation of X-ray levels have been synchronized with the help of positive feedback due to the inverse effect of shock waves in the wall volume on the parameters of the diaphragm at the entrance to the chamber. The loop of this positive feedback is shown in Fig. 30.1. The period of this feedback loop (repetition frequency is $\omega \approx 22$ kHz) is determined by three time–space factors: (a) the time of movement of the oil jet with cavitation bubbles from the exit of the diaphragm; (b) the duration of the shock wave movement to the chamber wall; (c) the duration of the acoustic shock wave propagation along the wall to the entrance to the diaphragm. This system is "bubble" analog of electronic Van der Pol generator [8].

There are at least three physical mechanisms, by which the action of shock wave in the volume of the cavitation chamber may lead to X-ray generation on the outer surface of the same chamber.

The first mechanism is associated with a well known in quantum mechanics methods of sudden acceleration of atom with a nuclear charge Z to the velocity v during reflection of the shock wave front. Consider for example the excitation of $1s_0 \rightarrow 2p_0$ radiation transition at sudden acceleration. The wavefunction of atom electron in the initial $1s_0$ state has the form

$$\Psi_{100}(\vec{r},t) = \sqrt{Z^3/4\pi a^3}\, e^{-Zr/a} \exp(-iE_1 t/\hbar)$$

The electron $2p_0$ wavefunction in atom moving with velocity $\vec{v} = \vec{e}_z v$ along z-axis is the following [6]:

$$\Psi_{210}(\vec{r},t) = r' \sqrt{Z^5/32\pi a^5} \cos\theta e^{-Zr'/2a} e^{im_e v z'/\hbar} e^{-i(E_2 + m_e v^2/2)t/\hbar}, \quad \vec{r}' = \vec{r} + \vec{e}_z vt.$$

The probability of electron transition $1s_0 \rightarrow 2p_0$ at fast change in the atom velocity is the following [6]:

$$W_{100,210} = \left| \int_V \Psi^*_{100}(\vec{r},0)\Psi_{210}(\vec{r},0) e^{imvz/h} dV \right|^2 = \frac{(9/32)(v/v_{100})^2}{[(v/v_{100})^2 + 9/4]^6},$$

where $v_{100} = Ze^2/\hbar \approx 2.3 * 10^8 Z$ cm/s. At typical velocity $v \equiv v_{sw} \approx 4 \cdot 10^6$ cm/s of the atomic motion in the shock wave front, we find the excitation probability

$$W_{100,210} \approx 2.2 * 10^{-3} (v/v_{100})^2 \approx 7 \cdot 10^{-7}/Z^2.$$

In a similar way, one can calculate the probability of excitation of other levels of energy. The spontaneous decay of these excited levels in atoms with $Z \gg 1$ leads to X-ray emission.

The second mechanism of X-ray generation at reflection of shock waves from the outer surface is associated with a strong deformation of the electronic shells of atoms in the shock wave front due to a large pressure gradient $\nabla P \approx P_{max}/\Delta L \approx 10^{11} - 10^{12}$ bar/cm within a very narrow front [6]. Such the deformation leads to the emergence of vacancies in the inner electronic shells of the atoms and to X-ray generation.

The third mechanism is connected with the possible short-term formation of coherent correlated states and ionization of atoms on the outer surface of the wall at reversible changing of interatomic distance by shock waves [9–12].

30.3 Investigation of Abnormal X-Ray at Cavitation of Free Water Jet

The second part of our research was carried out using a KMT water jet system [7] that can form a high-speed caviting water jet at P = 250–2000 atm (Fig. 30.2a). It was found that this jet after the output from the channel generates X-ray radiation with energy $E_x \approx 0.9$ keV at $P = 600$ atm and $E_x \approx 1.1$ keV at $P = 1000$ atm (Fig. 30.2b, water). It was found also that the outer surface of the thick (35 mm) metal wall of the rod with a cylindrical inner channel generates X-rays with a peak at $E_x \approx 1.7$–1.8 keV at $P = 600$ atm (Fig. 30.2b, Fe). When a very thin layer of fine lead powder is deposited to outer rod surface by medical gel for ultrasound, maximum of X-ray energy shifts to $E_x \approx 4.8$–5.0 keV (Fig. 30.2b, Pb). The mechanisms of X-ray generation is connected with the similar action of shock wave on surface atoms (see Sect. 30.1).

To increase the reliability of measurements and to investigate the spatial distribution of radiation, X-ray films were used. For this aim, two sheets of film were

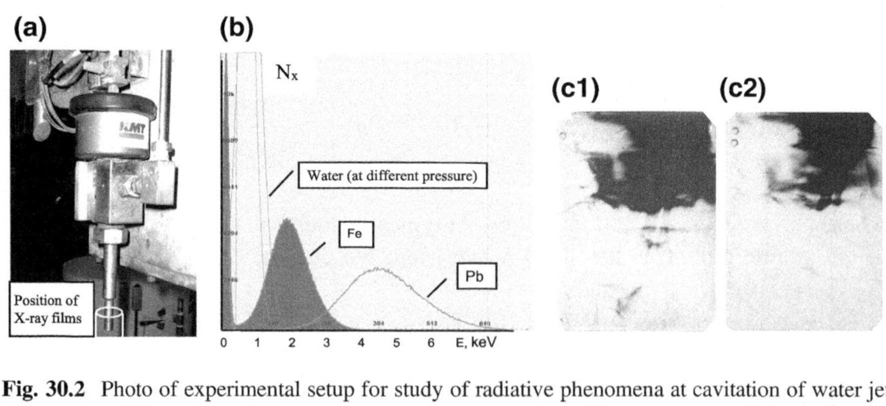

Fig. 30.2 Photo of experimental setup for study of radiative phenomena at cavitation of water jet (**a**); spectra of X-ray from different surface atoms at cavitation of fast water jet at 600 atm (**b**); the images on X-ray films, rolled into a cylinder coaxially with the rod: **c1**—the film was closer to the rod, **c2**—the film was behind the first film

placed without a gap in a common package of light-protective black paper. The package was rolled in the form of a cylindrical surface with a radius of $R \approx 3$ cm and was situated over the entire length of the rod. The distance from the outer surface of the rod to the film was $\Delta R_1 \approx 2.3$ cm and from the center of the water jet was $\Delta R_2 \approx 3$ cm. Figure 30.2c1, c2 shows the view of the two coaxial films after 15-min exposure at water pressure $P = 600$ atm. It can be seen that the spatial distribution of X-rays on both films is the same. The maximum radiation intensity was at the top of the images, which were close to the surface of the metal. The radiation energy can be clearly estimated by comparing the weaker darkening of the second film, which is associated with the radiation absorption in the volume of the first film. Estimations show that the X-ray energy corresponded to $E_x \approx 2$ keV. The lower part of both films was located below the end of the metal rod (see Fig. 30.2a) and softer X-ray radiation from water jet ($E_x \approx 1$ keV) did not reach the films, but was absorbed in air. The results obtained by X-ray detector and X-ray films are identical. This confirms that the results obtained by X-ray spectrometer are reliable!

30.4 Summary

Results of experiments and theoretical analysis show that the intensive acoustic shock waves connected with the cavitation processes in liquid jet are a source of intensive X-ray with a frequency that depends on kind of surface atoms. Relatively soft radiation with energy $E_x \approx 1$ keV is generated by a surface of the fast water jet in the area of cavitation and by surface of cavitation chamber made of Plexiglas. The more high X-ray ($E_x \approx 2$ keV) is generated by a surface of a cylindrical steel rod with caviting water jet. At the presence of heavy atoms on the rod, surface energy increases to $E_x \approx 5$ keV. It is important to note that this X-ray radiation is characterized by large transverse coherence, which is due to the synchronized excitation of surface atoms by a quasi-planar shock wave. Similar systems can be used as a safe laboratory and industrial sources of quasi-coherent X-ray with an activity up to 0.1–1 curie [13].

References

1. Barber, B.P., Hiller, R.A., Löfstedt, R., Putterman, S.J., Weninger, K.R.: Defining the unknowns of sonoluminescence. Phys. Rep. **281**, 65–70 (1997)
2. Adamenko, S.V., Vysotskii, V.I.: Evolution of annular self-controlled electron-nucleus collapse in condensed targets. Found. Phys. **34**, 1801–1831 (2004)
3. Adamenko, S.V., Vysotskii, V.I.: Neutronization and protonization of nuclei—two possible ways of the evolution of astrophysical objects and the laboratory electron-nucleus collapse. Found. Phys. Lett. **19**, 21–36 (2006)

4. Adamenko, S.V., Vysotskii, V.I.: Mechanism of synthesis of superheavy nuclei via the process of controlled electron-nuclear collapse. Ibid **17**, 203–233 (2004)
5. Kornilova, A.A., Vysotskii, V.I., et al.: Generation of intense directional radiation during the fast motion of a liquid jet through a narrow dielectric channel. J. Surf. Invest. **1**, 167–171 (2007)
6. Kornilova, A.A., Vysotskii, V.I., et al.: Generation of X-rays at bubble cavitation in a fast liquid jet in dielectric channels. Ibid **3**, 275–283 (2009)
7. Kornilova, A.A., Vysotskii, V.I., et al.: Generation of intense X-rays during ejection of a fast water jet from a metal channel to atmosphere. Ibid **4**, 1008–1017 (2010)
8. Ott, E.: Chaos in Dynamical Systems. Cambridge University Press, New York (1993)
9. Vysotskii, V.I., Adamenko, S.V., Vysotskyy, M.V.: Acceleration of low energy nuclear reactions by formation of correlated states of interacting particles in dynamical systems. Ann. Nucl. Energy **62**, 618–625 (2013)
10. Vysotskii, V.I., Vysotskyy, M.V.: Coherent correlated states and low-energy nuclear reactions in non stationary systems. Eur. Phys. J. A **49**, 99 (2013)
11. Vysotskii, V.I., Vysotskyy, M.V.: Correlated states and transparency of a barrier for low-energy particles at monotonic deformation of a potential well with dissipation and a stochastic force. J. Exp. Theor. Phys. **118**, 534–549 (2014)
12. Vysotskii, V.I., Vysotskyy, M.V.: The formation of correlated states and optimization of the tunnel effect for low-energy particles under non monochromatic and pulsed action on a potential barrier. Ibid **121**, 559–571 (2015)
13. Kornilova, A.A., Vysotskii, V.I., et al.: Patent WO201001906 A1 (2008)

Chapter 31
Laser-Plasma X-Ray Source Based on Cryogenic Targets

S. Amano

Abstract We developed a laser-plasma source based on cryogenic targets that generates continuous repetitive X-ray pulses. This source has a translating substrate system with a closed cryostat of He gas that can continuously supply solid Ar, Kr and Xe targets. The plasma emissions showed stable, continuous output power at frequencies of 1–10 Hz. At a laser intensity of 5×10^{12} W/cm^2, the Ar plasma emitted strongly in the water window region (2.3–4.4 nm), and its average X-ray power was 140 mW at a laser power of 1 W.

31.1 Introduction

Laser-plasma X-ray (LPX) sources that can generate high average X-ray power with continuous repetitive pulses, rather than a single shot, are required in most of their applications, especially in industrial use. Also, such applications require longer operational lifetimes. For this reason, we have been studying such a continuous LPX source that uses a cryogenic target of solid rare gas. Rare gas is considered to be an ideal deposition-free target because an inert gas, and its chemically inactive debris, will vaporize instantly, rather than deposit on optics near the plasma. This is advantageous during long lifetime continuous operation with repetitive pulses. We also chose to use a cryogenic solid target to increase the conversion efficiency and brightness, resulting from the solid target's higher density. Additionally, a smaller gas load for evacuation by the exhaust pump system was also expected in the solid state when compared with gas and liquid jets. In the present work, we developed a system to continuously supply cryogenic targets, and we report the characteristics of our developed X-ray source based on cryogenic Ar, Kr and Xe targets.

S. Amano (✉)
Laboratory of Advanced Science and Technology for Industry,
University of Hyogo, Ako, Hyogo, Japan
e-mail: sho@lasti.u-hyogo.ac.jp

T. Kawachi et al. (eds.), *X-Ray Lasers 2016*, Springer Proceedings
in Physics 202, https://doi.org/10.1007/978-3-319-73025-7_31

31.2 Experimental Setup

Figure 31.1 shows a side view of the developed LPX source, including the system that continuously supplied the cryogenic targets [1]. In our source, a copper substrate was attached to the tip of a cryostat head with a closed loop of He gas, which chilled the substrate surface to 15 K. The target gas was blown onto the substrate where it condensed into a solid layer with a thickness of 500 μm. The substrate coated with the solid layer translates up and down, moving in only one dimension, supplying a fresh target surface for each laser shot. The vacuum chamber was evacuated using a 190 l/s turbo molecular pump to a pressure of less than 1 Pa. The pump source was a conventional Q-switched Nd:YAG rod laser (Spectra-Physics, PRO-230), which could deliver pulses at a wavelength of 1ω (1064 nm) with a pulse width of 10 ns. The maximum pulse energy and repetition rate were 1.2 J and 10 Hz, respectively. The pulses were expanded with a beam expander before they passed through a window, and were then focused perpendicularly onto the target with a spot diameter of 50 μm by a lens with a focal length of 500 mm; this process produced plasma and emitted X-ray radiation. X-ray spectra were measured by a spectrometer with a resolution of 0.2 nm, which consisted of a transmission grating (TG) and a back-illuminated two-dimensional charge-coupled device (CCD). The TG was 1000 line/mm with a circular aperture, whose diameter was 46 μm. The spectrometer was located at 45° from the laser incident axis.

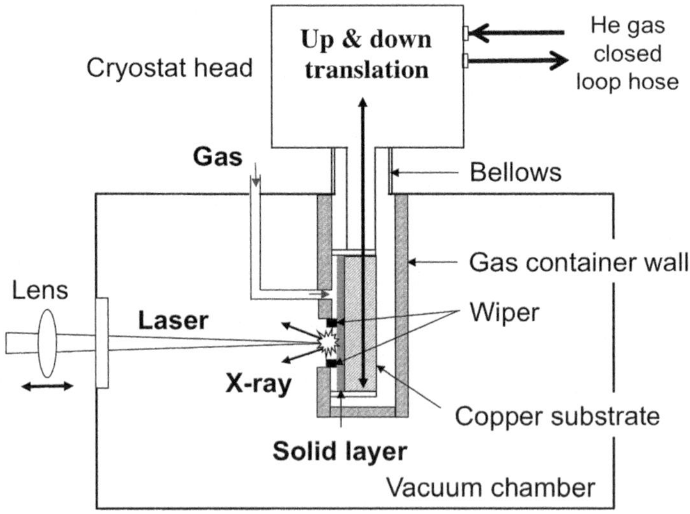

Fig. 31.1 Side view of the LPX source

31.3 Results and Discussion

31.3.1 Continuous Operation

To operate the LPX, generating continuous repetitive soft X-ray pulses, we determined the required laser frequency f, translation speed V and gas flow rate F. For example, the operational parameters for the Ar target were as follows: $V = 4$ mm/s and the gas flow rate $F = 250$ ml/min at the laser energy $E = 1$ J and $f = 1$ Hz, or $V = 11$ mm/s and $F = 400$ ml/min at $E = 0.1$ J and $f = 10$ Hz. Under the two sets of conditions, the laser pulses were focused on the Ar target. Figure 31.2 shows the results monitored using an X-ray diode (Opto Diode, SXUV-100 Mo/Si). These results show that the output power was stable and did not change as the target moved up and down, both at 1 and 10 Hz. This result indicates that each continuous laser shot had a fresh target surface, as designed. We also generated continuous repetitive X-ray pulses from the Kr and Xe targets, again as designed.

31.3.2 X-ray Spectra and Power

When the target was irradiated with pulse at 1 J and 1 Hz, the laser intensity was 5×10^{12} W/cm^2. Figure 31.3 shows the spectra measured at this intensity from the solid Ar, Kr and Xe. The X-ray powers were roughly estimated to be more than 10 mW in 10% bandwidth at 3.0 nm for the Ar target, at 9.0 nm for the Kr target, and at 10.8 nm for the Xe target, respectively [1].

Figure 31.3 shows that the Ar spectrum exhibited strong emissions across the water window region (2.3–4.4 nm). We measured the power in the water window region accurately. An absolutely calibrated X-ray diode (Opto Diode, AXUV100Ti/C2) measured the accurate energy and angular distribution by scanning it on a rotating stage around the plasma. The measured distribution could be fitted by

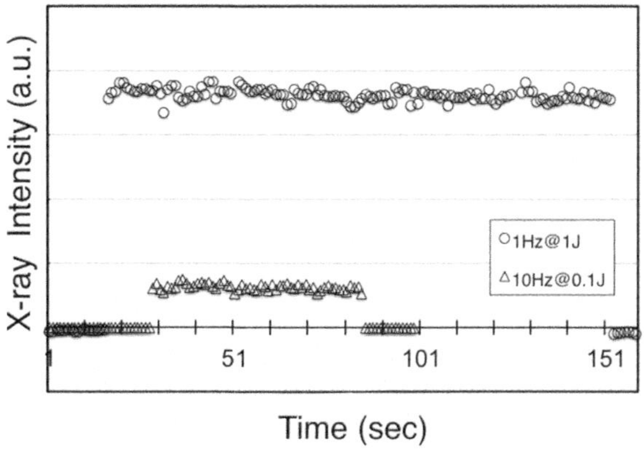

Fig. 31.2 EUV intensity from the Ar targets, monitored by an X-ray diode

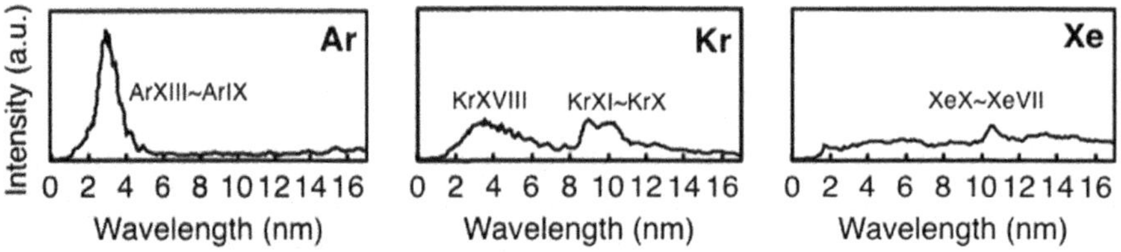

Fig. 31.3 Measured spectra from the solid Ar, Kr and Xe targets

$(\cos\theta)^{0.41}$. Accounting for this distribution, we conclude that the spatially integrated conversion efficiency was 14% in the water window (2.3–4.4 nm) at a laser energy of 1 J. Therefore, we achieved a high power of 140 mW (1 J × 1 Hz × 14%) in the water window [2].

Acknowledgements This work was supported by JSPS KAKENHI (Grant No. 23654204 & 26390113) and the Hyogo Science and Technology Association (Grant No. 26049).

References

1. Amano, S.: Laser plasma cryogenic target on translating substrate for generation of continuously repetitive EUV and soft X-ray pulses. Rev. Sci. Instrum. **85**, 063104 (2014)
2. Amano, S.: High-average-power water window soft X-rays from an Ar laser plasma. App. Phys. Express **9**, 07620 (2016)

Chapter 32
Photoionization of Atomic Neon Induced Using Nanosecond Pulses of Extreme Ultraviolet (EUV)

I. Saber, A. Bartnik, P. Wachulak, W. Skrzeczanowski, R. Jarocki and H. Fiedorowicz

Abstract We report on spectral investigations of Neon-photoionized plasma in the EUV and ultraviolet/visible (UV/VIS) ranges generated by nanosecond pulses of EUV. The EUV source is based on a gas puff target irradiated with Nd:YAG laser. The radiations from the EUV sources were collected and focused using different EUV collectors. The EUV radiations were focused on the Ne gas stream, injected into a vacuum chamber synchronously with the EUV pulses. Irradiation of the Ne gas resulted in photoionized plasmas formation consisted of atoms and ions with different ionization states. Differences in spectral line strengths for Ne ions were observed. The strongest spectral lines are from the neutral and singly ionized Ne emission lines. Using the Boltzmann plot method and with an assumption of a partial local thermodynamic equilibrium, the plasma electron temperature was estimated from the characteristic emission lines based on Ne II and Ne I species. A spectral analysis code PrismSPECT has been used to reproduce the experimental spectra theoretically. A summary of the results and the basic atomic processes dominated in the plasma is briefly discussed.

32.1 Introduction

Irradiation of a gas medium with a photon beam will ionize the gas. This ionized gas is called plasma, contains electrons, atoms, and ions. Estimating the electron temperature in the plasma and other plasma parameters is one of the basics plasma diagnostics. Investigation on the emission spectra of such plasma represents a key knowledge for photoionized plasma constituents and clarifies atomic kinetics

I. Saber · A. Bartnik (✉) · P. Wachulak · W. Skrzeczanowski · R. Jarocki
H. Fiedorowicz
Institute of Optoelectronics, Military University of Technology, Warsaw, Poland
e-mail: andrzej.bartnik@wat.edu.pl

T. Kawachi et al. (eds.), *X-Ray Lasers 2016*, Springer Proceedings
in Physics 202, https://doi.org/10.1007/978-3-319-73025-7_32

processes governed by it [1]. And also, increases further understanding of the fundamental aspects of photoexcitation and photoionization mechanisms during the plasma formation. For the time being, the invention of X-ray lasers has increased the future research domain of atomic and plasma physics. Especially, the invention of X-ray free-electron laser (FLASH) has enhanced the demand of studying the light–matter interaction in the soft X-ray region [2, 3]. Together with accelerator at Spring-8 in Japan [4] and the Linac Coherent Light Source (LCLS) in Stanford [5], intense and brilliant X-ray sources are state-of-the-art devices for the plasma physics studies, in particular for photoionization experiments. Differently, it is possible to ionize the gas using laser-produced plasma EUV sources or discharge-produced plasmas. There were considerable experiments reported previously, for instance, for laser-produced plasma [6] and discharged-produced plasma [7] as well dielectric-barrier discharge [8, 9]. Depending on the radiation source, intense EUV/SXR or electric discharge can be used to ionize a gas stream. To further explore the spectral behavior in atomic gases, we present in this work a preliminary result of the investigation on the photoionized plasma of Ne gas in the EUV and UV/VIS spectral range. The experimental sources were based on Nd: YAG laser with different pulse energies ranged and pulse durations. The laser plasmas were created by irradiation of a double-stream gas puff target with a Xe gas as the working gas. The EUV radiation pulses emitted from the sources were collected and focused onto a Ne gas stream injected into a vacuum chamber, through an auxiliary nozzle, synchronously with the EUV pulses. The irradiation of the gas by a focused nanosecond EUV pulses resulted in producing low-temperature photoionized plasma. This photoionized plasma emits radiation in the EUV and UV/VIS spectral ranges. The emitted radiation was measured using either a grazing incidence, flat-field spectrograph (McPherson Model 251), equipped with a 450 lines/mm toroidal grating or an Echelle Spectra Analyzer ESA 4000. The collected radiation from such plasma contains various emission lines corresponding to the various ionization degrees of Ne ions. Mostly observed emission lines were from the neutral and singly ionized charged states. The electron temperature in the plasma was estimated using the Boltzmann plot method, assuming the photoionized plasma is in a partial local thermodynamic equilibrium (pLTE).

32.2 Experimental Arrangement

The main elements of the experimental setup are a commercial Nd:YAG laser, a gas puff target, an EUV collectors, the Echelle spectrograph equipped with an intensified charge-coupled device (ICCD) and optionally a TGS EUV spectrograph. Schematic view of the setup is shown in Fig. 32.1. This source was based on a double-stream gas puff target; irradiated with the 10 Hz repetition rate and 10 J/10 ns Nd:YAG laser. The target was created by pulsed injection of xenon (8 bar) into a hollow stream of helium by employing electromagnetic valve system equipped with a double nozzle setup. After irradiation of such gas puff target with

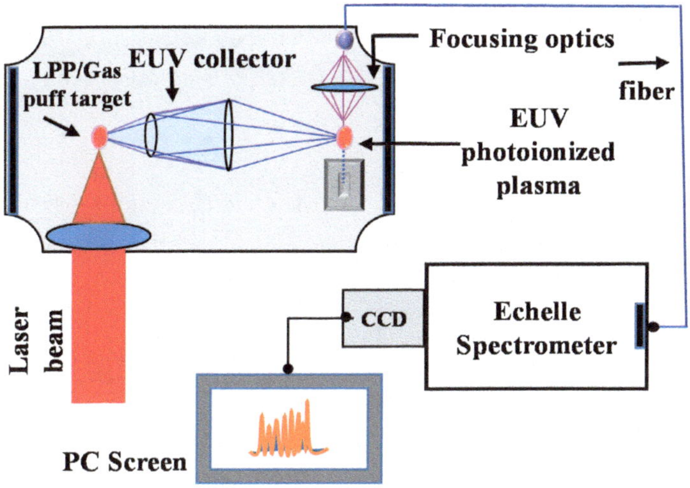

Fig. 32.1 Schematic view of the experimental setup of laser-produced plasma EUV source for photoionization experiments

Nd:YAG laser pulses a plasma is formed. The plasma radiation was focused into an interaction region. An atomic Ne gas was injected into the interaction region from an auxiliary electromagnetic valve. The backing pressure of the Ne gas was 5 bar. Irradiation of the Ne gas injected into the interaction region synchronously with EUV pulses leads to ionization and excitation of atoms and ions. The focusing conditions and photoionized plasma parameters were adjusted in the way to obtain maximum radiation intensity in the interaction region. The maximum power density in the region is around 10^8 W/cm^2 with the EUV fluence or power density of 300 mJ/cm^2. More information about the source description and focusing condition of the EUV radiation presented in [10]. The EUV radiation was focused using a multifoil collector. The spectral analyzing instrument was an Echelle Spectra Analyzer ESA 4000 spectrograph, equipped with the ICCD Kodak KAF 1001 Camera. The spectrometer system is capable of simultaneous measurements of complex spectra within the wide UV/VIS spectral range (200–780 nm) and a spectral resolution of $\lambda/\Delta\lambda \approx 20000$.

32.3 Experimental Results and Discussion

In Figs. 32.2 and 32.3 results of measured spectral lines in the UV/VIS region obtained from photoionized of Ne plasma were shown. Figure 32.2 shows the shorter wavelength spectral range while the longer spectral region is shown in the Fig. 32.3. The spectra are accumulated for 10 pulses integrated over 3 s with the pulse energy of 6 J. Different emission lines from Ne ions were observed.

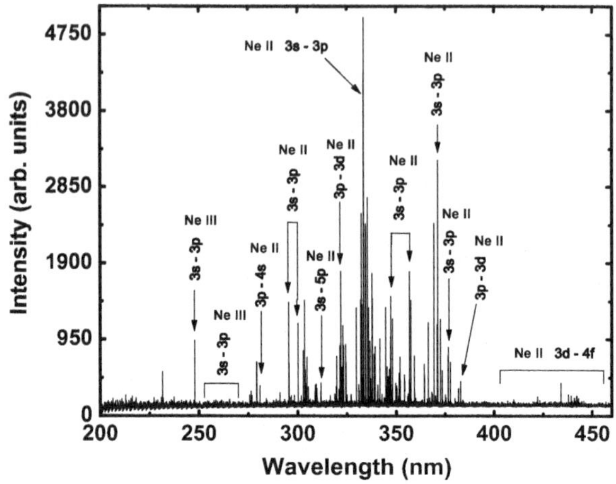

Fig. 32.2 UV/VIS emission spectrum lines for Ne-photoionized plasma obtained using laser-produced plasma EUV source

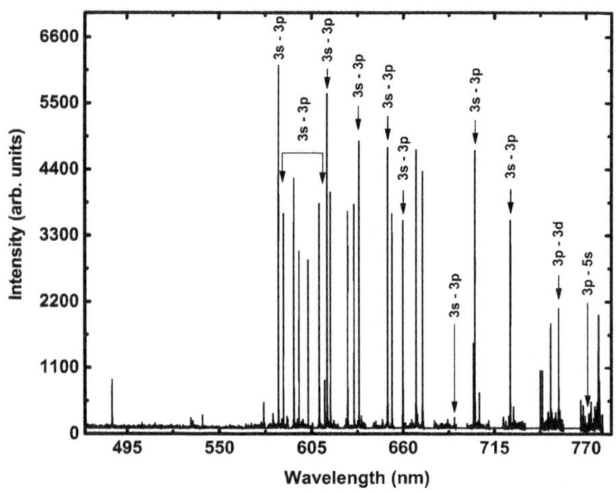

Fig. 32.3 UV/VIS emission spectrum lines for Ne-photoionized plasma obtained using laser-produced plasma EUV source

The detected transition lines are belong to neutral and up to double ionized Ne. The most intense emission lines were expanded over the entire spectral region, mostly dominated by neutral and singly ionized Neon. For the singly ionized Ne lines, Fig. 32.2, the strongest intensity lines corresponded to the UV spectral region, ranged from 300 to 380 nm.

In case of neutral Ne, Fig. 32.3, the optical spectral region approximately 585–730 nm showed significantly higher intensities. These lines, for Ne I and Ne II, mostly are emitted from radiative transitions $2s^22p^53s$-$2s^22p^53p$. Of course, there are also many lines correspond to the doubly ionized Ne III with relatively weak intensities comparing to the Ne I and Ne II spectral lines intensities.

In Fig. 32.4, the calculated emission lines using PrismSPECT [11] were put together with the measured spectra. This a steady-state spectral simulation,

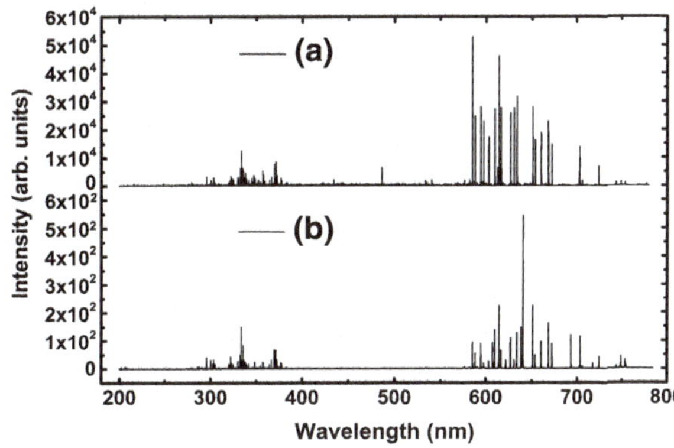

Fig. 32.4 Measured (**a**) and calculated (**b**) Ne-photoionized plasma in the UV/VIS spectral range obtained using the laser-produced plasma EUV source

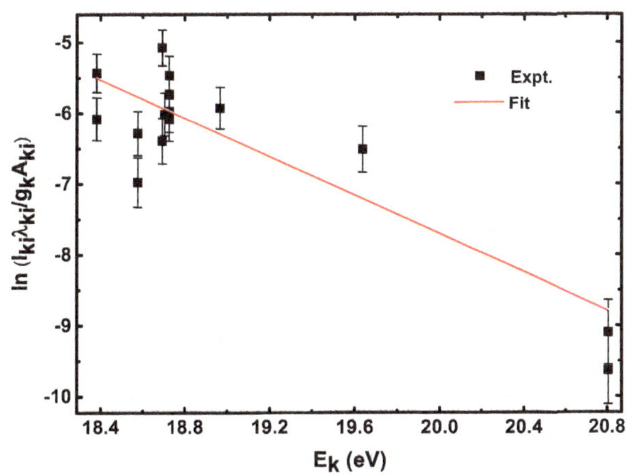

Fig. 32.5 Evaluation of the electron temperature using Boltzmann plot based on spectral lines originating from Ne I in the UV/VIS spectral ranges. The estimated electron temperature is $T_e = 0.74 \pm 0.19$ eV

the electron temperature that used in the simulation was of 1.99 eV with ion density of 2.778×10^{16} cm^{-3}.

Figure 32.5 shows the calculation of electron temperature for spectral lines in neutral Ne using Boltzmann plot method. It is important to mention a Boltzmann plot represents $\ln(I_{ki}\lambda_{ki}/g_k A_{ki})$, where I_{ki}, λ_{ki}, g_{k} and A_{ki} are integrated line intensity, transition wavelength, statistical weight and spontaneous transition probabilities, respectively, as a function of the transition energy of the upper levels E_{k} in eV. All possible lines with the strongest radiation transition probabilities were taken into account. Those overlapped and with less accuracy measurements in emission lines were avoided. Taking such data for evaluating electron temperature may affect the certainty of the result. Even so, it can be seen from the figure the plotted lines showed somehow a significant spread at the lowest transition energy, although the spectrum is accumulated for 10 EUV pulses. A fitted line to the experimental points gives rise an approximate electron temperature of 0.74 ± 0.19 eV with an estimating error of less than 20%.

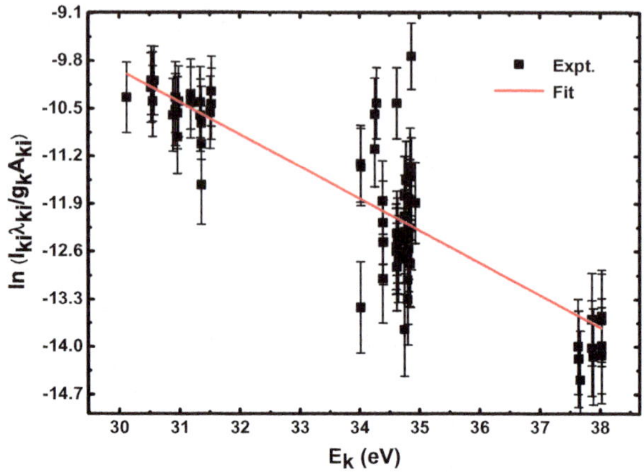

Fig. 32.6 Evaluation of the electron temperature using Boltzmann plot based on spectral lines originating from Ne II in the UV/VIS spectral ranges. The estimated electron temperature is $T_e = 2.11 \pm 0.08$ eV

There were many spectral lines corresponding to the singly ionized Ne. The Boltzmann plot for the characteristic emission lines, in this case, is shown in Fig. 32.6. The estimated electron temperature of these lines is about an order of three higher than that of the neutral atom, $T_e = 2.11 \pm 0.08$ eV. Considering stronger emission lines and covering a broad transition energy of the upper states in constructing the Boltzmann plot will give a better accuracy.

Part of the present experimental spectrum was recorded in the EUV region. A sample spectrum of Neon in the EUV range is presented in Fig. 32.7, measured spectrum (top) and simulated (bottom) with the spectral analysis code, Prism-SPECT. The experimental spectrum is acquired using the second source parameters; a 0.8 J, 4 ns and 10 Hz repetition rate. The spectra were accumulated over 100 pulses within 15 s using a grazing incident flat-field spectrometer (McPherson, Model 251) equipped with 450 lines/mm toroidal grating. A spectral resolution of the spectrograph $\lambda/\Delta\lambda \approx 500$. The modeled spectrum was obtained with an electron temperature of 1.95 eV and ion density of 5.0×10^{15} cm^{-3}. Known and observed emission lines are identified with [12]. There are two most intense emission lines in Ne I ion from $2s^22p^5$–$2sp^6$ transitions were observed. Most of the spectral lines in the spectra are mainly from singly ionized Ne and correspond to the $2s^22p^5$–$2s^2p^4nl$ transition lines.

Enhancement of the EUV radiation intensities in laser-produced plasmas (LPP) is essential from the points of view possible future applications in science and technology. A potential way to obtain efficient EUV emission in the interaction region, is to increase a laser pulse energy and ensure a proper focusing and absorption the laser energy by a target gas. This incident laser energy on the target gas is responsible for creating the ionization and excitation states for observed emission lines in a specific wavelength region. It is not required or necessary to reach the laser irradiance intensities the threshold conditions of the target material to produce efficient emission lines in laser-produced plasma EUV sources.

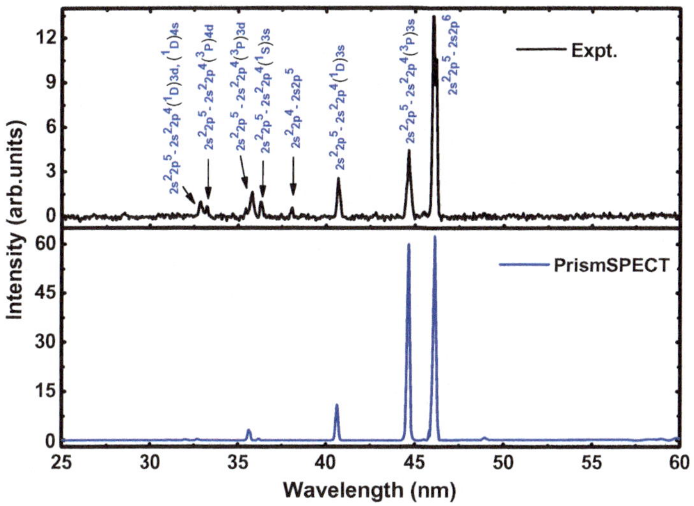

Fig. 32.7 Measured spectrum (top) and calculated (bottom) of ionized neon gas with EUV radiation pulses obtained using the laser-produced plasma source, accumulated 100 pulses with the pulse energy of 0.8 J

32.4 Conclusions

Overall, emission lines from Ne-photoionized plasmas produced are presented. The photoionized plasmas were produced by irradiation of the Ne gas using the EUV radiation produced from Xe/He gas puff target irradiated with a Nd:YAG laser of various parameters. The Neon gas was injected into the vacuum chamber synchronously with the exciting EUV pulses. The EUV beam was focused onto the gas stream employing different EUV collectors. The photoionized plasmas were analyzed using the Echelle Spectra Analyzer and a grazing incident flat-field spectrometer (McPherson, Model 251) equipped with 450 lines/mm toroidal grating. Different lines with relatively different intensities from Ne ions were detected. The strongest intensity lines dominated by neutral and singly ionized ions. For the higher ionization states, for example, Ne III, the overlapping of many spectral lines is significant. Also, the intensities of these lines are rather weak comparing to the neutral or singly ionized emission lines.

Theoretical spectra were calculated using PrismSPECT based on data from experimental emission lines. Analysis of the modeled spectrum revealed that main atomic processes contributed to the Ne spectra were from bound–bound and bound–free transitions. However, the contributions from the free–free transitions were almost negligible. This may explain the discrete and continues bands in Ne-photoionized plasma emission lines. A comparison between the measured and simulated spectra is showed reasonable accuracy within the given wavelength range. Boltzmann plots corresponding to radiative transitions lines in Ne I and Ne II were constructed and the electron temperature was estimated. An increased electron temperature from the neutral atom to the singly ionized Ne is shown. Justifying that at the earlier times of plasma formation there has been a high electron density attributed to a few collisional processes between photoionized plasma species and hence a strong interactions between electrons-atoms or electrons-ions in the plasma. More investigations on plasma-based EUV laser is needed by enhancing the intensity of EUV light sources to study the higher ionization states which give rise to the inner-shell transitions and other applications in EUV lithography as well. These also give other advantages related to higher ionization states from LPP EUV sources are needed and benchmarking the spectral model.

Acknowledgements The authors acknowledge the financial support from the EU FP7 Erasmus Mundus Joint Doctorate Programme EXTATIC under framework partnership agreement FPA-2012-0033 and partially with the support of the European Union's Horizon 2020 Programme (LASERLAB-EUROPE project grant agreement No. 654148). The work is also supported by the National Centre for Science, award No. UMO-2016/23/B/ST7/00949.

References

1. Kallman, T.R., McCray, R.: X-ray nebular models. Astrophys. J. Suppl. Ser. **50**, 263 (1982)
2. Ayvazyan, V., et al.: First operation of a free-electron laser generating GW power radiation at 32 nm wavelength. Eur. Phys. J. D **37**, 297–303 (2005)
3. Ackermann, W., et al.: Operation of a free-electron laser from the extreme ultraviolet to the water window. Nat. Photonics **1**, 336 (2007)
4. Shintake, T.: A compact free-electron laser for generating coherent radiation in the extreme ultraviolet region. Nat. Photonics **2**, 555–559 (2008)
5. Bostedt, C., Boutet, S., Fritz, D.M., Huang, Z., Lee, H.J., Lemke, H.T., Robert, A., Schlotter, W.F., Turner, J.J., Williams, G.J.: Linac coherent light source: the first five years. Rev. Mod. Phys. **88**, 015007 (2016)
6. Bartnik, A., Fiedorowicz, H., Fok, T., Jarocki, R., Szczurek, M., Wachulak, P.: Low temperature photoionized Ne plasmas induced by laser-plasma EUV sources. Laser Part. Beams **33**, 193–200 (2015)
7. Mahmood, S., Shaikh, N.M., Kalyar, M.A., Rafiq, M., Piracha, N.K., Baig, M.A.: Measurements of electron density, temperature and photoionization cross sections of the excited states of neon in a discharge plasma. J. Quant. Spectrosc. Radiat. Transf. **110**, 1840–185 (2009)
8. Sewraj, N., Merbahi, N., Marchal, F., Ledru, G., Gardou, J.P.: VUV spectroscopy and post-discharge kinetic analysis of a pure xenon mono-filamentary dielectric barrier discharge (MF-DBD). J. Phys. D Appl. Phys. **42**, 045206 (2009)
9. Carman, R.J., Kane, D.M., Ward, B.K.: Enhanced performance of an EUV light source (λ = 84 nm) using short-pulse excitation of a windowless dielectric barrier discharge in neon. J. Phys. D Appl. Phys. **43**, 025205 (2010)
10. Bartnik, A.: Laser-plasma extreme ultraviolet and soft X-ray sources based on a double stream gas puff target: interaction of the radiation pulses with matter. OPTO-Electron. Rev. **23**(2), 172–186 (2015)

11. Prism Computational Sciences, Inc., PRISMSPECT, 'Spectral Analysis Code', 455 Science Drive, Suite 140, Madison, WI, 53711. http://www.prism-cs.com/Software/PrismSPECT/overview.html
12. Kelly, R.L.: Atomic and Ionic spectrum lines below 2000 a Angstroms; hydrogen through Krypton (supplement No. 1 to, Vol. 16), Part I (H-Cr). J. Phys. Chem. Ref. Data **16**, 1–649 (1987)

Chapter 33
Effects of Equation of State on Fluid Simulations for Laser-Produced Plasmas

C. Ikeda, A. Sunahara and S. Namba

Abstract We developed a new fluid simulation code for laser-produced plasmas that applied HLL-HLLC scheme for equations of conservation for mass and momentum and upwind difference method for electron temperature and ion temperature. We applied new equation of state library named BADGER v1.0 to our simulation code and compared with the case of use of SESAME library. Our simulation results show the sensitivity of fluid dynamics for the equation of state.

33.1 Introduction

We have developed a fluid simulation code for laser-produced plasmas (LPP) that have two characteristics. First, easily becomes non-equilibrium between ion and electron, because relaxation time between ion and electron in the ablated plasma can be much longer than laser pulse duration, in which it is needed that electron temperature and ion temperature are calculated independently. Second, if we applied the differential form to the Euler equation for LPP that is typically compressible, conservation low of physical conserved values such as mass, momentum and total energy easily fail with accompanying large numerical errors. On the other hand, conservation scheme hardly deals with two independent energy equations of ion and electron that are derived from the first law of thermodynamics. Therefore, we developed a new scheme for the one-fluid, two-temperature model, which is combined with both conservation scheme and non-conservation scheme, in which mass and momentum of the fluid is solved by the conservation scheme. For ion and electron energy equations, we adopted the non-conservative scheme, which easily treats the two-temperature model. Because effects of equation of state (EOS) on

C. Ikeda · S. Namba (✉)
Department of Mechanical Science Engineering, Hiroshima University, Hiroshima, Japan
e-mail: namba@hiroshima-u.ac.jp

A. Sunahara
Institute for Laser Technology, Osaka, Japan

T. Kawachi et al. (eds.), *X-Ray Lasers 2016*, Springer Proceedings in Physics 202, https://doi.org/10.1007/978-3-319-73025-7_33

fluid dynamics are considered to be important, we should introduce appropriate EOS, which corresponds to target material. After the confirmation of conservation property of our code, we carried out the fluid simulations of laser ablation process and investigated effects of EOS on the simulation results. Here, we introduced new Equation of State library: BADGER v1.0 and compared it with the SESAME library.

33.2 Proposed Scheme for LPP Simulation

In order to improve the conservation of mass, momentum and total energy, we have developed a new scheme for LPP simulation.

Figure 33.1 shows the flow chart of our simulation code. First, mass and momentum (ρ, $\rho\mathbf{u}$) are calculated from equations of conservation for mass and momentum with pressure ($\overleftrightarrow{P}$) given by EOS with HLL (Harten–Lax–van Leer)-HLLC (HLL contact) approximate Riemann solver [1], which ensures conservation for physical values and compatible conservative values. Second, ion temperature and electron temperature are calculated by upwind difference method. Here, the heat capacity (c_V) and pressure term $p_{th} = T\left(\frac{\partial p}{\partial T}\right)$ are obtained from EOS. Finally, electron thermal conduction equation is calculated with Spitzer–Harm electron conductivity coefficient (κ) and electron-ion temperature relaxation equation is calculated.

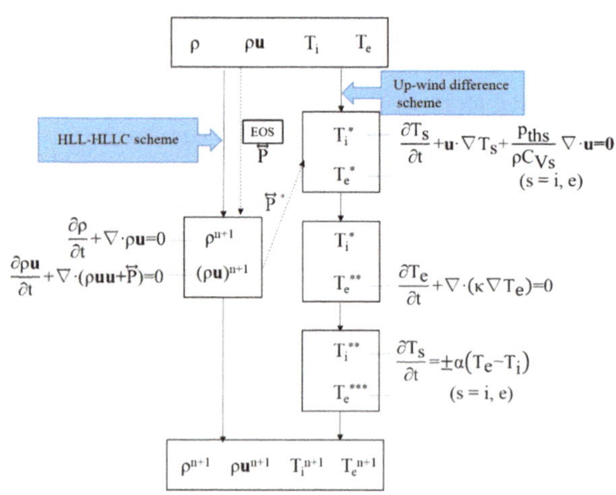

Fig. 33.1 The flow chart of our simulation code

33.3 New Equation of State Library: BADGER v1.0

BADGER v1.0 is library developed by Heltemes et al. (2012) [2]. Ion equation of state is based on Cowan model introducing scaled binding model. This model ensures positive pressure and switches to ideal gas at low density. Electron EOS and ionization degree are based on Thomas–fermi model taking into account ionization energy and ionization potential lowering contribution. Furthermore, the zero-temperature term in Helmholtz free energy is always subtracted from the result to correct positive pressure at zero temperature. Thermodynamic variables such pressures can be obtained from differencing the Helmholtz free energy.

33.4 Simulation Results

33.4.1 Simulation Condition

Our calculation simulates the condition of irradiation of 1.064 nm YAG laser on the carbon solid target (2.26 g/cm^3) with 10 ns pulse duration (FWHM) and 1×10^{11} W/cm^2 peak intensity up to time t = 5 ns. Computational region has 2D cylindrical coordinate and 1000 μm × 1000 μm (1000 × 1000 meshes) area. Time step was determined by Courant values of 0.1.

33.4.2 Conservation Ratio for Total Mass

We examined the ratio of total mass from the sum of initial values as shown in Fig. 33.2. Here, as a reference, we compared our result with that by Combined Unified Procedure (CIP) scheme, which is a non-conservative scheme and widely used for the LPP simulations. The solid line shows the mass conservation ratio calculated by our new code, and dot line shows the result obtained by CIP method.

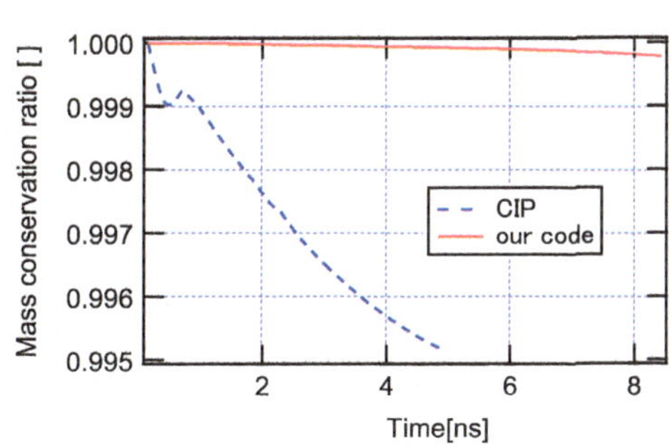

Fig. 33.2 Time evolution of mass conservation rate

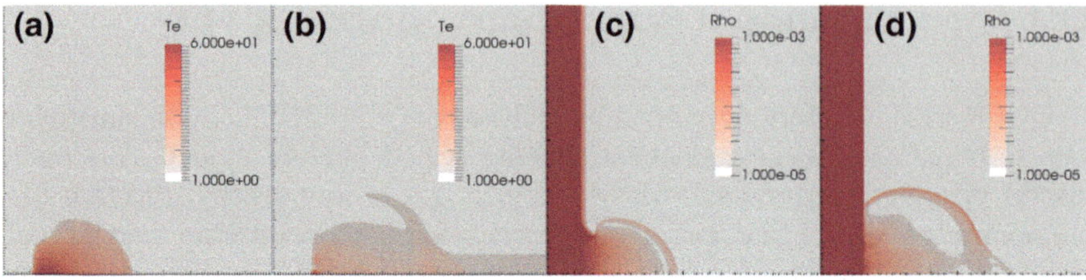

Fig. 33.3 At t = 5 ns, **a**, **b** electron temperature profile with BADGER v1.0 and SESAME, **c**, **d** mass density profile with BADGER v1.0 and SESAME

Both codes used BADGER v1.0 EOS and measured up to 10000 time steps. Figure 33.2 shows that the conservation of mass is significantly improved from CIP.

33.4.3 Effect of Equation of State

In order to investigate effects of EOS on profiles of LPP, we carried out simulations with BADGER v1.0 EOS and SESAME EOS library under the same condition. Figure 33.3a and c shows electron temperature and mass density profiles at the time t = 5 ns in the use of BADGER v1.0 EOS, and Fig. 33.3b and d shows ones in the use of SESAME EOS, respectively. We see the higher electron temperature shown in Fig. 33.3a for BADGER v1.0 EOS compared to that shown in Fig. 33.3b for SESAME EOS. In BADGER 1.0 EOS, average ionization degree that took into account ionization energy and ionization potential lowering is more realistic (Fig. 33.4a, b), and its effects on laser energy deposition and electron temperature. On the other hand, we see the smaller plasma expansion shown in Fig. 33.3a for BADGER v1.0 EOS compared to that of SESAME EOS shown in Fig. 33.3b. In the case of BADGER v1.0 EOS, electron sound speed is lower in the region of low temperature solid (Fig. 33.4c) than SESAME EOS (Fig. 33.4d), so plasma expansion is smaller.

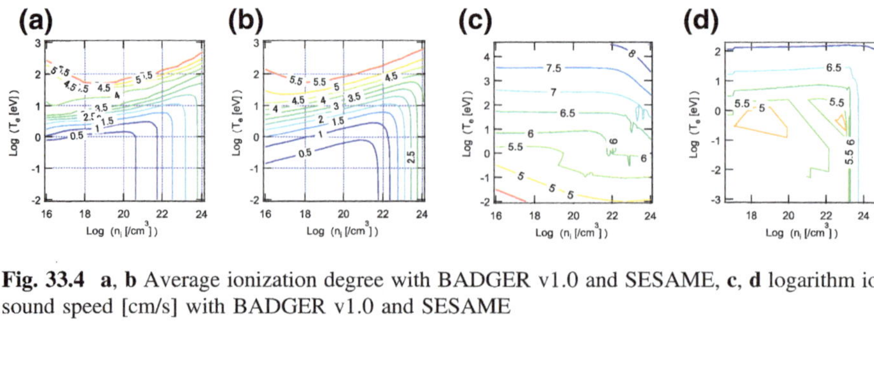

Fig. 33.4 **a**, **b** Average ionization degree with BADGER v1.0 and SESAME, **c**, **d** logarithm ion sound speed [cm/s] with BADGER v1.0 and SESAME

33.4.4 Conclusion

Improvement of conservation for mass shows the better conservation property of our new code with conservation and non-conservation hybrid scheme. In conclusion, our new code is useful when we are emphasizing on LPP profile.

Also, we compared two EOS models of BADGER v1.0 and SESAME for the carbon laser ablation. BADGER v1.0 showed relatively lower sound velocity in solid state, resulting in smaller plasma expansion, because more realistic ionization energy and potential lowering were incorporated in the code.

References

1. Kim, S.D., et al.: J. Comput. Phys. **228**, 7634 (2009)
2. Heltemes, T.A., Moses, G.A.: Comput. Phys. Commun. **183**, 2629 (2012)

Part V
X-Ray Imaging

Chapter 34
Soft X-Ray Laser Ablation Mass Spectrometry for Chemical Composition Imaging in Three Dimensions (3D) at the Nanoscale

C. S. Menoni, I. Kuznetsov, T. Green, W. Chao, E. R. Bernstein, D. C. Crick and J. J. Rocca

Abstract Analytical probes capable of assessing the molecular composition and mapping it in 3D at the nanoscale will transform materials research, biology, and medicine. Herein, we describe recent advances of soft X-ray laser ablation mass spectral imaging that highlight the unique capabilities of the method to identify intact molecular content in dielectrics and fullerenes and to map the molecular composition of single microorganisms in 3D at the nanoscale.

34.1 Introduction

Table-top soft X-ray (SXR) lasers producing pulses with energies from 1 to 100 μJ at repetition rates of up to 400 Hz, and with wavelengths ranging from 8.8 to 46.9 nm [1–9] have enabled the implementation of photonic application testbeds that are paving the way to scientific research and technological innovation. Nanoscale zone plate and holographic imaging, nanopatterning, nanoablation,

C. S. Menoni (✉) · I. Kuznetsov · T. Green · J. J. Rocca
Department of Electrical & Computer Engineering, Colorado State University, Fort Collins, CO, USA
e-mail: carmen.menoni@colostate.edu

W. Chao
Lawrence Berkeley Lab, Center for X-Ray Optics, Berkeley, CA, USA

C. S. Menoni · E. R. Bernstein
Department of Chemistry, Colorado State University, Fort Collins, CO, USA

D. C. Crick
Department of Microbiology, Immunology and Pathology, Colorado State University, Fort Collins, CO, USA

J. J. Rocca
Department of Physics, Colorado State University, Fort Collins, CO, USA

T. Kawachi et al. (eds.), *X-Ray Lasers 2016*, Springer Proceedings in Physics 202, https://doi.org/10.1007/978-3-319-73025-7_34

nanofabrication, and nanoscale chemical imaging all exploit the nanometer-scale wavelength to achieve superior resolution, to image nanostructures and their dynamics with a single laser shot and print over areas, a fraction of a millimeter, with high throughput [10–20]. This paper describes advances in SXR laser ablation mass spectrometry, an analytical method first demonstrated by our group using a compact 46.9 nm SXR laser [20].

Pulsed laser ablation in combination with mass spectrometry forms the basis of one of the most broadly used analytical tools to assess chemical composition of the solid phase [21–23]. The chemical makeup of a specimen in the probed region is obtained by mass-discriminating the ions generated in the laser-created plasma. A mass spectrum typically contains numerous peaks that identify atoms and molecules within the probed region of the sample.

In pulsed laser ablation mass spectrometry, the mechanisms of ablation and ionization are dictated by the laser–matter interactions, which are a function of the laser wavelength, pulse duration and laser fluence. These interactions are also responsible for shaping molecular fragmentation which in some cases introduces interferences in a mass spectrum.

There are numerous works that have studied SXR laser ablation of different materials including metals, polymers, and dielectrics [24–32]. These works have shown that SXR laser ablation has distinct traits. In SXR laser ablation, the energetic SXR photons are absorbed by direct photoionization on most solids within 10–100 nm from the surface due to strong absorption. The SXR photons interact with the sample throughout the laser pulse duration, even for nanosecond pulses, due to reduced inverse bremsstrahlung. Bond breaking leads to the formation of plasmas with electron temperatures of the order of a few electron volts and density higher than optically created plasmas [33, 34]. The high degree of localization of the SXR laser energy and strong absorption in large bandgap dielectrics made it possible to ablate craters micron deep at threshold fluences of ~0.1 J/cm^2 (or equivalently $6 \cdot 10^8$ W cm^{-2} for 1.7 ns pulse duration at $\lambda = 46.9$ nm), 20–40× smaller than when using UV lasers [35]. The ablation threshold of dielectrics is further reduced by 10× when using EUV laser pulses of picosecond time duration [25]. The mechanisms for picosecond SXR laser ablation have been associated with thermomechanical processes following the 100% absorption of the laser energy by the solid [25].

SXR photons have also sufficient energy to ionize atoms, molecules, and fragments in the plasma plume. Studies of photoionization of gas-phase metal and oxide clusters using 26.4 eV photons ($\lambda = 46.9$ nm, 1.5 ns pulse duration) in combination with mass spectrometry have shown that the SXR photons efficiently ionize inorganic clusters [36]. Studies of SXR laser-created plasmas in solids show that plasmas created using nanosecond pulses at $\lambda = 46.9$ nm are significantly colder and less ionized than those created by 120 ps pulses at $\lambda = 800$ nm [33].

SXR laser pulses can also be focused to ~80 nm diameter spots using a zone plate lens to ablate polymers [37]. This coupled with the nm-penetration depth of SXRs made it possible to demonstrate three-dimensional (3D) nanoscale composition imaging in organic thin film samples for the first time [20].

In this chapter, we show applications of SXR laser ablation mass spectrometry to the chemical analysis and composition imaging of different solid samples. We selected SiO_2 as an example to highlight differences in the mass spectrum depending on the polarity of the ions detected. We use C_{60} to show that the instrument has high sensitivity and large mass range detection capability. Lastly, we show an example of the 3D imaging capabilities of the method.

34.2 Experimental Details

The SXRL ablation mass spectrometry system, shown in the Fig. 34.1, has been previously described [20]. Pulses from a compact 46.9 nm discharge pumped SXRL are guided into the interaction chamber and focused onto the sample's surface using a zone plate lens. The 200 nm outer zone width zone plate focuses the SXR laser beam to spots of ~200 nm. The zone plate contains a central opening 50 μm in diameter that allows for ion extraction. The sample and the zone plate are mounted on precision-controlled stages that allow precise micrometer and nanometer positioning. A set of electrostatic plates, appropriately biased, is used to extract positive or negative ions into the time of flight (TOF) analyzer, which is operated in reflectron mode. Firing of the laser, sample motion, and ion detection are all computer-controlled.

We have previously shown that it is possible to ablate spots in a polymer resist with volumes varying from 66 to 2 atto-liters by reducing the laser fluence from 3 to 0.2 J/cm^2 [20]. This is an important capability for achieving nanometer-scale resolution in chemical imaging applications. SXR laser ablation mass spectrometry has achieved a sensitivity of 0.04 atto-mole in the detection of alanine, a detection sensitivity 40× larger than that of secondary ion time-of-flight mass spectrometry [20].

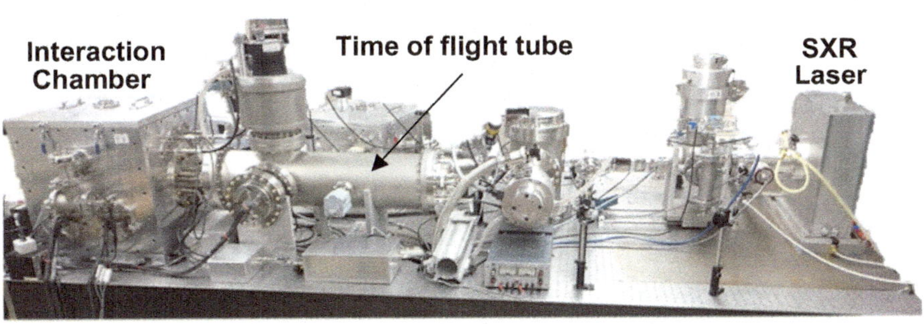

Fig. 34.1 Photograph of the SXR laser ablation mass spectrometry instrument showing the compact capillary discharge laser (λ = 46.9 nm), the interaction chamber, and the time-of-flight mass detector

Thin film samples are either deposited or spun over an indium tin oxide (ITO) coated substrate. In some cases, a thin, ~20 nm metal layer is added onto the sample's surface to prevent surface charging. The fullerene samples consisted of a 15–30 nm thick layer of commercially available C_{60} (Sigma-Aldrich, Milwaukee, WI, USA) deposited onto the ITO-coated glass slide.

34.3 Results

34.3.1 Mass Spectral Analysis

The SXR laser–material interactions are different depending on the specific sample. We have shown here two examples, SiO_2 and C_{60}, to highlight the distinct features obtained in their mass spectra. Figure 34.2 shows mass spectra of SiO_2 obtained with positive and negative ion extraction. The positive ion spectrum only contains elemental signatures from the glass matrix, Si^+ and O^+, and from impurities, Na^+ and K^+. In contrast, the negative ion spectrum contains molecular signatures that are identified as SiO_2^- and $[SiO_2 + O]^-$.

Fullerenes are comprised entirely of carbon acting as both electron donors and electron acceptors. The ionization energy (IE) of C_{60} is 7.54 eV and the fragmentation energy is 40.8 eV ($C_{60}^+ \rightarrow C_2 + C_{58}^+$) [38]. Their high order of symmetry along with high degrees of freedom allow for charge delocalization and the ability to withstand large amounts of excess internal energy above the IE [39]. The ablation of C_{60} with 26.5 eV photons exploits fullerene's high photoionization cross section (>500 Mb at 26.5 eV) to enhance ion yield while minimizing doubly charged ions [40]. The unfragmented C_{60} fullerene peak is observed at 720 m/z. Formation of "magic fullerenes" at C_{70} (840 m/z) and C_{84} (1008 m/z) is observed.

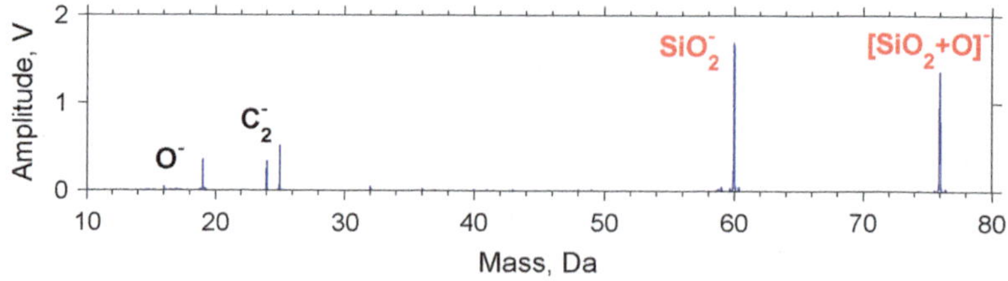

Fig. 34.2 Mass spectra of SiO_2 obtained with: (top) positive ion extraction and (bottom) negative ion extraction

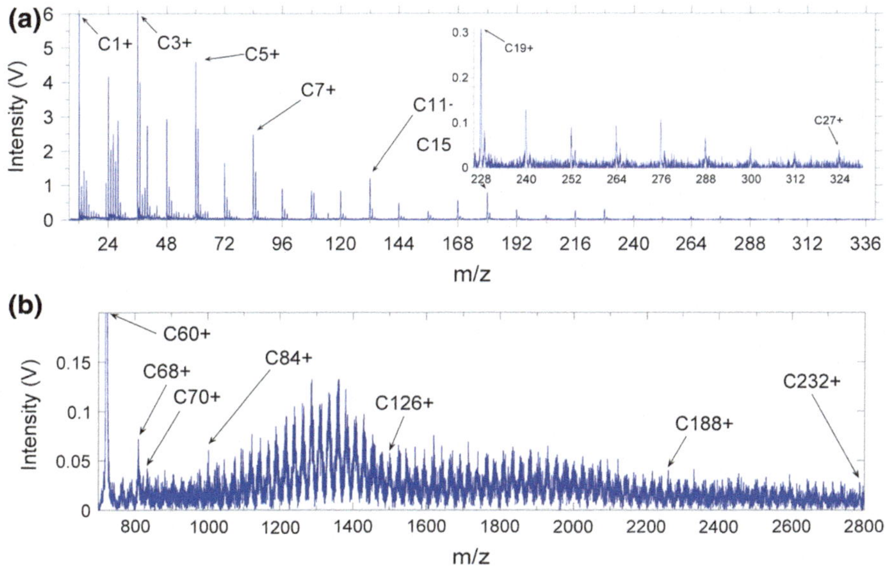

Fig. 34.3 **a** Mass spectrum of C_{60} fullerene, low-mass range (0–336 m/z), highlighting the presence of carbon series C_1–C_{28} with C_1 separation. No doubly charged species were present. **b** Mass spectrum of C_{60} fullerene, high mass range (700–2800 m/z), highlighting the formation of stable giant fullerenes with C_2 separation. Magic fullerenes are highlighted at C_{60}, C_{70}, and C_{84}

Consistent C_2 separation of even carbon clusters from C_{68} to C_{232} with distributions centered near 1300 and 1900 m/z is observed in the mass spectrum shown in Fig. 34.3b. Envelopes of distribution occur from ~C_{84} to C_{126} and C_{126} to C_{188}. Previous work has shown similar results using laser desorption/ionization (LDI) methods [41]. We speculate that the formation of higher order fullerenes occurs during plasma expansion.

The observation of carbon fragments is fluence-dependent. The spectra shown in Fig. 34.4 are plotted on the same scale and normalized to fluence. Fragmentation is highest at 70 mJ/cm^2 and decreases at lower fluences resulting in an increase in the C_{60} parent ion signal. Parent ion is prominent while fragments are nearly absent at 15 mJ/cm^2. Fragments include C_3–C_{24} at intervals of C_1. Similar to other work [42], the presence of coalesced fragments above C_{60} is observed only in the presence of low-mass fragments.

34.3.2 *Mass Spectral Imaging*

The SXR laser ablation mass spectrometry system shown in Fig. 34.1 can map chemical composition in 3D with nanoscale resolution [20]. Composition images are constructed from the analysis of single-shot mass spectra obtained from

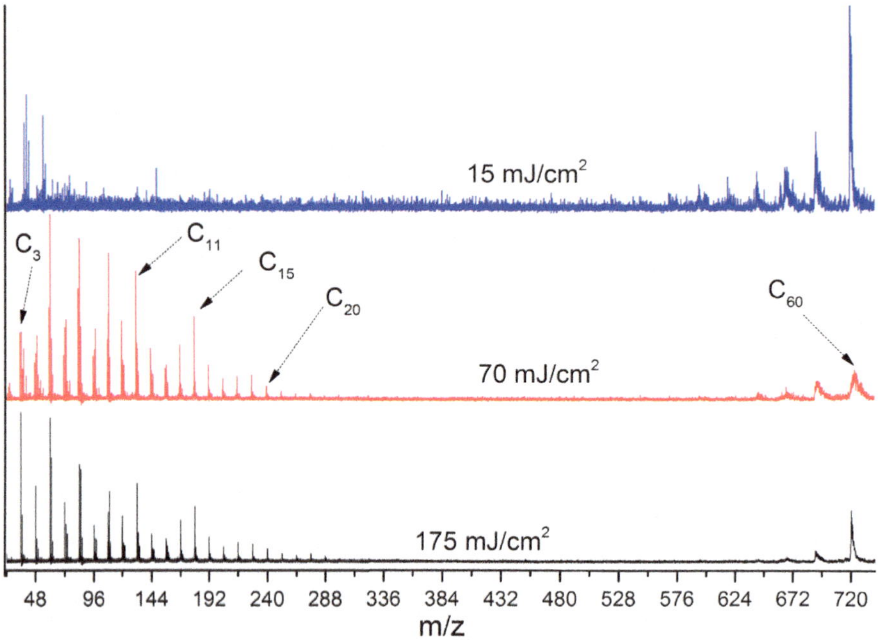

Fig. 34.4 Mass spectrum of C_{60} fullerene at different fluences. Fragments include C_3–C_{24} at intervals of C_1. The C_{60} parent is observed at 720 m/z in all spectra

atto-liter ablated craters (voxels) as the sample is displaced laterally on a plane perpendicular to and parallel to the optical axis defined by the SXR laser beam. An example of 3D composition maps of specific molecular components in an isolated *M. smegmatis* bacterium is shown in Fig. 34.5. In these images, contours of constant amplitude, isolines, of an ion peak in the mass spectrum are plotted. The lateral and depth displacements are 0.3 μm and 0.2 μm, respectively. The resultant 3D data array, with a full mass spectrum in each voxel, is then postprocessed to produce smoothed profiles of each molecular distribution inside the bacterium. For example, ion 91 m/z (blue color plot) shows similar distribution in all five data layers assessed, while the 86.1 m/z ion (green color plot) distribution shows isolines extending throughout the bacterium the most among three analyzed molecular maps. The best lateral and depth resolutions obtained from the analysis of organic materials are 75 nm and 20 nm, respectively [20]. A similar lateral resolution has been obtained in the analysis of inorganic samples [43]. Resolution is affected by the laser fluence, the sensitivity of the detection system, and noise in the spectrum which is sample-dependent. Nevertheless, the ability to probe molecular composition from atto-liter volumes with the level of sensitivity demonstrated by SXR laser ablation mass spectrometry is unparalleled.

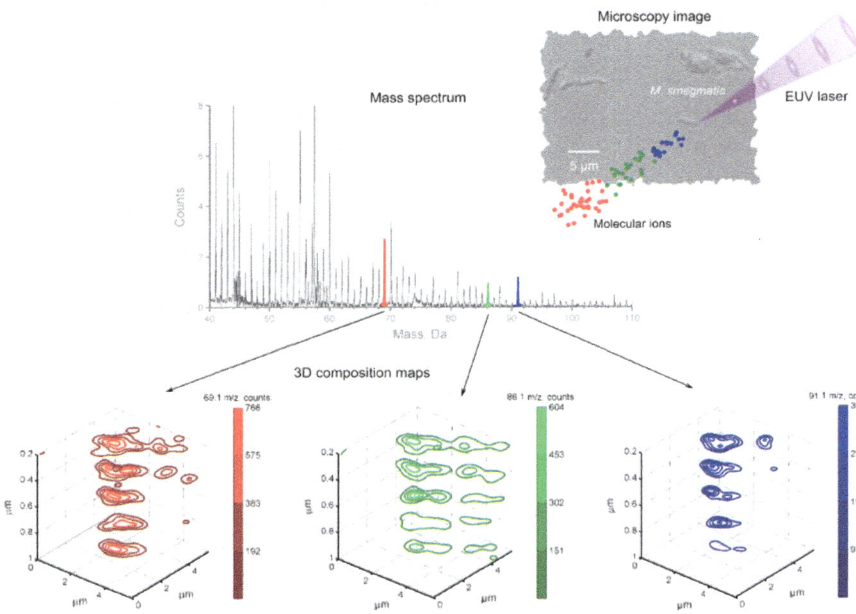

Fig. 34.5 Schematics showing how composition maps are formed by plotting the intensity of a salient single peak in the mass spectrum in 3D. Each voxel produces a full mass spectrum as shown in the figure. The colored molecular ions in the microscopy image represent how masses separate in the time-of-flight chamber

34.4 Conclusions

SXR laser ablation mass spectrometry is a highly sensitive method with the capability to analyze and map the chemical composition of solid samples in three dimensions with nanoscale resolution. Its nanoscale resolution is a combination of the high localization of the SXR laser beam into atto-liter volumes, the distinct interaction of the SXR laser radiation and materials, and the ability of the high energy photons to ionize most atoms and molecules.

The analysis of the SiO_2 and C_{60} mass spectra showed intact molecular ions. This is the case for SiO_2 when using negative ion extraction. The mass spectra of C_{60} contain a wealth of information, in terms of fragmentation as well as coalescence mechanisms, which requires further studies. Advances in understanding the physics of the interaction of SXR lasers and materials will be critical for SXR laser ablation mass spectrometry to achieve its highest potential as a nanoscale diagnostic tool. Improved sensitivity and extended mass range are expected to provide increasingly detailed information that will enable, for example, the understanding of subcellular chemical organization in biological systems, surface chemistry on nanostructures, and their interaction within heterogeneous materials at a level not previously possible.

References

1. Heinbuch, S., Grisham, M., Martz, D., Rocca, J.J.: Demonstration of a desk-top size high repetition rate soft X-ray laser. Opt. Express **13**, 4050 (2005)
2. Martz, D.H., Alessi, D., Luther, B.M., Wang, Y., Kemp, D., Berrill, M., Rocca, J.J.: High-energy 13.9 nm table-top soft-X-ray laser at 2.5 Hz repetition rate excited by a slab-pumped Ti:sapphire laser. Opt. Lett. **35**, 1632 (2010)
3. Benware, B.R., Macchietto, C.D., Moreno, C.H., Rocca, J.J.: Demonstration of a high average power tabletop soft X-ray laser. Phys. Rev. Lett. **81**, 5804 (1998)
4. Curtis, A.H., Reagan, B.A., Wernsing, K.A., Furch, F.J., Luther, B.M., Rocca, J.J.: Demonstration of a compact 100 Hz, 0.1 J, diode-pumped picosecond laser. Opt. Lett. **36**, 2164 (2011)
5. Macchietto, C.D., Benware, B.R., Rocca, J.J.: Generation of millijoule-level soft-X-ray laser pulses at a 4-Hz repetition rate in a highly saturated tabletop capillary discharge amplifier. Opt. Lett. **24**, 1115 (1999)
6. Reagan, B.A., Wernsing, K.A., Curtis, A.H., Furch, F.J., Luther, B.M., Patel, D., Menoni, C. S., Rocca, J.J.: Demonstration of a 100 Hz repetition rate gain-saturated diode-pumped table-top soft X-ray laser. Opt. Lett. **37**, 3624 (2012)
7. Baumgarten, C., Pedicone, M., Bravo, H., Wang, H.C., Yin, L., Menoni, C.S., Rocca, J.J., Reagan, B.A.: 1 J, 0.5 kHz repetition rate picosecond laser. Opt. Lett. **41**, 3339 (2016)
8. Rocca, J.J., Wang, Y., Larotonda, M.A., Luther, B.M., Berrill, M., Alessi, D.: Saturated 13.2 nm high-repetition-rate laser in nickellike cadmium (30, 2581, 2005). Opt. Lett. **31**, 129 (2006)
9. Furch, F.J., Reagan, B.A., Luther, B.M., Curtis, A.H., Meehan, S.P., Rocca, J.J.: Demonstration of an all-diode-pumped soft X-ray laser. Opt. Lett. **34**, 3352 (2009)
10. Vaschenko, G., Brizuela, F., Brewer, C., Grisham, M., Mancini, H., Menoni, C.S., Marconi, M.C., Rocca, J.J., Chao, W., Liddle, J.A., Anderson, E.H., Attwood, D.T., Vinogradov, A.V., Artioukov, I.A., Pershyn, Y.P., Kondratenko, V.V.: Nanoimaging with a compact extreme-ultraviolet laser. Opt. Lett. **30**, 2095 (2005)
11. Vaschenko, G., Brewer, C., Brizuela, F., Wang, Y., Larotonda, M.A., Luther, B.M., Marconi, M.C., Rocca, J.J., Menoni, C.S.: Sub-38 nm resolution tabletop microscopy with 13 nm wavelength laser light. Opt. Lett. **31**, 1214 (2006)
12. Wachulak, P.W., Capeluto, M.G., Marconi, M.C., Menoni, C.S., Rocca, J.J.: Patterning of nano-scale arrays by table-top extreme ultraviolet laser interferometric lithography. Opt. Express **15**, 3465 (2007)
13. Wachulak, P.W., Marconi, M.C., Bartels, R.A., Menoni, C.S., Rocca, J.J.: Volume extreme ultraviolet holographic imaging with numerical optical sectioning. Opt. Express **15**, 10622 (2007)
14. Brewer, C.A., Brizuela, F., Wachulak, P., Martz, D.H., Chao, W., Anderson, E.H., Attwood, D.T., Vinogradov, A.V., Artyukov, I.A., Ponomareko, A.G., Kondratenko, V.V., Marconi, M. C., Rocca, J.J., Menoni, C.S.: Single-shot extreme ultraviolet laser imaging of nanostructures with wavelength resolution. Opt. Lett. **33**, 518 (2008)
15. Wachulak, P., Grisham, M., Heinbuch, S., Martz, D., Rockward, W., Hill, D., Rocca, J.J., Menoni, C.S., Anderson, E., Marconi, M.: Interferometric lithography with an amplitude division interferometer and a desktop extreme ultraviolet laser. J. Opt. Soc. Am. B-Opt. Phys. **25**, B104 (2008)
16. Brizuela, F., Wang, Y., Brewer, C.A., Pedaci, F., Chao, W., Anderson, E.H., Liu, Y., Goldberg, K.A., Naulleau, P., Wachulak, P., Marconi, M.C., Attwood, D.T., Rocca, J.J., Menoni, C.S.: Inspection 13.2 nm table-top full-field microscope. In: Schellenberg, F.M., LaFontaine, B.M. (eds.) Alternative Lithographic Technologies (2009)
17. Brizuela, F., Wang, Y., Brewer, C.A., Pedaci, F., Chao, W., Anderson, E.H., Liu, Y., Goldberg, K.A., Naulleau, P., Wachulak, P., Marconi, M.C., Attwood, D.T., Rocca, J.J.,

Menoni, C.S.: Microscopy of extreme ultraviolet lithography masks with 13.2 nm tabletop laser illumination. Opt. Lett. **34**, 271 (2009)
18. Urbanski, L., Isoyan, A., Stein, A., Rocca, J.J., Menoni, C.S., Marconi, M.C.: Defect-tolerant extreme ultraviolet nanoscale printing. Opt. Lett. **37**, 3633 (2012)
19. Malm, E.B., Monserut, N., Brown, C.G., Wachulak, P.W., Xu, H., Balakrishnan, G., Chao, W., Anderson, E.H., Marconi, M.C.: Tabletop single-shot soft X-ray holography with a large field of view. Opt. Express **21**, 9959 (2013)
20. Kuznetsov, I., Filevich, J., Dong, F., Woolston, M., Chao, W., Anderson, E.H., Bernstein, E. R., Crick, D.C., Rocca, J.J., Menoni, C.S.: Three-dimensional nanoscale molecular imaging by extreme ultraviolet laser ablation mass spectrometry. Nat. Commun. **6**, 6944 (2015)
21. Castner, D.G., Ratner, B.D.: Biomedical surface science: foundations to frontiers. Surf. Sci. **500**, 28–60 (2002)
22. Becker, J.S.: Inorganic Mass Spectrometry. Wiley (2007)
23. McDonnell, L.A., Heeren, R.M.A.: Imaging mass spectrometry. Mass Spectrom. Rev. **26**, 606 (2007)
24. Juha, L., Bittner, M., Chvostova, D., Krasa, J., Otcenasek, Z., Prag, A.R., Ullschmied, J., Pientka, Z., Krzywinski, J., Pelka, J.B., Wawro, A., Grisham, M.E., Vaschenko, G., Menoni, C.S., Rocca, J.J.: Ablation of organic polymers by 46.9-nm-laser radiation. Appl. Phys. Lett. **86**, 034109 (2005)
25. Faenov, A.Y., Inogamov, N.A., Zhakhovskii, V.V., Khokhlov, V.A., Nishihara, K., Kato, Y., Tanaka, M., Pikuz, T.A., Kishimoto, M., Ishino, M., Nishikino, M., Nakamura, T., Fukuda, Y., Bulanov, S.V., Kawachi, T.: Low-threshold ablation of dielectrics irradiated by picosecond soft X-ray laser pulses. Appl. Phys. Lett. **94**, 231107 (2009)
26. Inogamov, N.A., Faenov, A.Y., Khokhlov, V.A., Zhakhovskii, V.V., Petrov, Y.V., Skobelev, I.Y., Nishihara, K., Kato, Y., Tanaka, M., Pikuz, T.A., Kishimoto, M., Ishino, M., Nishikino, M., Fukuda, Y., Bulanov, S.V., Kawachi, T., Artisimov, S.I., Fortov, V.E.: Spallative ablation of metals and dielectrics. Contrib. Plasma Phys. **49**, 455 (2009)
27. Inogamov, N.A., Faenov, A.Y., Zhakhovskii, V.V., Skobelev, I.Y., Khokhlov, V.A., Kato, Y., Tanaka, M., Pikuz, T.A., Kishimoto, M., Ishino, M., Nishikino, M., Fukuda, Y., Bulanov, S.V., Kawachi, T., Petrov, Y.V., Anisimov, S.I., Fortov, V.E.: Interaction of short laser pulses in wavelength range from infrared to X-ray with metals, semiconductors, and dielectrics. Contrib. Plasma Phys. **51**, 361 (2011)
28. Inogamov, N.A., Zhakhovsky, V.V., Faenov, A.Y., Khokhlov, V.A., Shepelev, V.V., Skobelev, I.Y., Kato, Y., Tanaka, M., Pikuz, T.A., Kishimoto, M., Ishino, M., Nishikino, M., Fukuda, Y., Bulanov, S.V., Kawachi, T., Petrov, Y.V., Anisimov, S.I., Fortov, V.E.: Spallative ablation of dielectrics by X-ray laser. Appl. Phys. A-Mater. Sci. Process. **101**, 87 (2010)
29. Ishino, M., Faenov, A.Y., Tanaka, M., Hasegawa, N., Nishikino, M., Tamotsu, S., Pikuz, T. A., Inogamov, N.A., Zhakhovsky, V.V., Skobelev, I.Y., Fortov, V.E., Khohlov, V.A., Shepelev, V.V., Ohba, T., Kaihori, T., Ochi, Y., Imazono, T., Kawachi, T.: Nanoscale surface modifications and formation of conical structures at aluminum surface induced by single shot exposure of soft X-ray laser pulse. J. Appl. Phys. **109**, 013504 (2011)
30. Ishino, M., Faenov, A.Y., Tanaka, M., Tamotsu, S., Hasegawa, N., Nishikino, M., Pikuz, T. A., Kaihori, T., Kawachi, T.: Observations of surface modifications induced by the multiple pulse irradiation using a soft picosecond X-ray laser beam. Appl. Phys. A-Mater. Sci. Process. **110**, 179 (2013)
31. Norman, G., Starikov, S., Stegailov, V., Fortov, V., Skobelev, I., Pikuz, T., Faenov, A., Tamotsu, S., Kato, Y., Ishino, M., Tanaka, M., Hasegawa, N., Nishikino, M., Ohba, T., Kaihori, T., Ochi, Y., Imazono, T., Fukuda, Y., Kando, M., Kawachi, T.: Nanomodification of gold surface by picosecond soft X-ray laser pulse. J. Appl. Phys. **112**, 013104 (2012)
32. Tomita, T., Yamamoto, M., Hasegawa, N., Terakawa, K., Minami, Y., Nishikino, M., Ishino, M., Kaihori, T., Ochi, Y., Kawachi, T., Yamagiwa, M., Suemoto, T.: Experimental verification of femtosecond laser ablation schemes by time-resolved soft X-ray reflective imaging. Opt. Express **20**, 29329 (2012)

33. Berrill, M., Brizuela, F., Langdon, B., Bravo, H., Menoni, C.S., Rocca, J.J.: Warm photoionized plasmas created by soft-X-ray laser irradiation of solid targets. J. Opt. Soc. Am. B-Opt. Phys. **25**, B32 (2008)
34. Gartside, L.M.R., Tallents, G.J., Rossall, A.K., Wagenaars, E., Whittaker, D.S., Kozlova, M., Nejdl, J., Sawicka, M., Polan, J., Kalal, M., Rus, B.: Extreme ultraviolet interferometry of warm dense matter in laser plasmas. Opt. Lett. **35**, 3820 (2010)
35. Reichling, M., Sils, J., Johansen, H., Matthias, E.: Nanosecond UV laser damage and ablation from fluoride crystals polished by different techniques. Appl. Phys. A-Mater. Sci. Process. **69**, S743 (1999)
36. Dong, F., Heinbuch, S., Rocca, J.J., Bernstein, E.R.: Single photon ionization of van der Waals clusters with a soft X-ray laser: $(SO_2)(n)$ and $(SO_2)(n)(H_2O)(m)$. J. Chem. Phys. **125**, 154317 (2006)
37. Vaschenko, G., Etxarri, A.G., Menoni, C.S., Rocca, J.J., Hemberg, O., Bloom, S., Chao, W., Anderson, E.H., Attwood, D.T., Lu, Y., Parkinson, B.: Nanometer-scale ablation with a table-top soft X-ray laser. Opt. Lett. **31**, 3615 (2006)
38. Hertel, I.V., Steger, H., Devries, J., Weisser, B., Menzel, C., Kamke, B., Kamke, W.: Giant plasmon excitation in free c-60 and c-70 molecules studied by photoionization. Phys. Rev. Lett. **68**, 784 (1992)
39. Wurz, P., Lykke, K.R.: Multiphoton excitation, dissociation, and ionization of C-60. J. Phys. Chem. **96**, 10129 (1992)
40. Juranic, P.N., Lukic, D., Barger, K., Wehlitz, R.: Multiple photoionization and fragmentation of C-60 in the 18–280-eV range. Phys. Rev. A **73**, 042701 (2006)
41. Raebiger, J.W., Alford, J.M., Bolskar, R.D., Diener, M.D.: Chemical redox recovery of giant, small-gap and other fullerenes. Carbon **49**, 37 (2011)
42. Beck, R.D., Weis, P., Brauchle, G., Kappes, M.M.: Mechanistic aspects of fullerene coalescence upon ultraviolet-laser desorption from thin-films. J. Chem. Phys. **100**, 262 (1994)
43. Green, T., Kuznetsov, I., Willingham, D., Naes, B.E., Eiden, G.C., Zhu, Z., Chao, W., Rocca, J.J., Menoni, C.S., Duffin, A.M.: Characterization of extreme ultraviolet laser ablation mass spectrometry for actinide trace analysis and nanoscale isotopic imaging. J. Anal. At. Spectrom. **32**, 1092 (2017)

Chapter 35
Coherent Diffraction Imaging with Tabletop XUV Sources

M. Zürch, A. Guggenmos, R. Jung, J. Rothhardt, C. Späth, J. Tümmler, S. Demmler, S. Hädrich, J. Limpert, A. Tünnermann, U. Kleineberg, H. Stiel and C. Spielmann

Abstract Coherent diffraction imaging (CDI) at wavelengths in the extreme ultraviolet range has become an important tool for nanoscale investigations. Employing laser-driven high harmonic sources allows for lab-scale applications such as cancer cell classification and phase-resolved surface studies in reflection geometry. The excellent beam properties support a spatial resolution below the wavelength, i.e., close to the Abbe limit. Unfortunately, the usually low photon flux of HHG sources limits their applicability. Recent advances in ultrafast fiber laser development cumulated in sources delivering average powers approaching the milliwatt level in the extreme ultraviolet. In comparison, a tabletop soft X-ray laser driven by moderate pump energies was recently employed for CDI featuring excellent temporal coherence and extraordinary high flux allowing for single-shot imaging.

M. Zürch (✉)
Chemistry Department, University of California, Berkeley, USA
e-mail: mwz@berkeley.edu

M. Zürch · C. Spielmann
Institute of Optics and Quantum Electronics, Jena University, Jena, Germany

M. Zürch · J. Rothhardt · S. Hädrich · J. Limpert · A. Tünnermann · C. Spielmann
Helmholtz Institute Jena, Jena, Germany

A. Guggenmos · C. Späth · U. Kleineberg
Ludwig-Maximilians-Universität München, Garching, Germany

A. Guggenmos · C. Späth · U. Kleineberg
Max-Planck-Institut für Quantenoptik, Garching, Germany

R. Jung · J. Tümmler · H. Stiel
Max Born Institute, Berlin, Germany

J. Rothhardt · S. Demmler · S. Hädrich · J. Limpert · A. Tünnermann
Institute of Applied Physics, Jena University, Jena, Germany

J. Limpert · A. Tünnermann
Fraunhofer Institute for Applied Optics and Precision Engineering, Jena, Germany

T. Kawachi et al. (eds.), *X-Ray Lasers 2016*, Springer Proceedings in Physics 202, https://doi.org/10.1007/978-3-319-73025-7_35

35.1 Introduction

The short wavelength radiation in the extreme ultraviolet (XUV) and soft X-ray range together with a high photon flux are the key elements for imaging nanoscopic structures. Coherent diffraction imaging (CDI) suits the needs for imaging in the XUV by omitting optical elements that would typically introduce high losses and limit the numerical aperture and thus the achievable resolution [1]. For the broader application of this technique, laboratory light sources of various kinds have been applied.

A prominent compact source increasingly applied for coherent diffraction imaging and holography at short wavelengths is based on high-order harmonic generation [2]. Such sources are driven by an amplified ultrafast laser system and feature high spatial coherence with sufficient narrowband emission lines. The latter is important for realizing temporal coherence that allows imaging down to the wavelength scale as has been recently demonstrated [3]. The spectral range covered with high photon flux by these sources is typically several 10 eV up to several 100 eV. Due to the low penetration depth at these photon energies, recent applications targeted the reflection geometry [4, 5] where CDI becomes a powerful technique yielding three-dimensional information of the surface. In addition, material specificity can be achieved [6]. The surface sensitivity becomes advantageous when nano- and microscale objects of a certain morphology are to be compared or classified. As an example, recently it was demonstrated that reflection geometry CDI at 35 eV photon energy can be used to classify different breast cancer cell expression profiles, solely by investigation of their diffraction patterns [7]. Other recent trends include using the full frequency comb of a harmonic source for gaining additional information comparable to Fourier transform spectroscopy [8] or raster scanning the sample, the so-called ptychography, for increasing the field of view and relaxing constraints in the image reconstruction [9].

In this contribution, we will focus on prospects for tabletop coherent diffraction imaging by improving established sources and applying new sources. A major bottleneck of HHG powered CDI experiments, thus far, are the limited photon flux and the resulting long integration times. In the second section, we will present recent progress in employing ultrafast high-power fiber lasers for a high harmonic generation. Due to the high repetition rate, an unprecedented average power near the milliwatt level per harmonic order is achieved. Another promising source is a laser plasma-based tabletop soft X-ray laser (SXRL), which we will introduce in Sect. 35.3 as a light source for CDI experiments. The extraordinary high photon flux and the narrow line width make this source interesting for CDI. The potential of these sources is evaluated in the closing section.

35.2 High Photon Flux XUV and Soft X-Ray Sources Based on High Harmonic Generation with Femtosecond Fiber Lasers

High harmonic generation provides short wavelength radiation with very good beam quality and spatial coherence. However, the conversion efficiency is typically not larger than 10^{-6} into a single harmonic in the XUV and even lower for the soft X-ray spectral region [10].

Femtosecond fiber lasers have significantly advanced during the recent years and nowadays deliver up to millijoule pulse energies at repetition rates of up to 1 MHz [11]. The corresponding average output power can be as high as 1 kW, which is two orders of magnitude higher compared to typical Ti:Sapphire-based femtosecond lasers. Key issues for efficient HHG are pulse-shortening via post-compression [12] and a high-density gas target to achieve phase matching in the typically required tight-focusing regime [13].

The combination of such a high average power femtosecond fiber laser with hollow-fiber-based post-compression and subsequent HHG in a continuous high-density gas jet recently demonstrated more than 10^{13} photons/s (>100 μW average power) within a single plateau harmonic at 30 eV [14]. A similar HHG source has already successfully been used for sub-wavelength imaging at high numerical aperture [3]. In CDI experiments, a higher spatial resolution requires shorter wavelengths (higher photon energies) once a numerical aperture near unity is reached. For that reason, the operation range of the fiber laser-driven HHG source has been shifted to 70 eV. In the following subsections, work toward high photon flux at the aforementioned photon energy and extensions up to the soft X-ray regime will be presented.

35.2.1 High Photon Flux 70 eV HHG Source

Higher cutoff energies can be achieved in HHG by focusing the driving laser to higher intensities, using longer driving wavelengths, and using a target gas with higher ionization potential (argon in this case). Here, a 35 fs, 0.5 mJ laser is focused to a peak intensity of $>2 \cdot 10^{14}$ W/cm^2 onto an argon gas jet to generate XUV radiation up to ~70 eV. The photon flux at 68.6 eV (H57) was particularly optimized by tuning the gas pressure, the jet position, and an aperture in front of the focusing lens. A spatiotemporal profile of the detected harmonics is shown in Fig. 35.1. At 100 kHz repetition rate this XUV source reached $>10^{11}$ photons/s in this single 1.3 eV bandwidth harmonic [15]. Very recently, this source has been used for CDI experiments demonstrating a record sub-20 nm resolution enabled by the combination of short wavelength, high spatial coherence, and high photon flux [16].

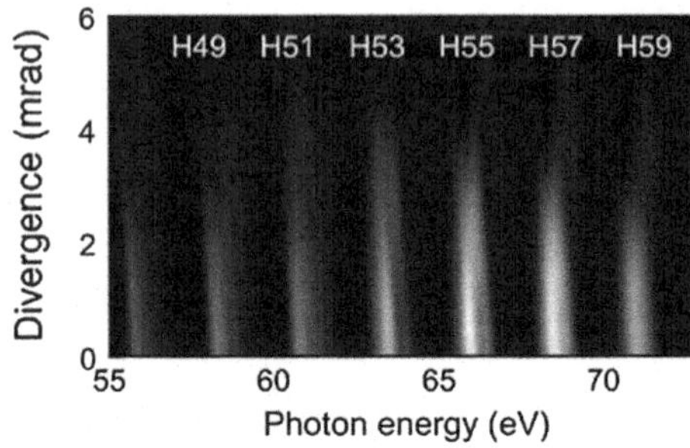

Fig. 35.1 Spatio-spectral distribution of high-order harmonics generated in argon

35.2.2 *High Photon Flux Coherent Soft X-Rays up to the Water Window*

HHG with a 1 μm wavelength laser into the soft X-ray region desires intensities on the order of 10^{15} W/cm^2 and very short (few-cycle) pulses to minimize the accumulated ionization of the target gas. These pulses can be generated efficiently by two-stage compression of a femtosecond fiber laser as described in [17]. Figure 35.2 displays typical HHG spectra that can be achieved with neon. Due to the few-cycle duration and a non-stabilized carrier-envelope phase, the harmonic lines are broadened and smeared out to a continuum. Nevertheless, a particular spectral region could be selected for imaging, e.g., by multilayer mirrors filtering a tailored central energy and bandwidth within the physical limits [18]. Although the conversion efficiency into the soft X-ray region is much lower compared to the XUV, still more than 10^9 photons/s/eV can be obtained up to the cutoff at ~150 eV (8 nm) [17].

HHG can be generated with this few-cycle laser in helium up to the so-called water window, but with a reduced photon flux due to poor phase-matching. Figure 35.2b displays measured spectra with and without an additional parylene filter clearly showing the carbon K-edge at 283 eV.

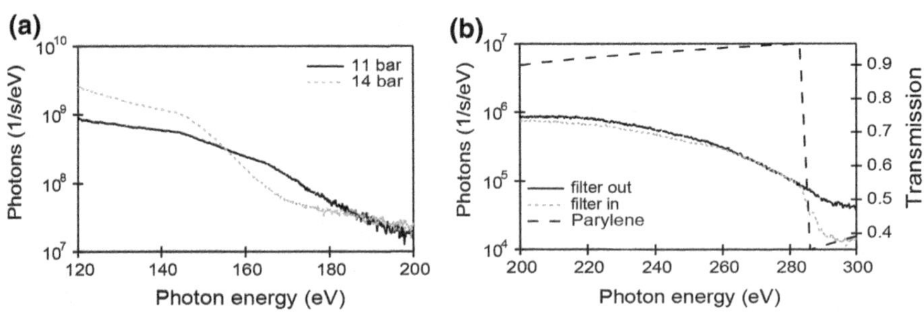

Fig. 35.2 **a** HHG spectra recorded with neon at different backing pressures of the gas jet. **b** HHG spectra ranging up to the water window with helium measured with and without an additional parylene filter

In the future, Tm-based fiber lasers delivering femtosecond pulses at 2 μm wavelength [19] will be employed to push the phase-matching cutoff into the water window and beyond. Thus, a new class of tabletop high photon flux soft X-ray sources will be available for tabletop imaging soon.

35.3 Solid-State Soft X-Ray Laser as Source for Coherent Diffraction Imaging

An alternative approach for generating coherent short wavelength radiation lies in employing plasma-based soft X-ray lasers (SXRL). These emit short pulses in the extreme ultraviolet XUV/soft X-ray range between 3 and 40 nm [20]. There are numerous schemes proposed for soft X-ray lasing, however, the transient collisional excitation scheme has proved to be the most reliable and promising for the development of compact laser-pumped SXRL [21]. It has been shown that combining this scheme with the grazing incidence (GRIP) pumping geometry a compact soft X-ray light source can be realized [22, 23]. Using about 1.5 J of pump energy in the GRIP scheme, strong XUV emission in the range between 10 and 20 nm with pulse energies of the SXRL up to 3 μJ were realized [24, 25]. Since the emission of the SXRL constitutes an atomic transition, the spectral bandwidth of the XUV emission is in the order of some picometer resulting in a very high temporal coherence [26]. Since these lasers typically operate in the self-amplified spontaneous emission (SASE) scheme, which is comparable to the operation scheme of most free-electron lasers, the transversal spatial coherence of the SXRL strongly depends on the geometry of the excitation scheme. The degree of spatial coherence ranges from few percents for a single-stage SXRL [27] up to 60% for a HHG-seeded X-ray laser [28]. For pump energies below 1 J, there are only very few data sets concerning the spatial coherence of a single stage laser. Nevertheless, these systems are worth further investigations, since they are attractive for applications requiring higher average photon flux, since pump lasers with pulse energies of 1 J or below are easily scalable to repetition rates up to 500 Hz [29].

Surprisingly, there are so far only very few examples of employing an SXRL for coherent diffraction imaging (CDI). For instance, in [30] an SXRL operating at 13.9 nm and pumped by a 1.5 J, 10 Hz Ti:Sa Laser was employed for coherent diffraction imaging. A spatial resolution of about 80 nm at 13.9 nm has been estimated at a numerical aperture of 0.08.

35.3.1 Molybdenum SXRL Operated with Moderate Pump Energy

Here, we report on recent experiments performed at an SXRL pumped in the GRIP geometry using moderate pump energies of below 0.5 J. The experimental layout is depicted in Fig. 35.3a. The soft X-ray laser was pumped by two pulses of a high repetition rate 100 Hz thin disk laser (TDL) chirped pulse amplification (CPA) system. The TDL system consists of a front-end with an Yb:KGW oscillator and stretcher delivering an output pulse of about 1 ns at 1030 nm. The stretcher is followed by a Yb:KGW regenerative amplifier. This output is subsequently divided into two pulses. Each of these pulses is amplified in a regenerative Yb:YAG thin disk amplifier to a level of about 100 mJ. The pulse from the first regenerative amplifier is compressed to a duration of about 200 ps using a grating compressor. The other pulse is fed to a thin disk multipass amplifier which amplifies the pulses to an energy up to 400 mJ. Subsequent compression in a grating compressor results in about 2 ps pulse duration. The long pulse (150 ps, $E \approx 70$ mJ) is focused by a combination of a cylindrical ($f = -500$ mm) and a spherical lens ($f = 380$ mm) onto the target at normal incidence. The resulting line focus has a width of about 30 μm. The generated plasma column is subsequently heated by the short pulse (2 ps, $E \approx 270$ mJ) focused according to the GRIP method by a spherical mirror ($f = 762$ mm) into the preformed plasma. In the presented case an optimal GRIP angle of 24° was found. The delay between the two pulses is a critical parameter [22, 23, 31], which can be set by the round trip time of the two amplifiers and a delay stage. Debris from the solid-state molybdenum target requires special attention at high repetition rate. In our experiments, we protected the optics against debris by thin glass plates or foils, and the SXRL output is guided through an aperture in order to reduce debris contamination on the following optical elements. The target itself is a Mo slab with a length of 50 mm and a width of 5 mm. An X-Y-Z scanning unit is synchronized with the laser and continuously moves the target such that a fresh spot is used every couple of laser shots. We found the most stable SXRL operation if the target surface was renewed after 5–10 laser shots. For a pulse combination of 70 mJ/150 ps in the long pulse and 270 mJ/2 ps in the short pulse, we estimated a lower limit for the energy of one single SXRL pulse of 100 nJ and a divergence of about 10 mrad. Typical beam profiles feature a rectangular-shaped mode being overlaid with higher frequency modulation. The latter is typical for SASE type lasers. The rectangular shape originates from the shape of the plasma column.

35.3.2 Coherent Diffraction Imaging Using the SXRL

In the CDI setup (Fig. 35.3a), the exit plane of the XUV source is imaged onto the sample with a demagnifying telescope consisting of two concave spherical mirrors

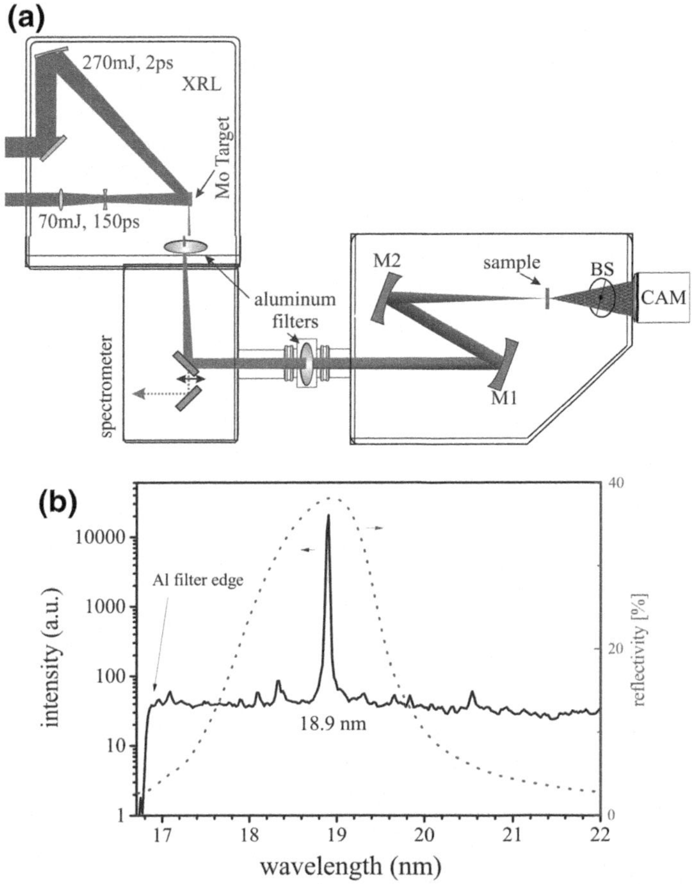

Fig. 35.3 **a** Typical experimental setup incorporating an SXRL and a CDI experiment. The molybdenum-based SXRL in this example emits at 18.9 nm (**b**). Aluminum filters suppress the stray light of the pump lasers and optically seal the experiment. Mirrors M1 and M2 are curved multilayer mirrors (**b**) that further spectrally clean up the SXRL beam and refocus the SXRL onto the sample. A back-illuminated cooled 16-bit CCD camera (CAM) measures the diffraction pattern

($ROC_1 = 2$ m, $ROC_2 = 1$ m) with a molybdenum/silicon multilayer coating. The mirrors are designed and optimized to have the highest reflectance coinciding with the SXRL emission line at 18.9 nm (Fig. 35.3b) [32]. Both multilayer mirrors have been realized with a dual ion beam deposition system described in [33]. From the model calculations and a reference measurement at the PTB synchrotron beamline at BESSY II a reflectivity of 14.6% (38.2% per single mirror, see Fig. 35.3b) under near normal incidence (2° off) at 18.9 nm was determined. In a CDI experiment a sample is placed in the rear focus of M2. The generated flux has been estimated prior to the imaging experiments with a calibrated XUV-sensitive photodiode. Under optimized conditions we obtained $(3.2 \pm 0.3) \times 10^{10}$ photons per shot, which leads to an overall capability of the system of producing up to 3×10^{12} photons per second in less than 0.01% bandwidth centered around 18.9 nm.

In the presented experiment, we used a 50 nm thick silicon nitride membrane covered with a 50 nm thick gold layer on top. With a focused ion beam we milled a shaped aperture (see STEM image in inset in Fig. 35.4a) into this slab. The sample was introduced into the refocused XUV beam and the diffracted light was recorded with large area XUV/X-ray-sensitive CCD (Andor iKon L, 27.6×27.6 mm^2 chip size) positioned 15 mm downstream. The resulting numerical aperture is 0.67 allowing for a half-pitch resolution of 14 nm. Within 300 laser shots a typical diffraction pattern exhibiting fringes extending to the edge of the detector could be recorded (Fig. 35.4a). Even with a single laser shot diffraction patterns covering the inner 30% of the detector around the central speckle were measured.

For the iterative phase retrieval, we employed a guided version of the hybrid input–output algorithm that proved to work well in previous CDI experiments [3]. Unfortunately, the phase retrieval algorithm was neither able to reconstruct the full field of view nor very sharp edges (Fig. 35.4c). We attribute this to a limited spatial coherence. While the diffraction pattern features fringes that should at least support a spatial resolution around 20 nm, we get at best a resolution of about 400 nm, while it is difficult to assign a resolution if the object is not fully resolved. From the radially decaying signal in real space, we conclude that further distanced features of

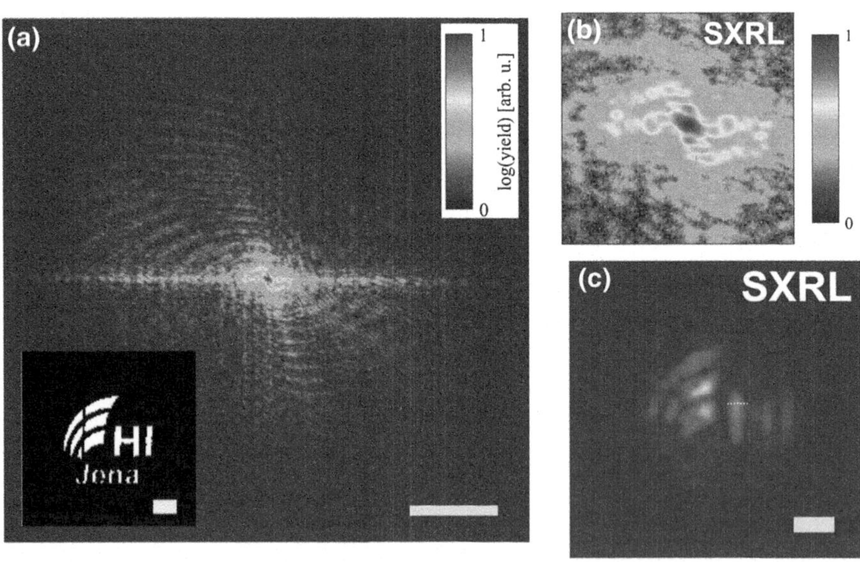

Fig. 35.4 Coherent diffraction imaging using a solid-state SXRL. **a** Measured diffraction pattern of an shaped aperture (STEM image inset) captured from 300 laser shots featuring fringes extending well to the edge of the detector. The scale bar in **a** is 10 μm^{-1} (inset 1 μm). **b** Detail of the diffraction pattern around zero momentum transfer. **c** The panel shows the modulus of the reconstructed real space (linear scale, scale bar 1 μm) for the SXRL measurement

the object produce speckles with reduced contrast which can consequently not be properly phased. The poor speckle contrast is also evident in Fig. 35.4b.

In this experiment we estimate the focal spot size to be roughly 15 μm full-width half maximum, while the object is 3 μm wide. A first analysis of the data suggests that for an object of about 1 μm width the spatial coherence would have been sufficient to phase out all speckles. The reconstruction quality should then be drastically improved since the decoherence observed in the presented experiment also compromises finding a proper support of the object which subsequently worsens the reconstruction. For smaller objects, however, the quality should be drastically improved.

Although the first measurements presented here resulted in low image quality due to the low spatial coherence, the speckles measured up to high momentum transfers in principle allowing sub-30 nm resolution within few laser shots justify more studies on these laser systems and more systematic investigations of the coherence properties [34].

35.4 Summary and Outlook

In this contribution, we have presented recent progress in the development of high-power ultrafast fiber lasers and discussed their capability of generating high harmonic radiation with high average power [35]. While the average power in the XUV approaches the milliwatt level, new developments allow scaling the photon energy into the energy range above 100 eV extending well into the water window. A combination of scaling toward higher photon energies and at the same time improving the flux should soon allow for coherent diffraction imaging at a spatial resolution at the 10 nm level. Short integration times will further allow for the application of these sources in tomographic and ptychographic scans in order to image three-dimensional and large volume specimen. To date, a spatial resolution down to 13 nm has been demonstrated using one of the presented fiber laser sources [16]. In contrast, SXRLs offer high pulse energies per shot confined in a very narrow linewidth making these lasers probably the best choice for single-shot imaging in a tabletop environment. The high average flux on the order of $>10^{12}$ photons/s in a picometer linewidth along with the scalability of the repetition rate of the presented thin disk pump laser hold promise for wide applicability in coherent diffraction imaging. One bottleneck experimentally observed is the limited spatial coherence of the SXRLs, which for now limits the usability to micron and sub-micron sized objects, e.g., bacteria, viruses, or particle clusters. Another possibility recently demonstrated to improve the flux and coherence property is a HHG-seeded SXRL [36] that essentially combines the best of two approaches.

Acknowledgements This work was partly supported by German Federal Ministry of Education and Research BMBF German–Korean collaboration program (01DR12011), the Korea–Germany program (Korean NRF 2010-00633, BK21), BMBF contracts 13N12082 "NEXUS" and 05P2015

"APPA," LaserLab Europe (INREX), EFRE (ProFit), the European Union's Seventh Framework Program (FP7/2007–2013)/ERC Grant agreement no [240460] "PECS," by the European Research Council under the ERC grant agreement no. [617173] "ACOPS" and the German Research Foundation (DFG) via the Excellence Cluster "Munich-Centre for Advanced Photonics" (MAP, EXC 158). M. Zürch acknowledges support from the Humboldt Foundation.

References

1. Sandberg, R.L., et al.: Studies of materials at the nanometer scale using coherent X-ray diffraction imaging. JOM **65**, 1208 (2013)
2. Krausz, F., Ivanov, M.: Attosecond physics. Rev. Mod. Phys. **81**, 163 (2009)
3. Zuerch, M., et al.: Real-time and sub-wavelength ultrafast coherent diffraction imaging in the extreme ultraviolet. Sci. Rep. **4**, 7356 (2014)
4. Zuerch, M., et al.: XUV coherent diffraction imaging in reflection geometry with low numerical aperture. Opt. Express **21**, 21131–21147 (2013)
5. Gardner, D.F., et al.: High numerical aperture reflection mode coherent diffraction microscopy using off-axis apertured illumination. Opt. Express **20**, 19050 (2012)
6. Shanblatt, E.R., et al.: Quantitative chemically-specific coherent diffractive imaging of reactions at buried interfaces with few-nanometer precision. Nano Lett. (2016). https://doi.org/10.1021/acs.nanolett.6b01864 (accepted manuscript)
7. Zuerch, M., et al.: Cancer cell classification with coherent diffraction imaging using an extreme ultraviolet radiation source. J. Med. Imaging **1**, 031008 (2014)
8. Witte, S., et al.: Lensless diffractive imaging with ultra-broadband table-top sources: from infrared to extreme-ultraviolet wavelengths. Light Sci. Appl. **3**, e163 (2014)
9. Thibault, P., et al.: High-resolution scanning X-ray diffraction microscopy. Science **321**, 379–382 (2008)
10. Winterfeldt, C., et al.: Colloquium: optimal control of high-harmonic generation. Rev. Mod. Phys. **80**, 117–140 (2008)
11. Müller, M., et al.: 1 kW 1 mJ eight-channel ultrafast fiber laser. Opt. Lett. **41**, 3439–3442 (2016)
12. Hädrich, S., et al.: 200 W average power energetic few-cycle fiber laser. In: High-Brightness Sources and Light-Driven Interactions, p. JT3A.1. Optical Society of America (2016)
13. Rothhardt, J., et al.: Absorption-limited and phase-matched high harmonic generation in the tight focusing regime. New J. Phys. **16**, 033022 (2014)
14. Hädrich, S., et al.: High photon flux table-top coherent extreme ultraviolet source. Nat. Photonics **8**, 779–783 (2014)
15. Rothhardt, J., et al.: High-repetition-rate and high-photon-flux 70 eV high-harmonic source for coincidence ion imaging of gas-phase molecules. Opt. Lett. **24**(16), 18133–18147 (2016)
16. Tadesse, G.K., et al.: High speed and high resolution table-top nanoscale imaging. Opt. Lett. **41**(22), 5170–5173 (2016)
17. Rothhardt, J., et al.: 53 W average power few-cycle fiber laser system generating soft X rays up to the water window. Opt. Lett. **39**, 5224 (2014)
18. Guggenmos, A., et al.: Chromium/scandium multilayer mirrors for isolated attosecond pulses at 145 eV. Opt. Lett. **40**, 2846–2849 (2015)
19. Stutzki, F., et al.: 152 W average power Tm-doped fiber CPA system. Opt. Lett. **39**, 4671–4674 (2014)
20. Suckewer, S., Jaegle, P.: X-ray laser: past, present, and future. Laser Phys. Lett. **6**, 411–436 (2009)
21. Nickles, P.V., et al.: Short pulse X-ray laser at 32.6 nm based on transient gain in neon-like titanium. Phys. Rev. Lett. **78**, 2748–2751 (1997)

22. Tuemmler, J., et al.: 10-Hz grazing incidence pumped Ni-like Mo X-ray laser. Phys. Rev. E **72**, 0374011–0374014 (2005)
23. Keenan, R., et al.: High-repetition-rate grazing-incidence pumped X-ray laser operating at 18.9 nm. Phys. Rev. Lett. **94**, 1039011–1039014 (2005)
24. Wang, Y., et al.: Demonstration of high-repetition-rate tabletop soft-X-ray lasers with saturated output at wavelengths down to 13.9 nm and gain down to 10.9 nm. Phys. Rev. A **72**, 7 (2005)
25. Cassou, K., et al.: Optimization toward a high-average-brightness soft-X-ray laser pumped at grazing incidence. Opt. Lett. **32**, 139–141 (2007)
26. Hasegawa, N., et al.: High-precision measurement of the spectral width of the nickel-like molybdenum X-ray laser. JCPS **163**, 012062 (2009)
27. Lucianetti, A., et al.: Transverse spatial coherence of a transient nickellike silver soft-X-ray laser pumped by a single picosecond laser pulse. Opt. Lett. **29**, 881–883 (2004)
28. Wang, Y., et al.: High-brightness injection-seeded soft-X-ray-laser amplifier using a solid target. Phys. Rev. Lett. **97**, 123901/1-4 (2006)
29. Baumgarten, C., et al.: 1 J, 0.5 kHz repetition rate picosecond laser. Opt. Lett. **41**, 3339–3342 (2016)
30. Kang, H.C., et al.: Single-pulse coherent diffraction imaging using soft X-ray laser. Opt. Lett. **37**, 1688–1690 (2012)
31. Luther, B.M., et al.: Saturated high-repetition-rate 18.9-nm tabletop laser in nickellike molybdenum. Opt. Lett. **30**, 165–167 (2004)
32. Guggenmos, A., et al.: Ion polished Cr/Sc attosecond multilayer mirrors for high water window reflectivity. Opt. Exp. **22**, 26526–26536 (2014)
33. Guggenmos, A., et al.: Aperiodic CrSc multilayer mirrors for attosecond water window pulses. Opt. Express **21**, 21728–21740 (2013)
34. Zürch, M., et al.: Transverse coherence limited coherent diffraction imaging using a molybdenum soft X-ray laser pumped at moderate pump energies. Sci. Rep. **7**, 5314 (2017)
35. Klas, R., et al.: Table-top milliwatt-class extreme ultraviolet high harmonic light source. Optica **3**(11), 1167–1170 (2016)
36. Depresseux, A., et al.: Table-top femtosecond soft X-ray laser by collisional ionization gating. Nat. Photonics **9**, 817–821 (2015)

Chapter 36
X-Ray Reflection Imaging of Inclined and Obliquely Illuminated Objects

I. A. Artyukov, A. S. Busarov, N. L. Popov and A. V. Vinogradov

Abstract It is well known that contrast high-resolution images are usually produced with normal incidence optics and samples observed normally to their surface. However, a number of recent X-ray experiments from 0.1 to 10 keV demonstrate a quest for imaging of slanted objects and the objects illuminated at grazing angles. A brief survey of previous studies, theoretical consideration, simulation, and application prospects of this new imaging technology is presented.

36.1 Introduction

In recent years, there have been published a number of experiments on imaging of obliquely observed objects. These experiments include investigations of surfaces and interfaces. In [1] a soft X-ray laser operating at 13.9 nm wavelength was used. The radiation was reflected off the multilayer mirror at an angle normal to the surface (Fig. 36.1). Then, it was reflected off the sample at 22.5° grazing angle. The zone plate imaged the sample onto CCD. The zone plate had a 10 mm focal length and 0.025 numerical apertures. Magnification was 79. The spatial resolution was 730 nm in the horizontal (the plane of the Fig. 36.1) and 360 nm in the vertical plane. The detector was positioned perpendicularly to the beam.

As the authors explained they wanted to create a versatile microscope for the study of dynamic surface morphology. The operation of such a scheme below 40 nm would be impossible without the use of oblique illumination (apart from thin objects fixed on a multilayer mirror). The Fig. 36.2 (also see [1]) shows the attenuation lengths of different chemical elements as a function of photon energy. The dashed line indicates the wavelength. In general, it is advisable to choose the smallest attenuation length for the study of a surface. At the same time, the wavelength should be relatively short to insure high-lateral resolution. For these

I. A. Artyukov · A. S. Busarov · N. L. Popov (✉) · A. V. Vinogradov
L.P. Lebedev Physical Institute, 53 Leninsky Prospekt, 119991 Moscow, Russia
e-mail: nick.l.popov@gmail.com

T. Kawachi et al. (eds.), *X-Ray Lasers 2016*, Springer Proceedings
in Physics 202, https://doi.org/10.1007/978-3-319-73025-7_36

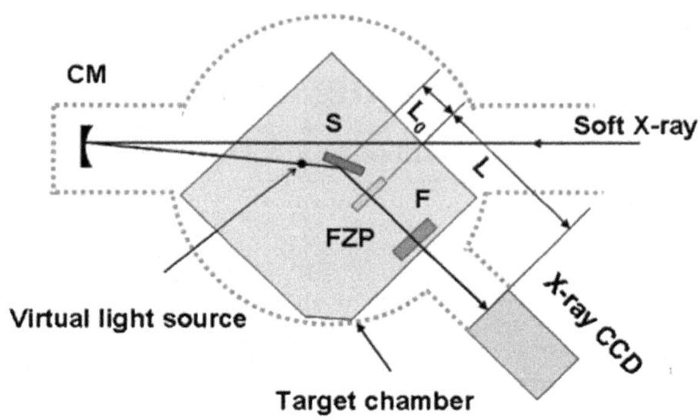

Fig. 36.1 Submicron scale image observation with a grazing incidence reflection-type single-shot soft X-ray microscope. $f \approx L0 = 10$ mm, $L = 500–1100$, $M = 79$, $NA = 0.025$, $\lambda = 13.9$ nm. Spatial resolution 730×360 nm (see [1])

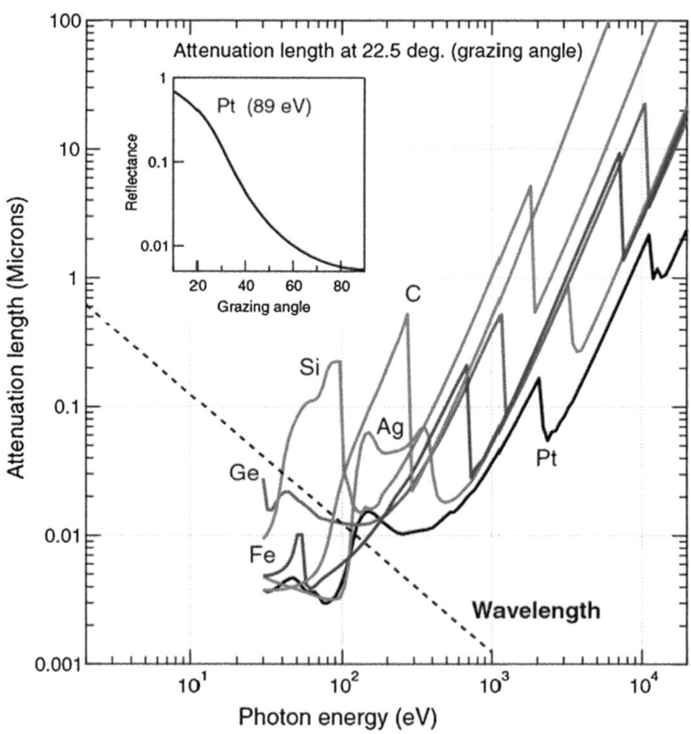

Fig. 36.2 Choice of wavelength λ and grazing angle θ

reasons, Pt and a wavelength of 13.9 nm were chosen for the study. However, at 13.9 nm, a sufficiently high reflection can only be achieved with grazing angles of incidence.

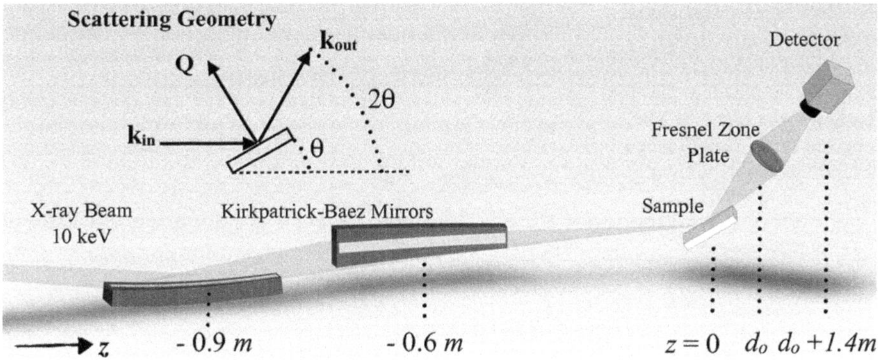

Fig. 36.3 X-ray beam modifies and images the calcite-water interface. $\lambda = 0.12$ nm, $\theta = 12°$, $f = 48.4$ mm, NA = 0.001, $d_0 = 50$ mm, M = 28, exposure < 1 s, spatial resolution 70 nm (see [2])

In [2], the authors studied the manner in which hard X-rays cause the dissolution of calcite. Simultaneously, the reflected radiation was used for real-time observation of calcite–water interface (Fig. 36.3). The reflection angle corresponded to the zero-order Bragg diffraction from atomic planes at (104). Namely, 12°. The reflected radiation was focused on the detector by the Fresnel Zone Plate. Magnification was 28. The spatial resolution was estimated at 70 nm from the vertical. The detector was positioned perpendicularly to the radiation. The radiation source was a synchrotron. A grazing angle was required in this case, since observation of a surface at atomic scale depth is impossible at a normal angle.

Generally, distinct, high-resolution images are produced through the use of normal incidence optics, with samples observed perpendicularly to their surface. However, besides of above examples, various applications demonstrate now a quest for X-ray imaging of slanted objects (see [1–5] and herein). The question arises: how can we maintain image sharpness in spite of object tilting? To solve this problem, we propose the theory of the extended Fresnel integral for tilted objects.

36.2 The Extended Fresnel Theory

An incident coherent beam interacting with object features produces at the surface of the object a wave field distribution $u_0(s)$ (see Fig. 36.4). Assuming that the scattered (reflected) wave remains to be paraxial we can find it as a solution of parabolic wave equation (PWE):

$$2ik\frac{\partial u(x,z)}{\partial z} + \frac{\partial^2 u(x,z)}{\partial x^2} = 0 \tag{36.1}$$

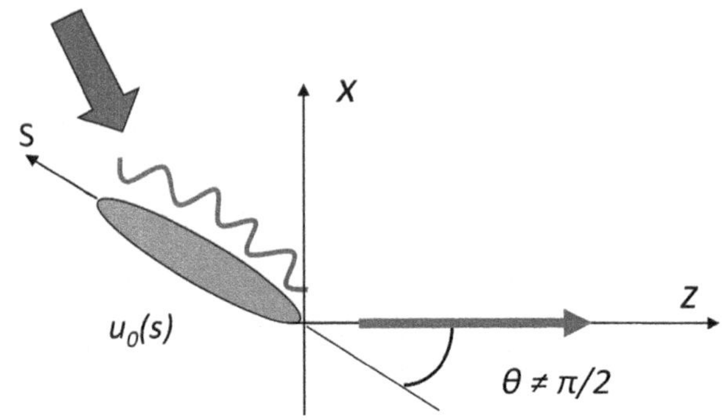

Fig. 36.4 Free propagation of coherent field reflected by a finite tilted object. An obliquely illuminated object is located in the plane S. The reflected beam is collimated around the Z-axis allowing application of PWE

with the boundary condition posed at the object surface:

$$u(x,z)|_{x=-ztg\theta} = u_0(s), \quad x = s\sin\theta, \quad z = -s\cos\theta \tag{36.2}$$

Basing on the approach (36.1), (36.2) the formula for wave field u(x, z) in the free space [3] is:

$$u(x,z) = \sqrt{\frac{k}{2\pi i}}(x\cos\theta + z\sin\theta) \times \int_{-z/\cos\theta}^{\infty} \frac{u_0(s)ds}{(z+s\cos\theta)^{3/2}} \exp\left[i\frac{k(x-s\sin\theta)^2}{2(z+s\cos\theta)}\right], \; x > -z\cos\theta. \tag{36.3}$$

The 3-D analog of (36.3) is [4]:

$$u(x,y,z) = \frac{k(x+ztg\theta)}{2\pi i} \times \int_{-\infty}^{\infty} dy' \int_{-\infty}^{z} \frac{u_0(y',z')dz'}{(z-z')^2} \exp\left\{ik\frac{(y-y')^2 + (x+z'tg\theta)^2}{2(z-z')}\right\}, \tag{36.4}$$
$$x > -z\cos\theta.$$

The wavefield of a tilted object (see (36.3), (36.4)) according to [4, 6] can be projected onto a vertical (normal to k-vector) plane and then propagated in the free space or through an optical system following the standard Fresnel theory. In this way, the field behind an ideal lens is figured out in analytical form. This easily allows being convinced that the image of an inclined object appears on the optically conjugate plane (Fig. 36.5) [6].

The amplitude of the object field at a given point is restored by the amplitude at the conjugate point:

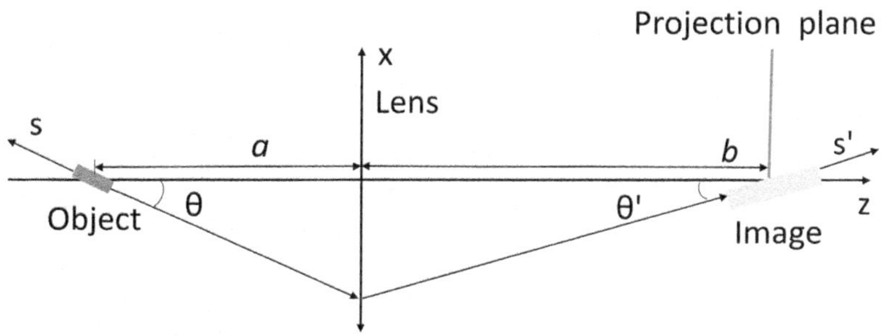

Fig. 36.5 The image of an inclined object appears on the optically conjugate plane

$$
\begin{aligned}
|u_0(s, y)| &= |u(s', y')|\left|\frac{f}{a + s\cos\theta - f}\right|;\\
s' &= \frac{1}{\cos\theta'}\left(\frac{f}{1 - \frac{f}{a + s'\cos\theta}} - b\right);\\
y' &= \frac{f}{f - a - s\cos\theta}y.
\end{aligned}
\tag{36.5}
$$

The equations (36.5) assume a lens of the infinite numerical aperture. The formulas (36.1)–(36.5) were implemented in our software, allowing simulation results of experiments on imaging of tilted objects.

36.3 The Simulation of a Microscope

To verify our theory, we simulated the experiment [1] to determine the limit of the microscope's resolution. For this purpose, we used as an object the meander shown in Fig. 36.6. Figure 36.7 shows image and projection (see Fig. 36.5) resolution in microns, where s is the coordinate on the object plane. The curves were constructed by 10 edges of the meander using 10–90% method.

The line with squares demonstrates the loss of resolution as one move away from the center of the object. Thus, the images of the object on the conjugate plane are of higher resolution than those on the projection plane (i.e., perpendicularly to the optical axis). The result is close to that obtained in [1], but only for the view field from −10 to 10 mkm.

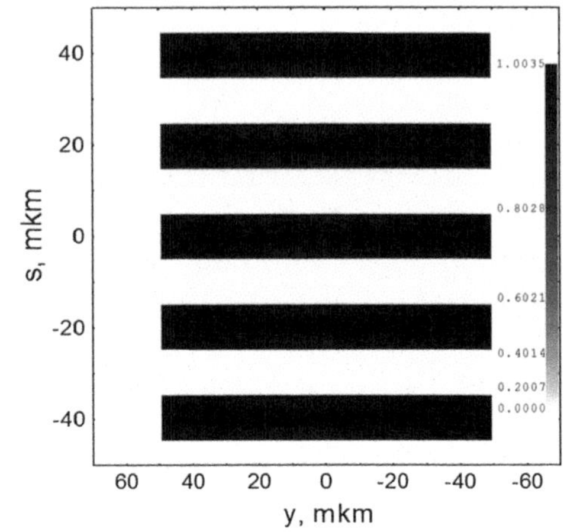

Fig. 36.6 The resolution was estimated by directing the microscope at the meander

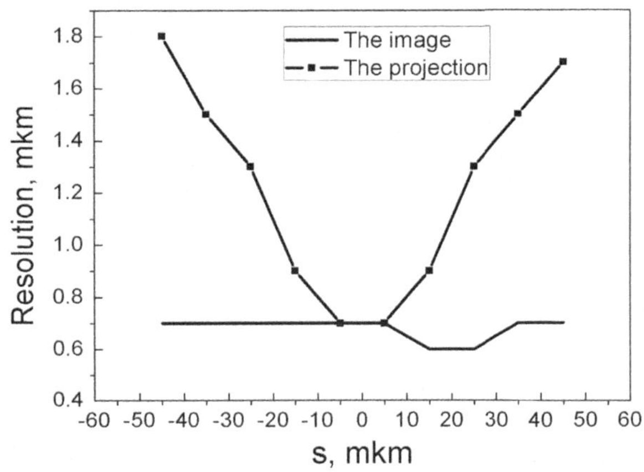

Fig. 36.7 The figure shows image and projection resolution in microns, where s is the coordinate on the object plane

36.4 Patterning

To test the above theory and to scale to shorter (subnanometer) wavelengths, the following experiment is suggested implying the usage of SXRL as an illuminating source (Fig. 36.8).

The experiment involves patterning of an inclined wafer and obliquely illuminated mask tilted to the optical axis. The wavelength is 13.5 nm, θ corresponds to the critical angle for gold, and lengths a and b have been chosen to produce a magnification of 1/500th. The Table 36.1 shows the resolution levels for different numerical apertures at two wavelengths 13.5 and 7 nm. An aperture of 0.3 corresponds to the width of the outermost Fresnel zone, which is equal to 12 nm. This is currently the highest resolution that has been achieved with a zone plate [7].

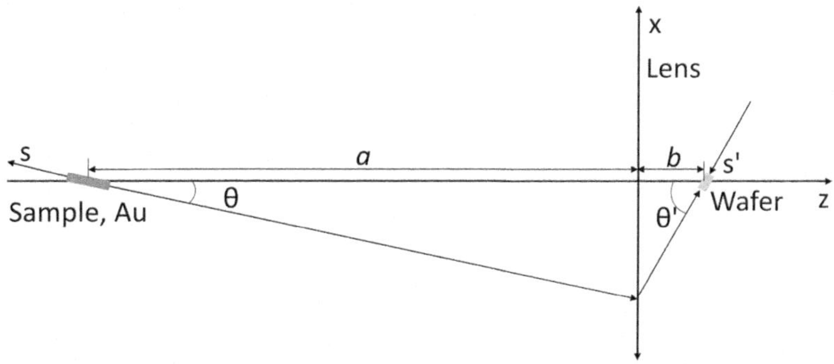

Fig. 36.8 The optical scheme of nanopatterning. $\lambda = 13.5$ nm, $\theta = 10^{\circ}$, $\theta' = 86^{\circ}$, $a = 8.80$ cm, b = 0.101 cm, f = 0.1 cm, M = 1/500

Table 36.1 The lens 10–90% resolution δ depending on the numerical aperture. f = 0.1 cm, M = 1/500

λ (nm)	θ	*a* (cm)	NA	δ (nm)	λ (nm)	θ	*a* (cm)	NA	δ (nm)
13.5	10°	8.8	0.3	16.1	7	5.2°	4.72	0.3	8.9
			0.2	24.9				0.2	13.2
			0.1	41.8				0.1	25.6

Apparently, the use of the wavelength of 7 nm enables to provide the print resolution of 10 nm or better.

36.5 Conclusion

The extended Fresnel theory provides a ground for simulation of imaging experiments with obliquely illuminated and obliquely observed samples. The simulation of an experiment on surface modification by 13.9 nm laser [1] was carried out. It showed that the theoretical resolution was achieved along the field of view of the microscope partially. It is expected that field of view, as well as resolution, can be improved by optical scheme optimization. To validate the extended Fresnel theory and to scale to shorter (subnanometer) wavelengths, the nanopatterning experiment is suggested.

The authors are indebted to T. Kawachi, C. Menoni, and J. Rocca for useful discussion of this work at ICXRL 2016 and after it.

The research has been supported by the Program of fundamental investigations of the Russian Academy of Science Presidium "Fundamental and applied problems of photonics and physics of new optical materials."

References

1. Baba, M., Nishikino, M., Hasegawa, N., Tomita, T., Minami, Y., Takei, R., Yamagiwa, M., Kawachi, T., Suemoto, T.: Submicron scale image observation with a grazing incidence reflection-type single-shot soft X-ray microscope, Jpn. J. Appl. Phys. **53**(8) (2014)
2. Laanait, N., Erika, B.R., Callagon, E., Zhang, Z., Sturchio, N.C., Lee, S.S., Paul, F.: X-ray–driven reaction front dynamics at calcite-water interfaces. Science **349**(6254), 1330–1334 (2015)
3. Artyukov, I.A., Popov, A.V., Vinogradov, A.V.: Wave field transformation at coherent imaging of a tilted reflection mask. In: SPIE, vol. 7451 (2009)
4. Artyukov, I.A., Feshchenko, R.M., Popov, N.L., Vinogradov, A.V.: Coherent scattering from tilted object. J. Opt. **16**(3) (2014)
5. Zhang, H., Liu, J., Zhai, M., Zhou, Y., Jia, D., Liu, T.: Theoretical analysis and experimental validation of sampling volume in tilted imaging system. IEEE Photonics J. **7**(6), 17669–17677 (2015)
6. Artyukov, I.A., Busarov, A.S., Popov, N.L., Vinogradov, A.V.: Optical transforms related to coherent imaging of inclined objects. In: X-Ray Lasers 2012, pp. 147. Springer Proceedings in Physics (2012)
7. Chao, W., Kim, J., Rekawa, S., Fischer, P., Anderson, E.H.: Demonstration of 12 nm resolution fresnel zone plate lens based soft X-ray microscopy. Opt. Express **17**(20), 17669–17677 (2009)

Chapter 37
Nanoscale Imaging Using a Compact Laser Plasma Source of Soft X-Rays and Extreme Ultraviolet (EUV)

P. Wachulak, A. Torrisi, M. Ayele, J. Czwartos, A. Bartnik, R. Jarocki, J. Kostecki, M. Szczurek, Ł. Węgrzyński and H. Fiedorowicz

Abstract Application of a compact laser plasma source of soft X-rays and extreme ultraviolet (EUV) in imaging with nanometer resolution is demonstrated. The source is based on a gas puff target irradiated with nanosecond laser pulses from a small commercial Nd:YAG laser. Soft X-ray radiation in the 'water window' spectral range and EUV near 10 nm are generated efficiently without production of target debris. Nanoscale imaging of biological samples as well as micro- and nanostructures using transmission soft X-ray and EUV microscopy based on Fresnel optics and soft X-ray contact microscopy is demonstrated.

37.1 Introduction

Soft X-ray and EUV microscopy provides information complementary to that obtained from optical, electron and atomic force microscopy techniques. Imaging of cellular structure and extended tissue in biological samples requires nanometer resolution and good sample penetration. It can be provided by current soft X-ray microscopic techniques operating in the 'water window' spectral range (wavelength: 2.3–4.4 nm; photon energy: 280–560 eV). The various techniques include transmission and scanning soft X-ray microscopy, 3D tomography and soft X-ray contact microscopy. Interest to nanoscale imaging in EUV spectral range is mainly connected with the development of tools for inspection of masks for EUV lithography, however, this technique can be also useful for investigations of micro- and nanostructures, including nanowires and magnetic nanostructures. Soft X-ray and EUV imaging with spatial resolution below 20 nm has been demonstrated using

P. Wachulak (✉) · A. Torrisi · M. Ayele · J. Czwartos · A. Bartnik · R. Jarocki · J. Kostecki · M. Szczurek · Ł. Węgrzyński · H. Fiedorowicz
Institute of Optoelectronics, Military University of Technology, 2, Kaliskiego Street, 00-908 Warsaw, Poland
e-mail: przemyslaw.wachulak@wat.edu.pl

T. Kawachi et al. (eds.), *X-Ray Lasers 2016*, Springer Proceedings in Physics 202, https://doi.org/10.1007/978-3-319-73025-7_37

synchrotron radiation, however, the limited access to synchrotrons is one of the main motivation of research focused on the development of imaging techniques based on laboratory sources of soft X-rays and EUV.

It was demonstrated that soft X-ray and EUV radiation can be produced from a high-temperature plasma generated by the interaction of high power laser pulses with matter and laser plasma light sources have been used in soft X-ray and EUV microscopy [1–3]. In this work, application of a compact laser plasma source in transmission soft X-ray microscopy based on Fresnel optics and contact soft X-ray microscopy are presented. The source is operating with a double-stream gas puff target irradiated with a nanosecond laser pulses from a small Nd:YAG laser (4 ns/ 0.8 J/10 Hz). The use of a double-stream gas puff target instead of a solid gives the possibility to develop an efficient and debrisless laser plasma EUV source [4]. The target is formed by injection of high-*Z* gas (xenon, krypton, argon, etc.) into a hollow stream from low-*Z* gas (hydrogen and helium) using a double nozzle setup [5]. The nozzle setup consists of a central nozzle in a form of a circular orifice surrounded with an outer nozzle in the form of a ring and is supplied with gases from two electromagnetic valves mounted in a common body. Strong EUV emission from the xenon gas puff target, exceeding the emission from solid targets, was demonstrated [4].

The double-stream gas puff target approach was used to develop a compact laser plasma EUV source for metrology applications [6]. The xenon target was irradiated with 4 ns/0.5 J pulses produced with repetition of 10 Hz from a commercial Nd: YAG laser. Conversion efficiency of the laser energy into EUV energy at 13.5 nm of about 1.6% was measured [6]. The source has been successfully used in the measurements of optical characteristics of Mo/Si multilayer mirrors at 13.5 nm [7]. In this chapter, we present application of this source in a newly developed stand-alone desk-top EUV microscope. The experiments on imaging of different objects were performed and the spatial resolution below 50 nm was demonstrated. It was shown that the EUV microscopy can provide structural information that is not achievable by conventional optical or scanning electron microscopy techniques.

37.2 Transmission Soft X-Ray Microscopy with a Fresnel Optics

The optical scheme of the transmission soft X-ray microscopy based on Fresnel optics using the compact laser plasma source is presented in Fig. 37.1.

Soft X-ray radiation in the 'water window' spectral range is produced by irradiation of a nitrogen gas puff target with nanosecond pulses from a Nd:YAG laser. The detailed description of the source and its parameters has been published in [8]. Radiation from the nitrogen plasma is collected by an axisymmetrical ellipsoidal grazing incidence mirror (Rigaku) and focused on a sample. The soft X-ray image is formed by a Fresnel zone plate with the outer zone width of 30 nm (Zoneplates Ltd.)

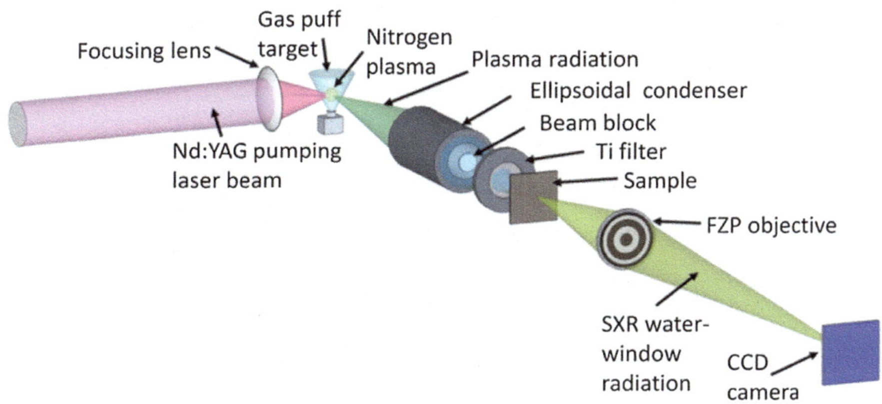

Fig. 37.1 Optical scheme of transmission soft X-ray microscopy with a laser plasma source

made in 400 nm thick silicon nitride. Quasi-monochromatic radiation that is required by Fresnel optics was obtained by selection of a spectral line at 2.88 nm from the emission of the nitrogen helium-like ions using 200 nm thick Ti absorption filter. The magnified image was registered with the use of a back-illuminated CCD camera, sensitive to soft X-rays (Andor, i-Kon). Schematic of the experimental setup of the transmission soft X-ray microscope based on Fresnel optics and a laser plasma source is shown in Fig. 37.2.

The entire setup fits on top of a single $1 \times 2\ \mathrm{m}^2$ optical table. It consists of a source vacuum chamber and a microscope chamber in which soft X-ray optical elements are mounted. An overall view of the system is shown in Fig. 37.3. The two cylindrical chambers are differentially pumped to provide sufficiently high vacuum in the microscope chamber (10^{-5} mbar).

Fig. 37.2 Schematic of the experimental setup of the transmission soft X-ray microscope based on Fresnel optics

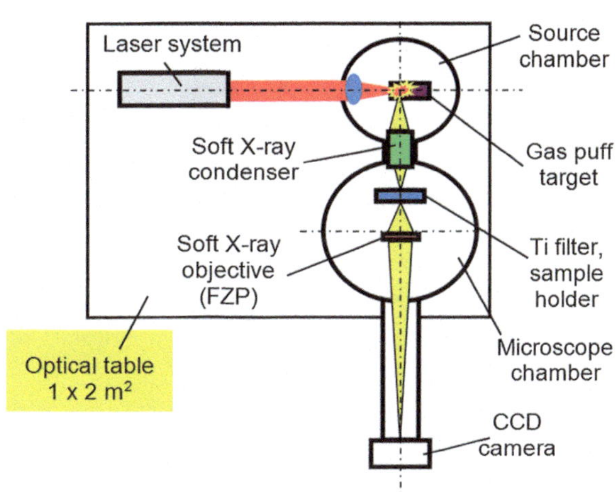

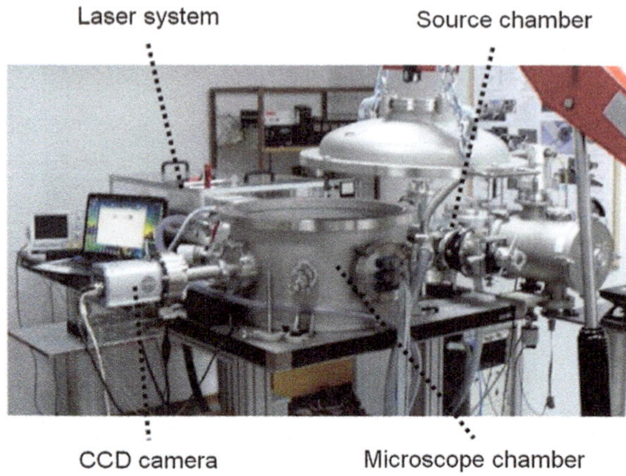

Fig. 37.3 Experimental setup for transmission soft X-ray microscopy based on Fresnel optics using a laser plasma source

This microscope was also successfully used to image a variety of real samples. Two examples of such study are depicted in Fig. 37.4. Figure 37.4a shows the SXR images of dehydrated CT 26 fibroblasts which were derived from colon carcinoma Mus musculus (strain BALB/c) and placed on top of 100 nm thick silicon nitride membrane.

Figure 37.4b depicts diatoms, placed on top of 30 nm thick silicon nitride membrane. The SXR exposure for both images was equal to 200 SXR pulses and lasted 22 s. The images of the cells and diatoms show their internal structure visible as modulation in the transmittance of the SXR light through the sample in the SXR image. The half-pitch spatial resolution of the SXR microscope measured employing a well-established knife edge test is equal to 60 nm in those images [9]. The SXR images show enhanced spatial resolution and high absorption contrast,

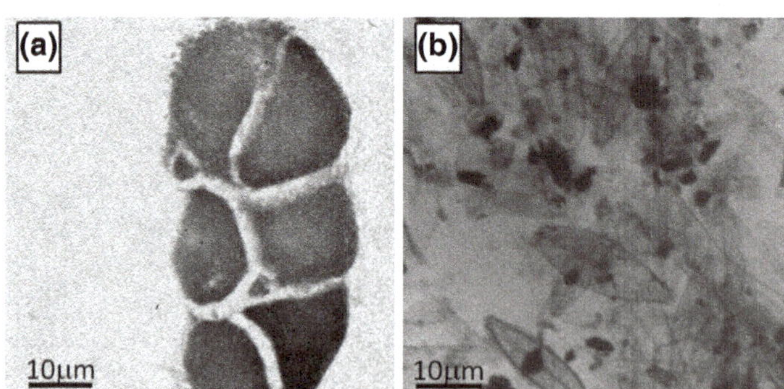

Fig. 37.4 SXR images of dehydrated CT 26 fibroblasts (**a**), derived from colon carcinoma Mus musculus (strain BALB/c) placed on top of 100 nm thick silicon nitride membrane, and diatoms placed on top of 30 nm thick silicon nitride membrane (**b**). The SXR exposure was 200 SXR pulses and lasted 22 s

which is possible to obtain in the 'water window' spectral region. Detailed results of the experiments have been published elsewhere [9, 10].

37.3 A Stand-Alone Compact EUV Microscope with Nanometer Spatial Resolution

Another example of the application of short wavelength radiation from the gas puff target sources is the EUV microscope developed recently in the frame of LIDER programme.

A scheme of the compact EUV microscope is shown in Fig. 37.5. A quasi-monochromatic EUV radiation is produced by irradiation of argon/helium double-stream gas puff target with nanosecond infrared pulses emitted from compact Nd:YAG laser, with subsequent spectral filtering provided by ellipsoidal Mo/Si multilayer condenser mirror and free-standing100 nm thick zirconium filter. The condenser has two roles here.

The first one is to focus the radiation, emitted from the argon plasma onto a sample, located in its second focal plane, as well as, as previously mentioned, to spectrally narrow down the broadband plasma emission. As a result, a cone of EUV light with a wavelength of 13.8 nm illuminates the sample.

A magnified image of that sample is then formed onto a CCD camera (Andor, i-Kon), using a zone plate made in 200 nm thick PMMA, with outer zone width of 50 nm (Zoneplates Ltd). The photographs of the compact stand-alone EUV microscope and the photograph of its internal components are shown in Fig. 37.6a, b respectively.

The EUV microscope was also used for imaging of real samples. Figure 37.7a shows EUV image of Chrysodidimus algae cells placed on top of 30 nm thick silicon nitride membrane acting as a support. The image of the object was acquired by integrating 100 EUV pulses over 10 s exposure time. The spatial resolution of this microscope was estimated applying a knife-edge test, which suggested a

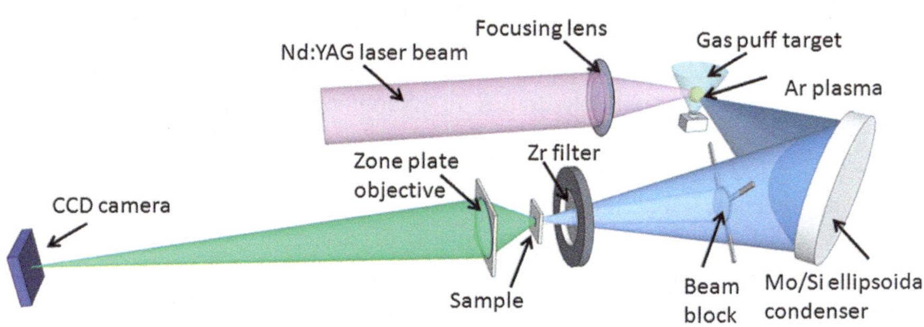

Fig. 37.5 Schematic of the experimental setup of the transmission EUV microscope based on Fresnel optics

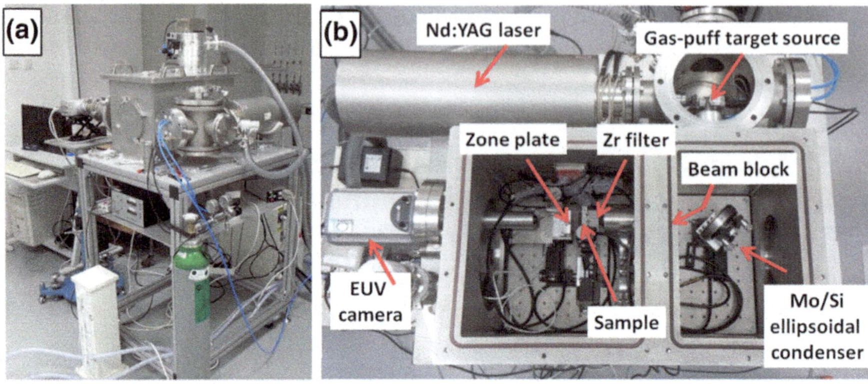

Fig. 37.6 Photograph of the compact stand-alone EUV microscope (**a**) and photograph of its internal components (**b**). All major components are indicated

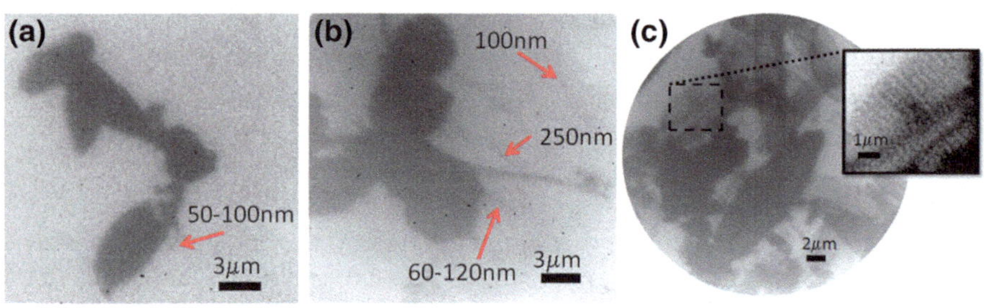

Fig. 37.7 The EUV image of Chrysodidimus cells (**a**) and CT 26 fibroplast cells (**b**) and diatoms (**c**) obtained with exposure of 100–200 EUV pulses and half-pitch spatial resolution of 48 nm, with visible features as small as 50–100 nm

half-pitch spatial resolution equal to 48 nm [11], almost an order of magnitude better than in case of classical full field visible light microscopes, limited by a wavelength to ~300 nm. The features, as small as 50–100 nm, are visible in the middle bottom part of the EUV image.

Another example is a sample of CT 26 fibroblast from Mus musculus colon carcinoma (strain BALB/c), prepared also on top of similar membrane. The sample was prepared with a gradual dehydration in ethanol series (final concentration 70% EtOH), without any fixation procedure. The EUV microscope image Fig. 37.7b was acquired with 200 EUV pulses, at a source repetition rate of 10 Hz, exposure time of 22 s and detector temperature of −20 °C. The image shows an improvement of the spatial resolution due to the employment of shorter wavelength, beyond the diffraction limit of the optical-visible light microscopes. It is possible to distinguish structures as small as 60 nm. Final samples are diatoms deposited on top of 30 nm thick Si_3N_4 membrane, depicted in Fig. 37.7c. More details about the EUV microscope can be found in [11, 12].

37.4 Soft X-Ray Contact Microscopy

Soft X-ray contact microscopy is a type of projection imaging technique, similar to radiography or shadowgraphy, in which the radiation which is illuminating the object is partially absorbed by it. The object is placed in contact with light-sensitive material (photoresist). The light that is locally transmitted by the object's structure illuminates the photoresist and changes its physical or chemical structure.

After the irradiation, the photoresist is chemically developed and modulation of the light intensity absorbed by the object is converted in this process to a modulation of the thickness of the resist. This creates a relief-like structure in its surface.

The height of the relief structure is proportional directly, however, not necessarily linearly, to the dose of radiation absorbed in the resist volume. In such process, a high-resolution imprint of the internal structure of the object can be stored in the surface of the photoresist and later converted to an image using, for example, AFM or SEM microscope.

The experimental setup for soft X-ray contact microscopy using the compact laser plasma EUV source and its overall view are presented in Fig. 37.8 and Fig. 37.9, respectively. The setup consists of Nd:YAG laser producing a laser beam, which is focused using an f = 25 mm focal length lens onto an argon/helium

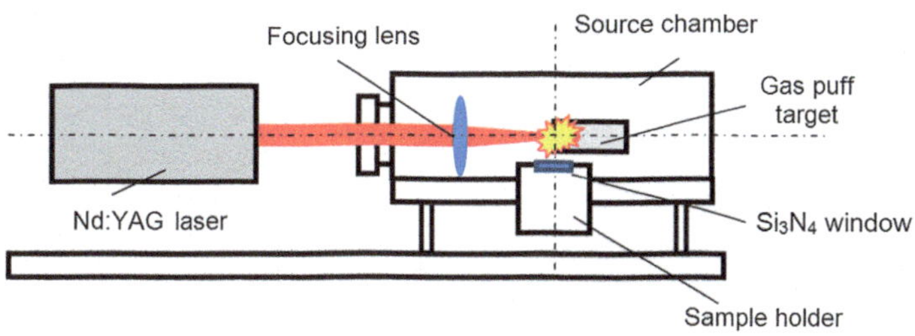

Fig. 37.8 Experimental setup for soft X-ray contact microscopy using a laser plasma source

Fig. 37.9 View of the setup for soft X-ray contact microscopy

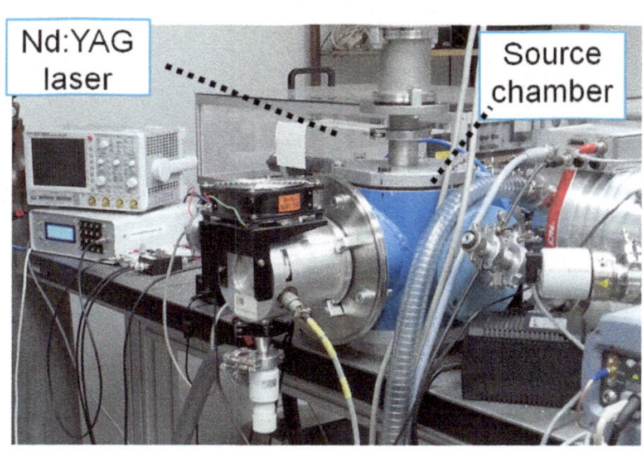

gas puff target produced using previously mentioned double electromagnetic valve system. The plasma produced as a result of laser interaction with gaseous target emits broad band radiation. Such radiation is spectrally filtered using silicon nitride membranes or titanium filters to produce radiation from the 'water window' spectral range. For quick exposures also a direct radiation from the plasma may be used.

The setup has been used in the preliminary experiments on soft X-ray contact microscopy. Figure 37.10 shows soft X-ray images obtained by the AFM scanning of the surface of 500 nm thick PMMA photoresist, of a HCV29 cell with 150 nm diameter gold beads obtained after 20 laser shots (2 s exposure time). In the SXR imprint at various scales of magnification cells and its internal structures can be seen. In the smallest scan, 150 nm in diameter gold beads can also be seen, used here as markers. Knife edge resolution test applied to those imprints suggests a spatial resolution of the order of 80 nm. The detailed description of the source and its parameters has been published in [13, 14].

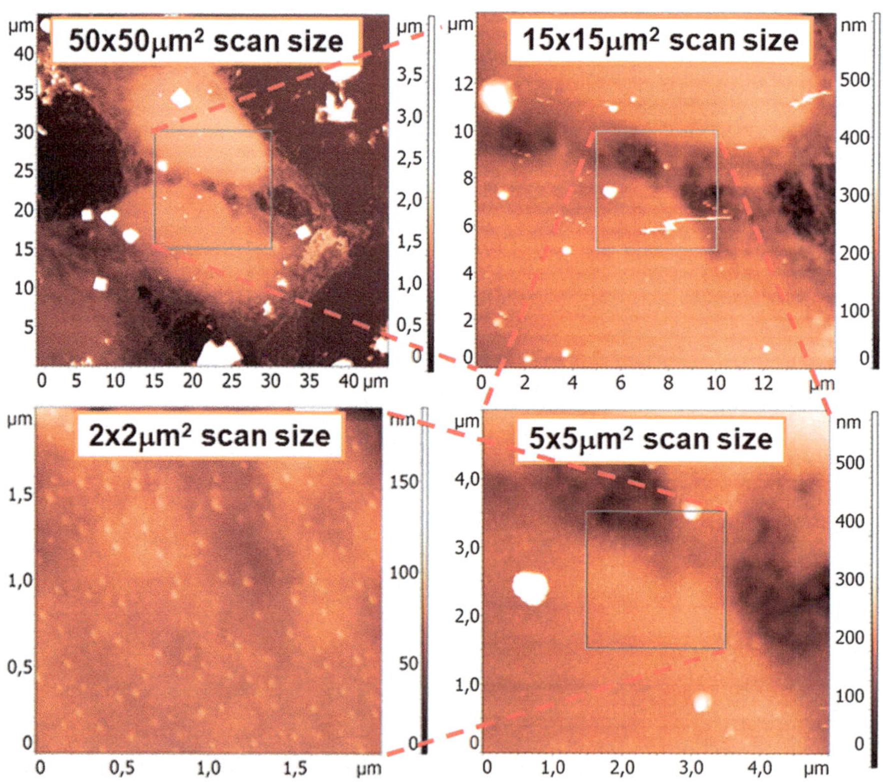

Fig. 37.10 Soft X-ray images of a HCV29 cell with 150 nm diameter gold beads obtained after 20 laser shots (2 s exposition time) at different scan sizes [14]

37.5 Summary and Conclusions

In summary, different imaging techniques producing high spatial resolution images of nanometer scale objects were briefly presented. The EUV/SXR microscopes, as well as contact microscope, are all based on a double-stream gas puff target source, debris free, compact, high flux and user friendly source, which was already employed in a variety of different experiments including also radiography, tomography, material processing photoionization experiments, radiobiology studies of doses absorbed in DNA strands, spectroscopy and metrology. Such compact, low maintenance and robust imaging systems may have potential applications in various fields of science, including biology, material science and nanotechnology.

Acknowledgements The research was supported by the Laserlab Europe Project (EU-H2020 654148), the Erasmus Mundus Joint Doctorate Programme EXTATIC (No. 2012-0033), the National Science Centre; award numbers UMO-2015/17/B/ST7/03718, UMO-2015/19/B/ST3/00435 and the National Centre for Research and Development, LIDER programme; award# LIDER/004/410/L-4/12/NCBR/2013. We thank to Šárka Vondrová, Jana Turňová and Prof. Miroslava Vrbová from Czech Technical University in Prague, Czech Republic, for preparing the cell samples investigated with the SXR full field microscope, to Dr. Andrzej Kowalik, from Institute of Electronic Materials Technology (ITME), Poland, for the preparation of the photoresist in their laboratory and to Prof. Małgorzata Lekka from the Department of Biophysical Microstructures, the Henryk Niewodniczański Institute of Nuclear Physics, Polish Academy of Sciences in Cracow, Poland for providing the cell samples investigated with the SXR contact microscope and Dr. E. Viaggiu, University of Tor Vergata, Rome (Italy), AlgaRes srl, for preparation of diatoms sample.

References

1. Michette, A.: Optical Systems for Soft X-rays. Plenum Press (1986)
2. Attwood, D.T.: Soft X-rays and Extreme Ultraviolet Radiation: Principles and Applications. Cambridge University Press (1999)
3. Bakshi, V.: EUV Sources for Lithography. SPIE Press Book (2006)
4. Fiedorowicz, H., Bartnik, A., Daido, H., Choi, I.W., Suzuki, M.: Strong EUV emission from a double-stream xenon/helium gas puff target irradiated with a Nd:YAG laser. Opt. Commun. **184**, 161–167 (2000)
5. Fiedorowicz, H., Bartnik, A., Jarocki, R., Kostecki, J., Rakowski, R., Szczurek, M.: Enhanced X-ray emission in the 1-keV range from a laser-irradiated gas puff target produced using the double nozzle setup. Appl. Phys. B **70**, 305–308 (2000)
6. Fiedorowicz, H., Bartnik, A., Jarocki, R., Kostecki, J., Krzywiński, J., Rakowski, R., Szczurek, A., Szczurek, M.: Compact laser plasma EUV source based on a gas puff target for metrology applications. J. Alloys Comp. **401**, 99–103 (2005)
7. Rakowski, R., Bartnik, A., Fiedorowicz, H., Jarocki, R., Kostecki, J., Krzywinski, J., Mikolajczyk, J., Pina, L., Ryc, L., Szczurek, M., Ticha, H., Wachulak, P.: Metrology of Mo/Si multilayer mirrors at 13.5 nm with the use of a laser-produced plasma extreme ultraviolet (EUV) source based on a gas puff target. Opt. Appl. **XXXVI**, 593–600 (2006)
8. Wachulak, P., Bartnik, A., Fiedorowicz, H., Rudawski, P., Jarocki, R., Kostecki, J., Szczurek, M.: Water window compact, table-top laser plasma soft X-ray sources based on a gas puff target. Nucl. Instrum. Methods B **268**, 1692–1700 (2010)

9. Wachulak, P.W., Torrisi, A., Bartnik, A., Adjei, D., Kostecki, J., Wegrzynski, L., Jarocki, R., Szczurek, M., Fiedorowicz, H.: Desktop water window microscope using a double-stream gas puff target source. Appl. Phys. B **118**, 573–578 (2015)
10. Wachulak, P., Torrisi, A., Nawaz, M.F., Bartnik, A., Adjei, D., Vondrová, Š., Turňová, J., Jančarek, A., Limpouch, J., Vrbová, M., Fiedorowicz, H.: A compact "water window" microscope with 60 nm spatial resolution for applications in biology and nanotechnology. Microsc. Microanal. **21**(5), 1214–1223 (2015)
11. Torrisi, A., Wachulak, P., Węgrzyński, Ł., Fok, T., Bartnik, A., Parkman, T., Vondrová, S., Turňová, J., Jankiewicz, B.J., Bartosewicz, B., Fiedorowicz, H.: A stand-alone compact EUV microscope based on gas-puff target source. J. Microsc. (under review)
12. Wachulak, P.W., Bartnik, A., Torrisi, A., Węgrzyński, Ł., Fok, T., Fiedorowicz, H.: A desk-top EUV nanoimaging system using a compact laser plasma light source. J. Measur. Sci. Technol. (submitted)
13. Ayele, M.G., Czwartos, J., Adjei, D., Wachulak, P., Ahad, I.U., Bartnik, A., Wegrzynski, Ł., Szczurek, M., Jarocki, R., Fiedorowicz, H., Lekka, M., Pogoda, K., Gostek, J.: Contact microscopy using a compact laser produced plasma soft X-ray source. J. Acta Phys. Pol. A **129**(2), 237–240 (2016)
14. Ayele, M.G., Czwartos, J., Adjei, D., Wachulak, P., Ahad, I.U., Bartnik, A., Wegrzynski, Ł., Szczurek, M., Jarocki, R., Fiedorowicz, H., Lekka, M.: Poster Presented on the 12th Multinational Congress on Microscopy (MCM 2015), Eger, Hungary (2015)

Chapter 38
ERL-Based Laser-Compton Scattering X-Ray Source for X-Ray Imaging

A. Kosuge, T. Akagi, Y. Honda, S. Araki, J. Urakawa, N. Terunuma, R. Nagai, T. Shizuma, M. Mori and R. Hajima

Abstract Nowadays, the generation of narrow-bandwidth, high brightness, and tunable X-rays via laser-Compton scattering (LCS) is expected for many scientific and industrial applications. In this paper, we will show that the energy recovery linac (ERL)-based LCS X-ray source can be applied to an X-ray imaging of biological specimens.

Keywords Laser-Compton scattering · Energy recovery linac · Enhancement cavity · Invertibility · X-ray absorption contrast image · Reversibility Refraction-enhanced image

38.1 Introduction

The Laser-Compton scattering (LCS) photons are generated by collision of laser photons and relativistic electrons. The LCS enables generation of narrow-bandwidth, high brightness, and tunable X-rays. Such a photon source is expected to bring breakthrough in fundamental researches [1], medical [2], and industrial applications [3]. The development of the Compact ERL (cERL) is ongoing at KEK to produce low emittance and high-current recirculating electron beams [4]. By combining a laser enhancement cavity with the cERL, the LCS X-rays with energy of 7 keV can be generated. In this paper, we will show the measurement results of the LCS X-rays and the X-ray absorption contrast imaging and refraction-enhanced imaging.

A. Kosuge (✉) · T. Akagi · Y. Honda · S. Araki · J. Urakawa · N. Terunuma
KEK, Oho, Tsukuba, Ibaraki 319-1195, Japan
e-mail: kosuge.atsushi@issp.u-tokyo.ac.jp

Y. Honda · N. Terunuma
Science of the Graduate University for Advanced Studies (SOKENDAI), Miura District, Kanagawa, Japan

R. Nagai · T. Shizuma · M. Mori · R. Hajima
National Institutes for Quantum and Radiological Science and Technology (QST), Tokai, Naka, Ibaraki 319-1106, Japan

T. Kawachi et al. (eds.), *X-Ray Lasers 2016*, Springer Proceedings in Physics 202, https://doi.org/10.1007/978-3-319-73025-7_38

38.2 Performance of the Seed Laser System and the Optical Cavity

As the seed laser of the enhancement cavity, the laser system delivers an average power of 45 W and pulse duration of 10 ps (FWHM). The pulse repetition rate is $f_{rep} = 162.5$ MHz which has an integer relation with the fundamental RF of the cERL. The laser beam ejected from the laser system is passed through a mode matching telescope in order to match the laser beam to the cavity mode. We employ a four-mirror cavity with two concave mirrors to produce a small spot laser beam inside a cavity. Since the LCS X-rays are generated by collision of laser photons and relativistic electrons, the laser beam is required to be well focused at the collision point where the spot size (rms) of the cERL electron beam at the experiment is 78 μm (horizontal) and 16 μm (vertical). The schematic and picture of the optical setup of our cavity locking laser system are shown in Fig. 38.1a, b. The optical cavity consists of an input coupler (M1 (R = 99.9%)), flat mirror (M2(R = 99.99%)), and two concave mirrors (M3 and M4(R = 99.999%)). The circling intracavity power P_{circ} is determined by measuring the power leaking from a concave mirror. From the leaking power measurement, when an injection power P_{in} was 24 W, a circulating power $P_{circ} = 10.4$ kW which corresponds to an enhancement factor of 433 was obtained. The enhancement factor can be further improved by suppressing the electrical noise of the cavity locking loop and the laser fluctuation caused by the mechanical vibration of the optical table. The optical cavity lock is realized with a Hänsch-Couillaud method [5], which utilizes polarization by monitoring changes in the polarization of the light field reflected from the cavity. In contrast to the original scheme, where a Brewster plate or a polarizer inside the cavity is needed, in our case the necessary polarization discrimination is given by the nonorthogonal incidence of the optical beam on the cavity mirrors. From the beam size measurement at the concave mirror (M3) together with a calculation by using the ABCD matrix, the waist sizes at the laser-electron collision point are estimated as $\sigma_x = \sigma_y = 30$ μm.

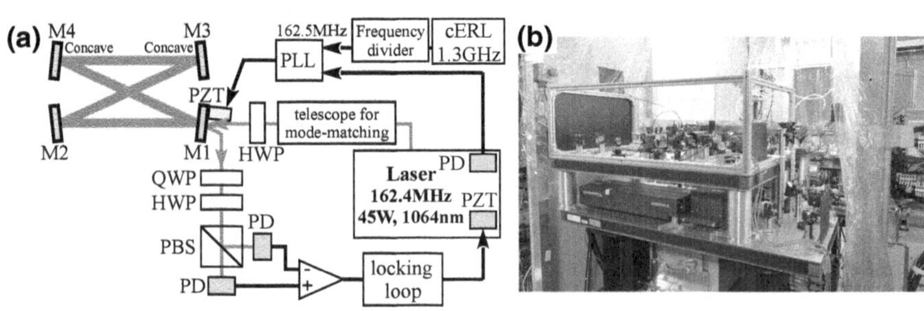

Fig. 38.1 **a** The schematic of the optical cavity and the cavity locking loop configuration. **b** The picture of the laser system and the optical cavity on the movable table in the cERL accelerator room

38.3 LCS X-ray Generation and X-ray Imaging Experiment

Around 7 keV X-ray which was generated by collision of 1064 nm laser photons and 20 MeV electrons at an angle of 18° was measured by a silicon drift detector (XR-100SDD, AMPTEK Inc.). Before the measurement, we adjusted spatial overlap and collision timing between the laser and electron beam. The LCS X-ray intensity evaluation was performed using the silicon drift detector, which was located 16.6 m away from the collision point. From the measured SDD spectrum, the central energy of 6.95 ± 0.01 keV, the FWHM spectrum width of 0.173 keV and detector count rate of 1370 cps was obtained within a detector area, ϕ4.66 mm. The measured spectrum width reflects the detector resolution. When we take into account the detector resolution, we estimate the actual spectrum width to be 30 ± 2 eV. The LCS photon flux at the collision point is estimated to be $(2.6 \pm 0.1) \times 10^7$photons/s from CAIN/EGS simulation with an electron beam current of 58 μA and an average laser power of 10 kW [6].

The LCS X-ray imaging was carried out by the setup shown in Fig. 38.2a. The X-ray images were recorded using a 2D photon counting X-ray detector (HyPix-3000, Rigaku, Pixel size: 100 μm × 100 μm). Since the transmittance of the 7-keV X-ray in the air ($T_{Air} \sim 1.4\%$ (L = 2.5 m)) is low, the tube filled with 70% the He gas was placed between the beryllium window and the detector. The transmittance of the tube was improved to 27%. With the hornet and detector at 19.1 m from the collision point, Fig. 38.2b shows the experimental result of the X-ray absorption contrast image. Figure 38.2c shows a refraction-enhanced image [7, 8] taken in the configuration with the hornet at 16.6 m and detector at 19.1 m

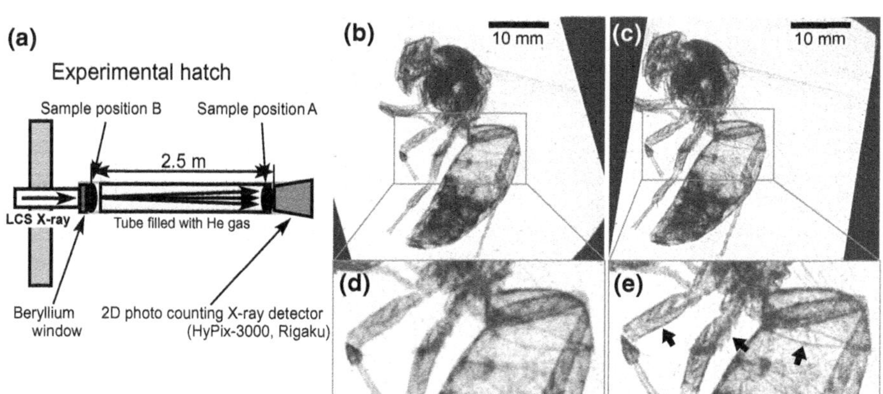

Fig. 38.2 **a** The schematic of LCS X-ray imaging experimental setup. Sample position A: absorption contrast, Sample position B: refraction-enhanced imaging. **b** X-ray absorption contrast image and **c** X-ray refraction-enhanced image of a hornet. **d** and **e** are the magnification of the red square of **b** and **c**, respectively

from the collision point. Due to the small source size of LCS X-rays, compared with Fig. 38.2b, c the refraction-enhanced image leads to enhanced edge contrast, revealing the fine structure of the legs and abdomen (Fig. 38.2d, e (red arrows)). The exposure time of Fig. 38.2b, c are 18 min (Averaged photon number: 0.08 photons/s/pixel) and 8 min (Averaged photon number: 0.11 photons/s/pixel), respectively.

38.4 Conclusion

We have demonstrated the generation of LCS X-ray at cERL in KEK. We also performed the X-ray absorption contrast imaging and refraction-enhanced imaging of a hornet. We confirmed that LCS X-rays from a small source size are useful for X-ray refraction-enhanced imaging. We plan to increase the photon flux by improving the electron beam current and the enhancement factor of the optical cavity to take an image in a short time.

Acknowledgements This work was supported by Photon and Quantum Basic Research Coordinated Development Program from the Ministry of Education, Culture, Sports, Science and Technology, Japan.

References

1. Hajima, R., et al.: Detection of radioactive isotopes by using laser Compton scattered γ-ray beams. Nucl. Instrum. Methods **608**, S57–S61 (2009)
2. Wiedemann, H., et al.: A compact radiation source for digital subtractive angiography. Nucl. Instrum. Methods **347**, 515–521 (1994)
3. Uberall, H., et al.: Channeling radiation as an X-ray source for angiography, X-ray lithography, molecular structure determination, and elemental analysis. In: SPIE Proceedings Short Wave Radiation Sources, vol. 1552, pp. 198 (1991)
4. Obina, T., et al.: Recent developments and operational status of the compact ERL at KEK. In: Proceedings of IPAC 2016, 1853 (2016)
5. Hänsch, T.W., Couillaud, B.: Laser frequency stabilization by polarization spectroscopy of a reflecting reference cavity. Opt. Commun. **35**, 441–444 (1980)
6. Nagai, R., et al.: Demonstration of high-flux photon generation from an ERL-based laser compton photon source. In: Proceedings of IPAC 2015, 1607 (2015)
7. Wilkins, S.W., et al.: Phase-contrast imaging using polychromatic hard X-rays. Nature **384**, 335–338 (1996)
8. Suzuki, Y., et al.: X-ray refraction-enhanced imaging and a method for phase retrieval for a simple object. J. Synchrotron Radiat. **9**, 160–165 (2002)

Chapter 39
2D and 3D Nanoscale Imaging Using High Repetition Rate Laboratory-Based Soft X-Ray Sources

H. Stiel, A. Blechschmidt, A. Dehlinger, R. Jung, E. Malm, B. Pfau, C. Pratsch, C. Seim, J. Tümmler and M. Zürch

Abstract In this contribution, we report about tomographic nanoscale imaging using a laser-produced plasma-based laboratory transmission X-ray microscope (LTXM) in the water window. The soft X-ray radiation of the LTXM is provided by a high average power laser-produced (1.3 kHz repetition rate, 0.5 ns pulse duration, 140 W average power) plasma source, a multilayer condenser mirror, an objective zone plate, and a back-illuminated CCD camera as a detector. In the second part of the contribution, we will present recent results on holography and coherent diffraction imaging using our high repetition rate X-ray laser. We will discuss advantages of these methods and its potential for nanoscale imaging.

H. Stiel (✉) · R. Jung · B. Pfau · C. Seim · J. Tümmler
Max-Born Institute, 12489 Berlin, Germany
e-mail: stiel@mbi-berlin.de

H. Stiel · A. Blechschmidt · A. Dehlinger · R. Jung · C. Seim
Berlin Laboratory for innovative X-ray Technologies (BLiX), 10623 Berlin, Germany

A. Blechschmidt · A. Dehlinger
Institut für Optik und Atomare Physik (IOAP), Technische Universität Berlin, 10623 Berlin, Germany

C. Pratsch
Helmholtz-Zentrum Berlin (HZB), 12489 Berlin, Germany

M. Zürch
Abbe Center of Photonics, Institute of Optics and Quantum Electronics, Jena, Germany

M. Zürch
Chemistry Department, University of California, Berkeley, USA

E. Malm
Synchrotron Radiation Research, Lund University, Lund, Sweden

T. Kawachi et al. (eds.), *X-Ray Lasers 2016*, Springer Proceedings in Physics 202, https://doi.org/10.1007/978-3-319-73025-7_39

39.1 Introduction

Laboratory-based laser-driven short-pulse X-ray sources like laser-produced plasmas (LPP), high harmonic generation (HHG), and plasma X-ray lasers (XRL) exhibit a great potential for imaging and spectroscopy in the soft X-ray range. These sources are complementary to large-scale facilities like synchrotrons or free electron lasers. LPP as well as XRL sources have already been successfully applied in nanoscale imaging in both life and material sciences. However, only few examples of high-resolution three-dimensional nanoscale images recorded with these sources exist. This is due to limited (in comparison with synchrotron sources) average photon flux and the lack of efficient X-ray optics. The first limitation can be overcome using high repetition rate (1 kHz and higher) pump lasers [1, 2]. The tremendous progress in the development of multilayer X-ray optics for the extreme ultraviolet (EUV) range has also led to the availability of multilayer coatings for wavelengths outside the EUV region allowing, e.g., efficiencies for a water window condenser optics of 4% or higher. For coherent laboratory-based X-ray sources (e.g., HHG or XRL), new imaging schemes such as coherent diffraction imaging or Fourier transform holography paved the way to nearly wavelength-limited resolution in 2D imaging [3–5].

In this contribution, we report about tomographic (3D) nanoscale imaging using a LPP-based laboratory transmission X-ray microscope (LTXM) in the water window. In the second part, we will present recent results on holography and coherent diffraction imaging using our high repetition rate XRL. We will discuss advantages of these methods and its potential for 2D and 3D nanoscale imaging.

39.2 Experimental

39.2.1 Laboratory Transmission X-Ray Microscope

Our LTXM, as already described in [1], is based on a laser-produced liquid nitrogen plasma source that provides the soft X-ray radiation. A Nd:YAG pulsed laser beam (1.3 kHz repetition rate, 0.5 ns pulse duration, FhG-ILT Aachen, Germany) with an average power of up to 140 W is focused onto a liquid nitrogen jet, producing a dense plasma which emits two strong lines located at 2.88 and 2.478 nm. A multilayer condenser mirror (OptiXfab, Jena) is used for collecting the radiation and selecting a specific wavelength. Due to a newly developed coating technology, the spherical multilayer mirror exhibits a reflectivity of up to 4% at 2.478 nm (500 eV). To prevent direct illumination by the plasma source, a central stop is placed in front of the sample. Furthermore, a 200 nm Al-filter blocks the infrared radiation from the laser. Behind the sample, a zone plate objective (Helmholtz-Zentrum Berlin) with an outermost zone width of 30 nm projects the image onto a cooled back-illuminated X-ray-sensitive CCD camera (SX-NTE 2048, Roper Scientific).

The magnification of the LTXM increases with the distance between the objective and the CCD camera. The sample prepared on standard grids for electron microscopy is placed on a stage allowing a rotation of 60° in both directions, resulting in a total of 120°. A cryogenic sample holder (GATAN 626) allows the transfer of frozen hydrated specimens to the sample stage.

39.2.2 High Repetition Rate X-Ray Laser

The plasma-based soft XRL operating in grazing incidence pump (GRIP) geometry was pumped by two pulses of a high repetition rate 100 Hz thin disk laser (TDL) chirped pulse amplification (CPA) system [6]. The TDL system consists of a front-end with a Yb:KGW oscillator, stretcher, and Yb:KGW regenerative amplifier followed by two regenerative amplifiers and one multipass amplifier. The output of the front-end is divided into two pulses. Each of them is amplified in a regenerative amplifier to a level of about 100 mJ. Whereas the pulse from the first regenerative amplifier is compressed to a duration of about 200 ps using a grating compressor, the output of the second regenerative amplifier is given to a thin disk multipass amplifier which amplifies the pulses to an energy of up to 400 mJ and compressed in the grating compressor to about 2 ps pulse duration at an energy of 270 mJ. The long pulse (150 ps, $E \approx 70$ mJ) is focused by a cylindrical and a spherical lens onto the target at normal incidence. The generated plasma column will then be heated by the short pulse focused by a spherical mirror into the preformed plasma. We have found for our Mo slab target (50 mm length, 5 mm width) an optimum GRIP angle of about 24°. The target was attached to a motorized stage with four degrees of freedom allowing the adjustment in three axes as well as the continuous renewing of the target surface.

39.3 Results

39.3.1 Water Window Tomography

The tomographic (three-dimensional) imaging of cells as known, e.g., from electron microscopy requires the collection of a so-called tilt series of the object imaged under several projection angles. As in electron tomography, the tilt series should be collected with an increment angle $\Delta\Theta$ as small as possible as well as a high maximum projection angle Θ. The tilt series is back-projected and so-called slices are generated. The coefficients of the Fourier transform upon which the back-projection algorithm is based can be exactly determined for the central region of the recorded projection images. For all other regions, the Fourier components have to be interpolated. Therefore, the maximum spatial frequency R for which unknown Fourier components can be interpolated depends on the dimension D of the object. Assuming that x is the tilting axis and z the optical axis, the Crowther criterion defines the maximum spatial resolution (as a first approximation) as $R = 1/D * \Delta\Theta$. This corresponds to a resolution in the y-direction of $d = \pi * D/N$, where N is the number of projections. Furthermore, due to geometrical and technical restrictions of the sample stage, the scanning range for the projection angle in our LTXM is restricted to ±60°. This issue known as the "missing wedge problem" leads to a reduced resolution in the z-direction defined by an elongation factor. For $\Theta = \pm 60°$, this factor amounts to 1.6. It should be noted that the resolution in the x-direction is not affected by the above-mentioned restrictions. It is defined only by the width of the outermost zone of the objective zone plate.

A practical example of tomographic imaging using our LTXM is the imaging of a diatom as shown in Fig. 39.1. Due to the limited DOF (depth of focus) of the zone plate (about 1 μm), the plane containing the highest number of nanoparticles, presumably the plane containing the Si_3N_4-window, was chosen to be focused on at each projection angle. As an example, two projection images at +30° and −30° are presented in Fig. 39.1a. First, insights about the actual position of the different structures inside and outside the diatom can already be gained from the tilted projections. Information that was not accessible from a single projection was for instance that the dark structure (circled in red) is clearly located outside the diatom. A tomographic reconstruction (cp. Fig. 39.1b) of the sample based on the projection series allows to receive more precise depth information.

39.3.2 Coherent Diffraction Imaging (CDI)

The setup for coherent diffraction imaging experiments is depicted in Fig. 39.2. The soft X-ray radiation (wavelength 18.9 nm) generated by the XRL is imaged onto the sample by means of a spherical (ROC = 1000 mm) and a flat multilayer mirror (30% reflectivity in normal incidence) resulting in a spot size of $0.45 \times 0.55\ mm^2$

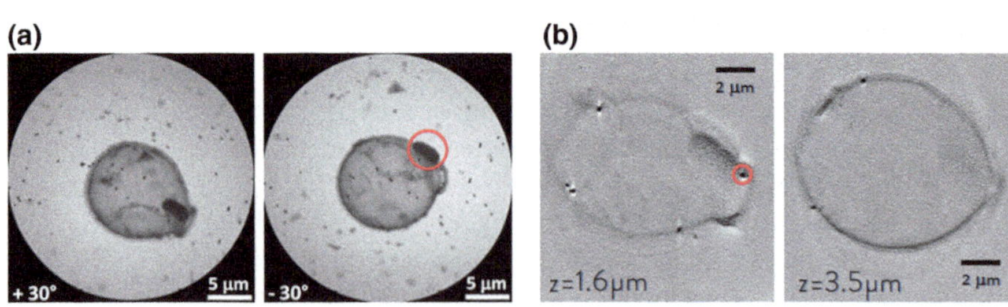

Fig. 39.1 **a** Left images: Examples for LTXM measurements of a diatom sample at different projection angles. Every image was recorded with an exposure time of 60 s. **b** Right images: Slices along the optical axis obtained from tomographic reconstruction. Nanoparticles with 250 nm size are clearly resolved. At z = 3.5 μm, the wall of the diatom is seen with a 150 nm resolution

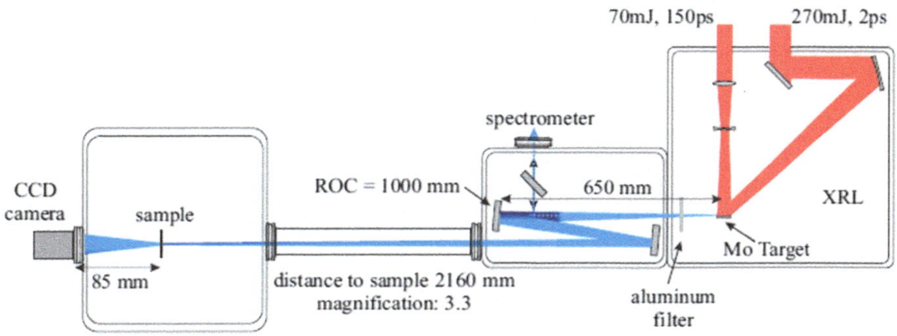

Fig. 39.2 Setup for the coherent diffraction imaging experiment using the XRL

(Fig. 39.3). Taking into account the limited spatial coherence (about 2%) of the unseeded XRL, an area of about 5000 μm^2 was coherently illuminated by a single pulse. As sample served an open aperture, shaped like a puzzle piece with dimensions of about 6×6 μm^2 and produced by focused-ion-beam milling into an Au film of 1.3 μm thickness (Fig. 39.3a). A cooled, back-illuminated, 16-bit CCD camera (Roper Scientific PI-MTE) placed at a distance of about 85 mm downstream of the sample captured the diffracted light with a numerical aperture of 0.16. The coherent diffraction pattern was recorded by accumulating 200 XRL shots (energy of each pulse $\approx$ 100 nJ) in a single exposure (Fig. 39.3b). In an iterative search, the sample's exit wave was retrieved from the diffraction pattern using the hybrid input–output (HIO) algorithm and applying the real-space support constraint. After 120 iterations, the search converged to the solution as presented in Fig. 39.3c. The reconstructed exit wave closely resembles the aperture's shape. The spatial resolution of approximately 200 nm is mainly limited by the low photon count and the incoherent summation of several XRL shots.

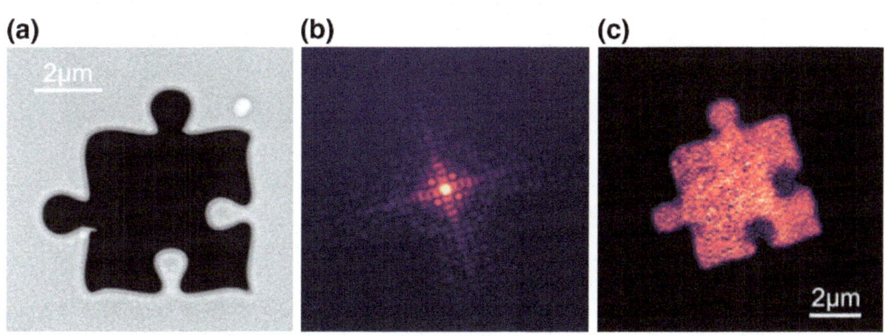

Fig. 39.3 **a** SEM image of the "puzzle piece" test sample, **b** diffraction pattern recorded with 200 XRL shots (logarithmic intensity scale), **c** reconstructed magnitude of the sample's exit wave

39.3.3 Fourier Transform Holography

The general setup for Fourier transform holography (FTH) consists of one or several small reference pinholes and a larger sample area [7]. In the far field, the interference signal between light scattered from the sample and light scattered from the reference is detected. The reconstruction is done via Fourier transform. The resulting image can be split into four components including a convolution between the image of the reference and the sample. In most experiments, the reference is created by milling small well-defined pinholes into an opaque membrane which also contains a transparent area for the sample. Since the size of the pinhole defines the resolution, it has to be relatively small. On the other hand, a small pinhole reduces the amount of photons in the reference and yields a bad signal-to-noise ratio (SNR). To circumvent this problem, setups with multiple pinholes or extended reference were proposed [8]. These types of milled references scatter the light inhomogeneously, resulting in a bright central spot and a fast declining signal for high spatial frequencies.

We, therefore, propose to use a single diffractive optical element (DOE) which generates simultaneously a homogeneous illumination profile for the sample area and a diffraction-limited reference. Figure 39.4 shows the setup for the FTH experiment as used in our proof of principle experiments. This approach has several

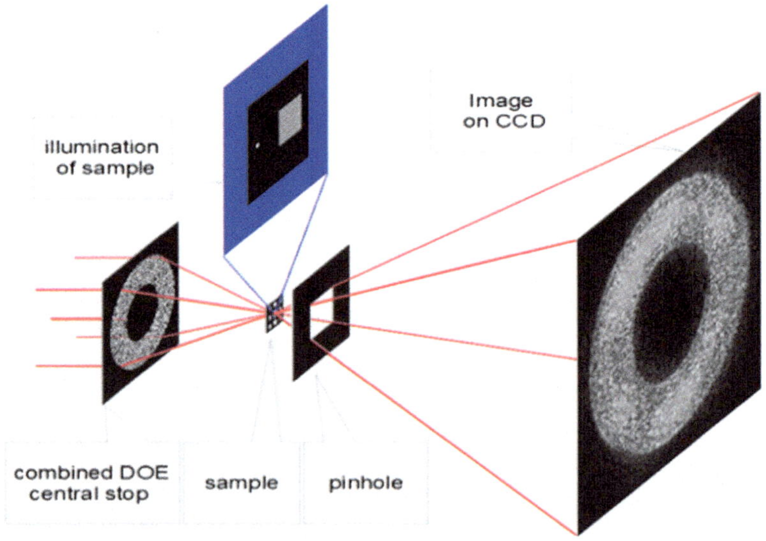

Fig. 39.4 Setup of the Fourier transform holography experiment. A DOE (diameter 400 μm, central stop of 200 μm) generates simultaneously reference and object illumination. The combination of central stop and pinhole selects the correct order of the DOE. For experimental reasons it was placed behind the sample. For a reduced dose and higher resolution, the pinhole should be placed between sample and DOE

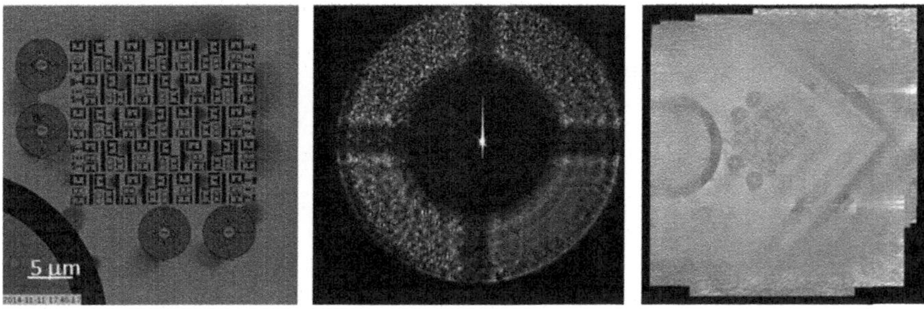

Fig. 39.5 Experimental results using the DOE and synchrotron radiation (324 eV) Exposure time: 100 ms. Photon flux on DOE: $2 * 10^{11}$ ph/s

advantages. The sample is freestanding and therefore much easier to prepare. Larger samples can be imaged via raster scanning and stitching of the resulting images. This also reduces speckles and a much larger part of the incident X-rays is used, which allows shorter exposure times. The intensity of the sample area and reference is a free design parameter and can be optimized for high SNR. A serious drawback is that the intrinsic stability between sample and reference of the simple FTH experiment is lost. Therefore, the sample has to be aligned with the focal plane of DOE.

To generate the DOE, we start with a random guess for its transmission function. It is assumed that the transmission function is purely phase shifting. The resulting field is propagated to the focal plane of the DOE. This field is then projected onto a field which satisfies the illumination conditions for the FTH. This modified field is back propagated to the DOE and a new transmission function is computed. The process is iterated until a sufficiently good match is found. The structure of the DOE is then generated by hard clipping the transmission function.

The capability of the method to image extended samples was tested using synchrotron radiation at the U41 beamline at the Helmholtz-Zentrum Berlin at 324 eV. The theoretically predicted best resolution for our first-order DOEs with outermost zone width of 50 nm is 80–100 nm. Figure 39.5 shows the structure under investigation (left image), the image on the CCD (middle), and the result of the stitching procedure (right).

References

1. Legall, H., Blobel, G., Stiel, H., Sandner, W., et al.: Compact X-ray microscope for the water window based on a high brightness laser plasma source. Opt. Express **20**, 18362–18369 (2012)
2. Martz, D.H., Selin, M., von Hofsten, O., Fogelqvist, E., et al.: High average brightness water window source for short-exposure cryomicroscopy. Opt. Lett. **37**, 4425–4427 (2012)
3. Sandberg, R.L., Song, C., Wachulak, P.W., Raymondson, D.A., et al.: High numerical aperture tabletop soft X-ray diffraction microscopy with 70-nm resolution. Proc. Natl. Acad. Sci. USA **105**, 24–27 (2008)

4. Kang, H.C., Kim, H.T., Kim, S.S., Kim, C., et al.: Single-pulse coherent diffraction imaging using soft X-ray laser. Opt. Lett. **37**, 1688–1690 (2012)
5. Zürch, M., Kern, C., Spielmann, C.: XUV coherent diffraction imaging in reflection geometry with low numerical aperture. Opt. Express **21**, 21131–21147 (2013)
6. Tümmler, J., Jung, R., Stiel, H., Nickles, P.V., Sandner, W.: High-repetition-rate chirped-pulse-amplification thin-disk laser system with joule-level pulse energy. Opt. Lett. **34**, 1378–1380 (2009)
7. Marchesini, S., Boutet, S., Sakdinawat, A.E., Bogan, M., et al.: Massively parallel X-ray holography. Nature **2**, 560–563 (2008)
8. Guizar-Sicairos, M., Fienup, J.R.: Holography with extended reference by autocorrelation linear differential operation. Opt. Express **15**, 17592–17612 (2007)

Chapter 40
The Observation of Transient Thin Film Structures During the Femto-Second Laser Ablation Process by Using the Soft X-Ray Laser Probe

N. Hasegawa, M. Nishikino, M. Ishino, N. Ohnishi, A. M. Ito, Y. Minami, M. Baba, A. Ya. Faenov, N. Inogamov, T. Kawachi, K. Kondo and T. Suemoto

Abstract We modified a soft X-ray laser (SXRL) interferometer synchronized with a Ti:sapphire laser to observe a single-shot image of the nanoscale structure dynamics of materials induced by an optical laser pulse. The lateral resolution on the sample surface was improved to 0.7 μm using precise imaging optics. Using this system, we succeeded in observing thin film structures above the solid (or liquid) surface in the femtosecond laser ablation process of metals (Au). The thin film worked as soft X-ray beam splitter. This result shows a thin film was smooth and dense (with a roughness of a few nanometers and near sold density). Fur-

N. Hasegawa (✉) · M. Nishikino · M. Ishino · T. Kawachi · K. Kondo
Kansai Photon Science Institute, National Institutes for Quantum and Radiological Science and Technology, Umemi-dai, Kizugawa-shi, Kyoto 619-0215, Japan
e-mail: hasegawa.noboru@qst.go.jp

N. Ohnishi
Department of Aerospace Engineering, Tohoku University, 6-6-01 Aramaki-AzaAoba, Aoba-ku, Sendai 980-8579, Japan

A. M. Ito
Department of Helical Plasma Research, National Institute for Fusion Science (NIFS), 322-6 Oroshi-cho, Toki City, Gifu 509-5292, Japan

Y. Minami · T. Suemoto
Institute for Solid State Physics, The University of Tokyo, Kashiwanoha, Kashiwa-shi, Chiba 277-8581, Japan

M. Baba
Saitama Medical University, 38 Moroyama-machi, Iruma-gun, Saitama 350-0495, Japan

A. Ya. Faenov
Osaka University, 1-1, Yamadaoka, Suita, Osaka 565-0871, Japan

N. Inogamov
Landau Institute for Theoretical Physics of Russian Academy of Sciences, 1-A, Akademika Semenova av., Chernogolovka, Moscow Region 142432, Russia

T. Kawachi et al. (eds.), *X-Ray Lasers 2016*, Springer Proceedings in Physics 202, https://doi.org/10.1007/978-3-319-73025-7_40

thermore, it gave rise to the possibility of generating novel transient soft X-ray optics.

40.1 Introduction

The dynamical processes of the formation of unique structures, such as submicron scale ripple and bubble structures [1], by the irradiation of ultrashort pulse lasers have attracted much attention in the area of novel laser processing. In order to understand the femtosecond laser ablation process, direct observation of the surface dynamics is required. In previous works, we developed a soft X-ray laser (SXRL) interferometer synchronized with a Ti:sapphire laser pulse [2, 3]. The Ni-like silver SXRL at a wavelength of 13.9 nm [4] is suitable for probing the initial process of surface morphological changes, because it has a small attenuation length (<10 nm), short duration (<10 ps), large photon number (>10^{10} photons/shot), and can penetrate the surface plasma; the critical electron density is 10^{24} cm^{-3}. The in-depth and lateral spatial resolutions on the sample surface were 1 nm and 2 μm, respectively. It was useful for the observation of the ablation process when the pump laser fluence was higher than the ablation threshold. However, it was insufficient to observe the dynamics of small structures around the ablation threshold such as the spallative ablation process predicted by molecular dynamics simulations [5].

In this study, we improved the lateral resolution of the system to 0.7 μm using precise imaging optics. Using this system, we succeeded in observing unique structures in the initial stage of the ablation process of Au pumped by an 80 fs Ti: sapphire laser pulse.

40.2 Improvement of the Optical Pump and Soft X-Ray Probe System

A schematic of the single-shot optical pump and SXRL probe system is shown in Fig. 40.1. This system consists of four parts: a Ni-like silver SXRL (13.9 nm, 7 ps), pump laser (795 nm, 80 fs), double time fiducial system, and soft X-ray imaging system with an interferometer. The timing accuracy between the SXRL and pump laser was 2 ps. The image of the illuminated area on the sample is transferred to the CCD surface by an imaging mirror. A double Lloyd's mirror divides the soft X-ray into objective and reference light and generates an interference pattern on the CCD surface. A fringe shift of one period corresponds to a 20 nm depth. Moreover, the depth resolution was 1 nm in the current experiment. This system can be switched between interferometry and reflective imaging easily by modifying the relative incline angle of the double Lloyd's mirror.

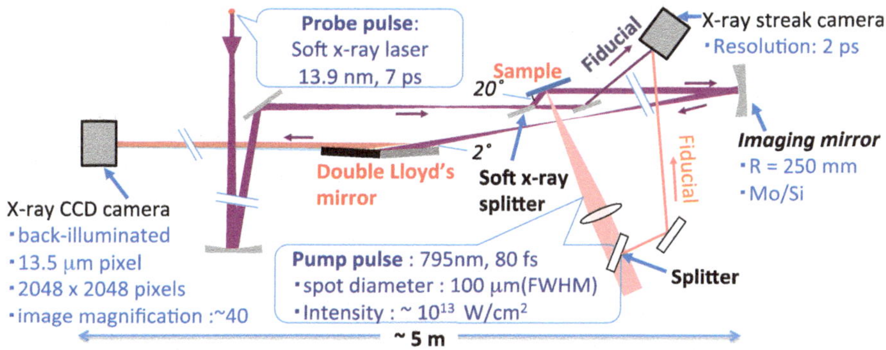

Fig. 40.1 Single-shot optical pump and soft X-ray laser probe system

In this study, the lateral resolution was improved using a high precision imaging mirror with a short focal length (f = 125 mm). It was evaluated using grooves fabricated on a Pt film using a focused ion beam (FIB). Figure 40.2a is the image of the test pattern observed by a scanning electron microscope (SEM). The rectangular dark areas show the grooves. Each groove has a width of 0.5–8 μm and a depth of 6 nm. The intervals of groove pairs from one edge to the other edge of the groove were the same as the width of the groove. The single-shot soft X-ray image and cross section at the area enclosed by the dotted line are shown in Fig. 40.2b and c, respectively. The longitudinal scale was 0.35 μm/CCD pixel. A pair of grooves 1 μm wide was clearly observed and that with a 0.5 μm width was not clearly observed. Furthermore, the soft X-ray intensity decreased to 2 pixels at the edge of

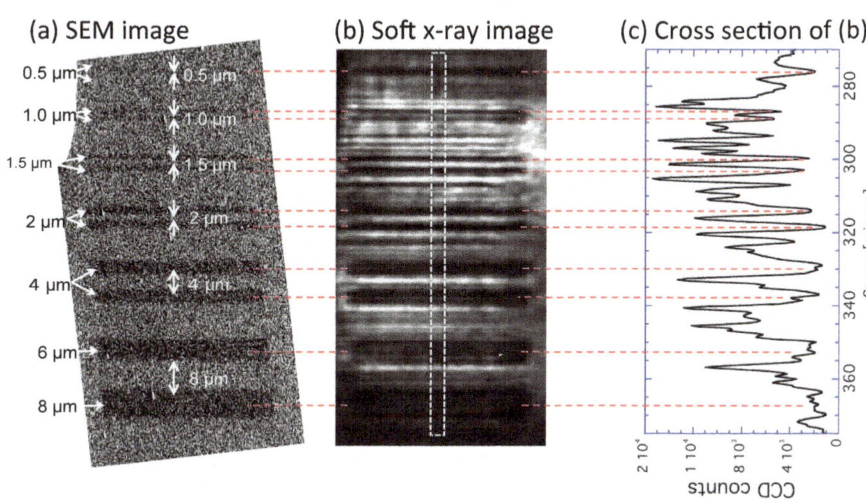

Fig. 40.2 Performance test of the soft X-ray imaging system. **a** SEM image of the test pattern. **b** Soft X-ray image. **c** Cross section of the soft X-ray image

most of the grooves. Therefore, the lateral resolution was evaluated to be 0.7 μm. The lateral resolution was within the area of 400 μm × 400 μm, and was sufficient to measure the ablation process of the 100-μm spot size.

40.3 Observation of the Transient Thin Film Structure in the Femtosecond Laser Ablation Process

We observed the ablation dynamics of Au (100 nm thick), which was irradiated by a pump beam with a peak fluence of about 1.1 J/cm^2. Figure 40.3a and b shows the temporal evolution of the interferogram of the ablated surface (= AF: ablation front). In Fig. 40.3a, 95 ps after the pump laser irradiation, the interference fringes were continuously smoothly. The height of the AF at the center was 20 nm. In Fig. 40.3b, at 389 ps, in addition to the interference fringes of the AF (height of ~40 nm), multiple concentric rings and thin interference fringes were observed around the dashed and solid arrows, respectively. Thin interference fringes imply the formation of another expanding structure above the AF. This is a thin film structure called an expansion front (EF). The height of the EF was over 100 nm. Figure 40.3c shows a schematic of (b). Because of the interference between the reflected SXRL from the AF and EF, multiple concentric rings (Newton's rings) can be generated. The observation of Newton's rings in the femtosecond laser ablation process has been reported using the visible probe beam [6]. The Newton's rings in the soft X-ray region imply that the EF was dense (near solid density), thin (<10 nm), and smooth (a roughness of a few nanometers) so as to function as a soft X-ray beam splitter, because the reflectivity of soft X-rays is quite sensitive to surface conditions. Detailed analyses of these results are underway.

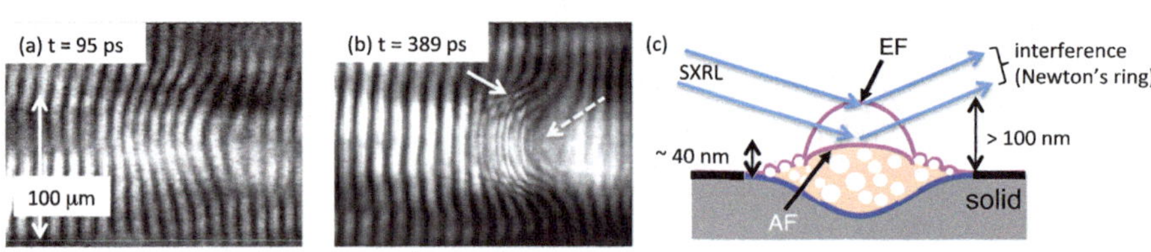

Fig. 40.3 **a**, **b** Snapshot of the interferogram of the femtosecond ablation process of Au. **c** Schematic of (**b**)

40.4 Summary

We modified the soft X-ray laser (SXRL) interferometer to observe the dynamics details of laser-induced materials. The lateral resolution on the sample surface was improved to 0.7 μm. Using this system, we succeeded in directly observing the expansion front (thin film structure) above the ablation front in the femtosecond laser ablation process of Au. The result indicated that the expansion front was dense (near solid density), thin (<10 nm), and smooth (with a roughness of a few nanometers). Further, it enables the possibility of developing novel transient soft X-ray optics.

References

1. Tomita, T., et al.: Effect of surface roughening on femtosecond laser-induced ripple structures. Appl. Phys. Lett. **90**, 153115-1–153115-3 (2007)
2. Suemoto, T., et al.: Single-shot picosecond interferometry with one-nanometer resolution for dynamical surface morphology using a soft X-ray laser. Opt. Express **18**, 14114–14122 (2010)
3. Hasegawa, N., et al.: Observation of the laser-induced surface dynamics by the single-shot X-ray laser interferometer. Proc. SPIE **8140**, 81400G-1–81400G-18 (2011)
4. Nishikino, M., et al.: Characterization of a high-brilliance soft X-ray laser at 13.9 nm by use of an oscillator-amplifier configuration. Appl. Opt. **47**(8), 1129–1134 (2008)
5. Perez D., Lewis L.J.: Molecular-dynamics study of ablation of solids under femtosecond laser pulses. Phys. Rev. B **67**, 184102-1–184102-15 (2003)
6. Temnov, V., et al.: Femtosecond time-resolved interferometric microscopy. Appl. Phys. A **78**, 483–489 (2004)

Chapter 41
Spectrally Resolved Spatial Interference for Single-Shot Temporal Metrology of Ultrashort Soft-X-Ray Pulses

Hugo Dacasa, Benoît Mahieu, Julien Gautier, S. Sebban and Philippe Zeitoun

Abstract Plasma-based soft-X-ray lasers seeded by high harmonics have only been temporally characterized using streak cameras, with a 1-ps temporal resolution. However, theoretical and experimental studies showed that sub-ps seeded soft-X-ray lasers are achievable. Existing methods do not provide enough information or require multiple shots, needing stable sources. In this work, we theoretically propose a single-shot method to characterize ultrashort soft-X-ray pulses, based on spectrally resolved spatial interference between an unknown pulse and a reference high-harmonic pulse. High harmonics have already been characterized, providing enough information to extract the unknown spectral phase from the 2D interferogram.

41.1 Introduction

In recent years, plasma-based seeded soft-X-ray lasers (PBSXRL) have become a useful tool for different applications requiring ultrashort, intense pulses in the soft-X-ray regime, such as biology [1], solid-state [2], and plasma physics [3].

In general, these sources consist of an elongated plasma, produced by an infrared laser on a solid or gaseous target, where population inversion is produced. This plasma is then seeded by a high-harmonic (HH) beam sent through it. The output pulse consists of a narrow part of the seed spectrum, greatly amplified [4]. While the plasma gain and its dynamics have been thoroughly measured in experiments, all which is currently known about the temporal properties of the resulting X-ray pulse comes solely from Maxwell-Bloch simulations [5, 6].

Experimental knowledge of the temporal properties of PBSXRL pulses would extend its range of application to fields where temporal resolution is key, while also confirming the validity of the related simulations.

H. Dacasa (✉) · B. Mahieu · J. Gautier · S. Sebban · P. Zeitoun
Laboratoire d'Optique Appliquée, UMR 7639 ENSTA-CNRS-Ecole Polytechnique,
Chemin de la Hunière, 91761 Palaiseau, France
e-mail: hugo.dacasa@ensta-paristech.fr

T. Kawachi et al. (eds.), *X-Ray Lasers 2016*, Springer Proceedings
in Physics 202, https://doi.org/10.1007/978-3-319-73025-7_41

41.2 Common Methods for Temporal Characterization of Ultrashort Pulses

In general, when trying to temporally characterize ultrashort pulses from a short-wavelength light source, one tries to adapt known methods from the visible and IR domain. However, this is not always possible, usually due to their reliance on non-linear crystals.

Autocorrelation measurements of soft-X-ray pulses can be made, as well as cross-correlation with pulses of longer wavelength. However, these methods obtain no information whatsoever about the spectral phase of the measured pulse. Additionally, assumptions must be made about its temporal shape, so these techniques are not enough for full temporal characterization. Another well-known technique in the IR range is FROG (Frequency-Resolved Optical Gating), which has successfully been adapted to the soft-X-ray domain for characterization of attosecond pulses obtained via high-harmonic generation, being known in this case as FROG CRAB (FROG for Complete Reconstruction of Attosecond Bursts) [7]. The main drawback of this technique, however, is its requirement of a high number of shots for a single measurement. This relies on the assumption that all shots are identical, or at least very similar, which is not necessarily the case, especially for a PBSXRL.

Ideally, one would prefer to use a single-shot technique, which could provide additional insight into the shot-to-shot stability of the source. Another well-known method for the IR range is SPIDER (Spectral Phase Interferometry for Direct E-field Reconstruction). It is a single-shot technique that relies on the spectral interference of the unknown pulse and a spectrally shifted, time-delayed replica. While it has already been adapted to HH pulses [8] and even seeded free-electron lasers [9], a spectrally shifted replica of a PBSXRL pulse cannot be produced due to the narrow bandwidth of the gain in the plasma.

Given that HH pulses have already been characterized, it would theoretically be possible to use referenced spectral interferometry, using the HH seed to obtain the phase of the PBSXRL pulse. In this case, the main requirement is an extremely high spectral resolution to be able to accurately resolve the interference fringes in such a narrow spectrum ($\Delta\lambda/\lambda = 10^{-6}$ or better). To solve this problem, we propose the use of a spatially encoded arrangement for referenced interferometry.

41.3 Sea Tadpole

An already-existing method for characterization of ultrashort IR pulses, SEA TADPOLE (Spatially Encoded Arrangement for Temporal Analysis by Dispersing a pair of Light E-fields) [10, 11] has never, to our knowledge, been implemented for the soft-X-ray range. It is based on spectrally resolved spatial interference between the unknown pulse and a reference pulse, whose spectral phase is known beforehand.

The two pulses are then sent at a small angle to a high-resolution 2D spectrometer (see Fig. 41.1).

By encoding the information in the spatial, rather than spectral, domain, the requirements for spectral resolution are not so strict as with conventional referenced spectral interference. The 2D interferogram essentially consists of spectrally resolved Young's fringes, following the equation:

$$S(\omega, x) = S_{unk}(\omega) + S_{ref}(\omega) + 2\sqrt{S_{ref}}\sqrt{S_{unk}} \cos\left(2kx \sin\theta + \varphi_{unk}(\omega) - \varphi_{ref}(\omega)\right),$$

where $S_{unk}(\omega)$, $S_{ref}(\omega)$, $\varphi_{unk}(\omega)$, and $\varphi_{ref}(\omega)$ are the spectra and spectral phases of the unknown and reference pulses, respectively and θ is half the angle between the pulses.

It is required that both pulses are synchronized so that spatial interference can occur. Additionally, the spectrum of the reference pulse must at least include that of the unknown pulse, not necessarily with the same central wavelength. To meet this condition for the case of a PBSXRL, the reference pulse can be a replica of the HH pulse used as a seed, whose phase can be measured by methods like HHSPIDER [8] or SEA SPIDER [12].

To show the use of the technique, we simulate the interference of an X-ray laser pulse with a high-harmonic pulse (see Fig. 41.2). Based on the parameters of the PBSXRL at Salle Jaune, in LOA [13], the unknown pulse has a central wavelength

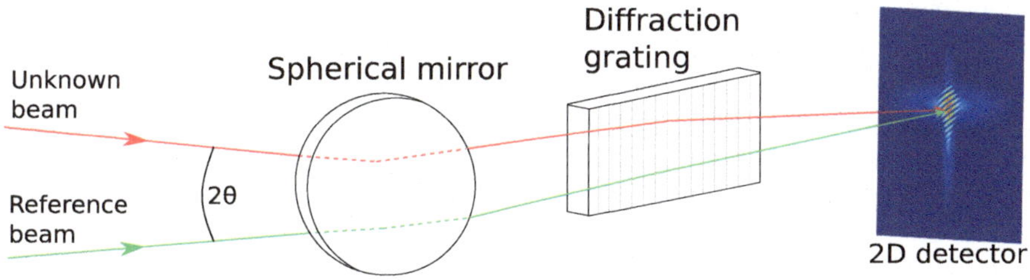

Fig. 41.1 Schematic drawing of the experimental setup for SEA TADPOLE

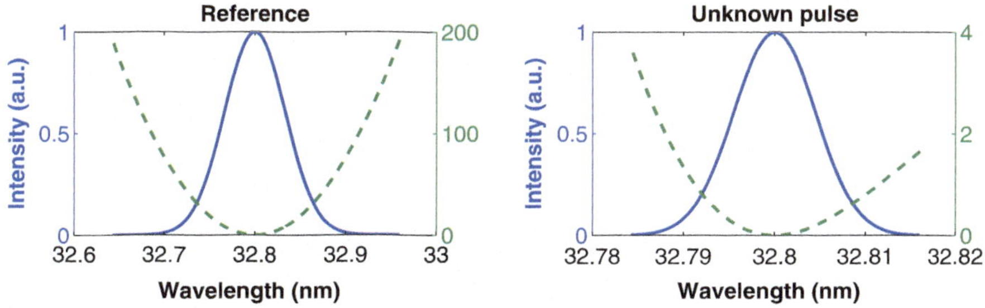

Fig. 41.2 Spectrum (solid lines) and spectral phase (dashed lines) of the high-harmonic reference pulse (left) and the PBSXRL unknown pulse (right)

of 32.8 nm, with its spectrum being 10 times narrower than that of the harmonic reference, which we assume to have the same central wavelength for the sake of simplicity. Both spectra are assumed to be Gaussian in shape, with arbitrary spectral phases. The reference pulse has a Group Delay Dispersion (GDD) of 5×10^3 fs^2 and no Third-Order Dispersion (TOD), whereas the unknown pulse has a GDD of 7×10^3 fs^2 and a TOD of 3×10^5 fs^3.

We assume that the 2D spectrometer uses a CCD with 2048×2048 square pixels, with a 13.5 µm side each, with a spectral resolution of $\Delta\lambda/\lambda = 10^{-5}$. The source-chip distance is assumed to be the same for both beams, however, the PBSXRL source is more divergent than the harmonic source, so we use a spatial FWHM of 1.7 mm on the spectrometer for the unknown pulse, and 0.6 mm for the reference. Both pulses are assumed to have the same intensity, to maximize the visibility of the interference fringes. The beams, forming an angle of 0.2°, produce the interferogram shown in Fig. 41.3. Note that the PBSXRL pulse has a narrower spectrum, but is spatially bigger, hence the cross-like shape.

Similar to referenced spectral interference, the first step to obtain the PBSXRL spectral phase from the interferogram is to perform a one-dimensional Fourier transform. In this case, however, this is done only in the x direction, not in ω. The resulting 2D image contains a central band and two sidebands at different positions of the K-axis, of which one is filtered. An inverse Fourier transform is then performed and the phase term $2kx \sin\theta + \varphi_{unk}(\omega) - \varphi_{ref}(\omega)$ is obtained. To retrieve the phase of the PBSXRL laser $\varphi_{unk}(\omega)$, it is just necessary to remove the reference phase and the spatial interference term. It is also interesting to note that the full spectral phase is obtained for every value of x where there are fringes, so it is possible to obtain and average several values for increased robustness of the result.

The obtained phase can be seen in Fig. 41.4, which compares it to the theoretical phase used as input in the code. This shows that the spectral phase of the PBSXRL can be accurately retrieved in a single shot if the conditions are met and the signal is strong enough.

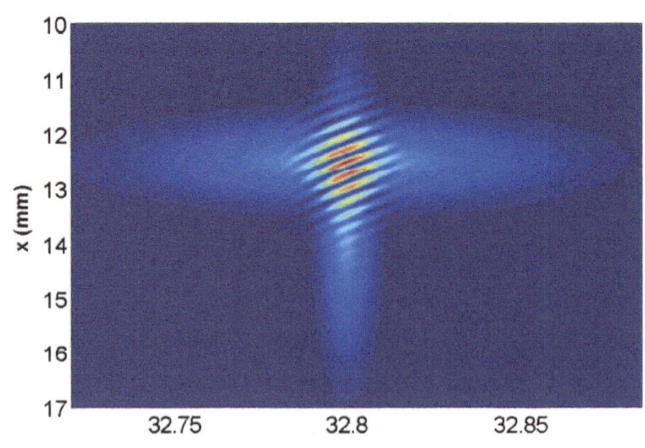

Fig. 41.3 Interferogram produced by the unknown and reference pulses in the 2D spectrometer (zoom)

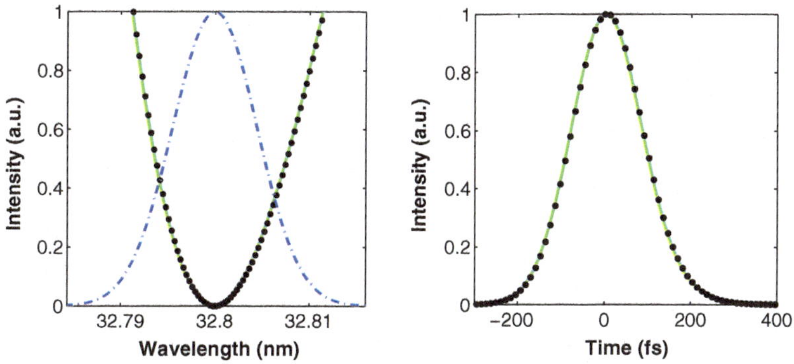

Fig. 41.4 Comparison of the input and calculated PBSXRL pulses. Left: input (solid line) and calculated (circles) spectral phases, with the pulse spectrum (dash-dot line). Right: input (solid line) and calculated (circles) temporal profiles of the PBSXRL pulses

41.4 Conclusions

SEA TADPOLE is a technique for full temporal characterization of ultrashort pulses, already used in the IR range but not yet experimentally transferred to the soft-X-ray range. It is a single-shot method based on spectrally resolved spatial interference between the unknown pulse and a reference pulse. It can allow for the study of the shot-to-shot stability of plasma-based soft-X-ray lasers, by using a replica of the high-harmonic seed pulse as a reference, whose spectral phase can be obtained with other known characterization techniques.

The experimental setup is relatively simple, consisting only of a 2D spectrometer, toward which the two beams are sent at a small angle. It is important that the spectral resolution is high enough, about $\Delta\lambda/\lambda = 10^{-5}$, to accurately measure the narrow spectrum produced by the PBSXRL, and the angle between both beams has to be kept small so that the produced interference fringes are well separated. Additionally, spectral phase retrieval is simple, performed by applying a non-iterative algorithm based on Fourier transform to extract the phase from the 2D interferogram and subtract the reference.

While we have focused solely to the case of soft-X-ray lasers, the technique can be applied to any source of ultrashort X-ray pulses, such as free-electron lasers, as long as a suitable reference can be found and characterized.

Acknowledgements The authors would like to thank COST Action MP1203 for their financial support.

References

1. Ros, D., et al.: Nucl. Instrum. Methods Phys. Res. A **653**(1), 76–79 (2011)
2. Mocek, T., Rus, B., Kozlová, M., Polan, J., Homer, P., Juha, L., Hájková, V., Chalupský, J.: Opt. Lett. **33**(10), 1087–1089 (2008)
3. Rossall, A.K., Tallents, G.J.: High Energy Density Phys. **15**, 67–70s (2015)
4. Depresseux, A., et al.: Nat. Photonics **9**, 817–821 (2015)
5. Oliva, E., Zeitoun, P., Sebban, S., Fajardo, M., Velarde, P., Cassou, K., Ros, D.: Opt. Lett. **34**(17), 2640 (2009)
6. Oliva, E., Zeitoun, P., Fajardo, M., Lambert, G., Ros, D., Sebban, S., Velarde, P.: Phys. Rev. A **84**, 013811 (2011)
7. Goulielmakis, et al.: Science **320**, 1614 (2008)
8. Mairesse, Y., Gobert, O., Breger, P., Merdji, H., Meynadier, P., Monchicourt, P., Perdrix, M., Salières, P., Carré, B.: Phys. Rev. Lett. **94**, 173903 (2005)
9. Mahieu, B., Gauthier, D., De Ninno, G., Dacasa, H., Lozano, M., Rousseau, J.P., Zeitoun, P., Garzella, D., Merdji, H.: Opt. Express **23**(14), 17667 (2015)
10. Bowlan, P., Gabolde, P., Shreenath, A., McGresham, K., Trebino, R.: Opt. Express **14**(24), 11892 (2006)
11. Bowlan, P.: Ph.D. thesis (2009)
12. Cormier, E., Walmsley, I.A., Kosik, E.M., Wyatt, A.S., Corner, L., DiMauro, L.F.: Phys. Rev. Lett. **94**, 033905 (2005)
13. Zeitoun, P., et al.: Nature **431**, 426–429 (2004)

Part VI
X-Ray Optics and Damage

Chapter 42
Multilayer Mirrors for Focusing Objective in 40-nm Wavelength Region

M. Toyoda, Y. Tamaru, S. Mori, K. Sawada, Y. Fu, E. J. Takahashi, A. Suda, F. Kannari, K. Midorikawa and M. Yanagihara

Abstract Isolated attosecond pulse (IAP) generation with pulse energy of a few μJ has been reported in the 40-nm wavelength region. For diffraction-limited focusing on the IAP, we are developing a Schwarzschild objective made of two-curved multilayer mirrors. To generate intense light fields with a maximum intensity of over 10^{16} W/cm^2 on the objective focus, we designed, fabricated, and tested multilayer mirrors for practical high reflectivity in the 40-nm wavelength region.

42.1 Introduction

Recently, isolated attosecond pulse (IAP) generation in the 40-nm wavelength region has been reported: an isolated intense pulse with pulse energy of a few μJ was demonstrated using a novel two-color gating method [1]. When an attosecond high-power extreme ultraviolet (EUV) pulse is focused using a diffraction-limited objective to produce a small focal spot with a diameter of a few handled nanometers, it is possible to generate extremely intense fields with a maximum intensity of over 10^{16} W/cm^2, thereby opening a new frontier of nonlinear optics. For the diffraction-limited focusing of the IAP, we are developing a Schwarzschild objective made of two-curved multilayer mirrors [2, 3]. This objective has two practical advantages: high spatial resolution resulting from the large numerical aperture optical design and spectral selectivity based on the Bragg reflection of

M. Toyoda (✉) · S. Mori · M. Yanagihara
Center for Advanced Microscopy and Spectroscopy, IMRAM, Tohoku University, 2-1-1 Katahira, Sendai 980-8577, Japan
e-mail: toyoda@tagen.tohoku.ac.jp

Y. Tamaru · K. Sawada · Y. Fu · E. J. Takahashi · K. Midorikawa
RIKEN Center for Advanced Photonics, 2-1 Hirosawa, Wako, Saitama 351-0198, Japan

Y. Tamaru · A. Suda
Tokyo University of Science, 2641 Yamazaki, Noda, Chiba 278-8510, Japan

K. Sawada · F. Kannari
Keio University, Yokohama 233-8522, Japan

T. Kawachi et al. (eds.), *X-Ray Lasers 2016*, Springer Proceedings in Physics 202, https://doi.org/10.1007/978-3-319-73025-7_42

multilayer mirrors. Producing this objective requires multilayer mirrors with practical high reflectivity. In this paper, we describe the design, fabrication, and testing of multilayer mirrors that are suitable for focusing in the 40-nm wavelength region.

42.2 Optical Design

To realize high reflectivity on multilayer mirrors, we applied the selection criterion for a coating material pair given by Yamamoto et al. [4]. According to the criterion, we expected high reflectivity from a material pair with both low absorption and a large Fresnel reflection coefficient at the interface, which corresponds to a large distance between the two materials on the complex plane plot of optical constants. After considering the criterion and availability of sputter targets, we chose several material pairs for the trial mirrors, as shown in Table 42.1. The period and thickness ratio of the mirrors were numerically optimized with IMD software [5] to give the maximum reflectivity at a wavelength of 40 nm.

Optical designs and calculated maximum reflectivity values for normal incident rays at a wavelength of 40 nm are tabulated in Table 42.1. We expected relativity high reflectivity from mirrors with an Mg spacer layer, while Si-based mirrors were expected to yield moderate reflectivity. For high reflectivity, the thickness of non-spacer layers, e.g., Mo, Cr, B_4C, and SiC, should be reduced, because the absorption of these materials is relatively large.

42.3 Experimental Results

The six trial mirrors in Table 42.1 were sputter deposited on Si wafers using a magnetron sputtering apparatus (SPL-500, Anelva Corp.) at Tohoku University. The period was examined with small-angle X-ray diffraction (XRD) and controlled by adjusting deposition time. We precisely controlled the period within 0.2 nm of the design values on Table 42.1. At-wavelength reflectivity was then measured with two different methods. First, we applied a reflectometer based on the higher

Table 42.1 Optical designs obtained with the numerical optimization procedure

Material	Period (nm)	Thickness ratio	Reflectivity
Mo/Mg	21.2	0.25	0.59
Cr/Mg	21.2	0.23	0.57
B_4C/Mg	21.2	0.25	0.54
SiC/Mg	21.2	0.25	0.53
Mo/Si	23.8	0.36	0.37
Sc/Si	21.0	0.33	0.31

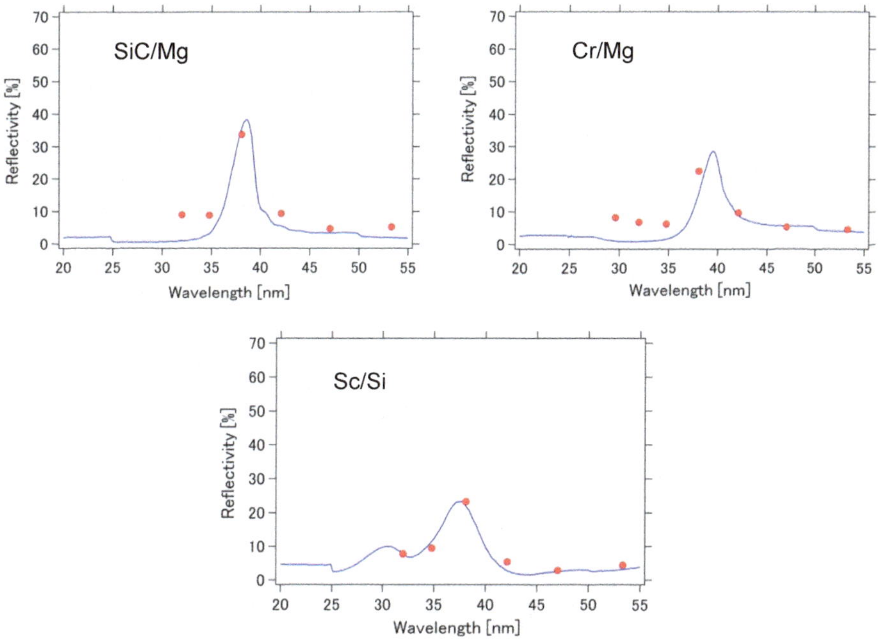

Fig. 42.1 Spectral reflectivity of trial mirrors. Solid curves and symbols represent data measured with synchrotron light and higher harmonics, respectively

harmonics of a Ti-sapphire laser at Riken. Symbols in Fig. 42.1 represent reflectivity for the higher harmonics at an incidence angle of 2°. We confirmed relatively high reflectivity (over 0.2) for the SiC/Mg, Cr/Mg, and Sc/Si multilayer mirrors. For the other samples, we observed reflectivity values less than half of those calculated. These degradations likely arose from the roughness due to heavy diffusion at the interface of the multilayer mirrors.

Finally, spectral reflectivity was confirmed by applying an EUV reflectometer equipped on the beamline BL5B of UVSOR at the Institute for Molecular Science. To suppress second- and higher-order lights from the monochromator, we applied a 300-nm thick Mg filter upstream of the reflectometer. The solid curves in Fig. 42.1 represent measured reflectivity as a function of wavelength. For reference, reflectivity data measured with the higher harmonics are indicated with symbols. The incident angle of the EUV rays on the trial mirrors was 11°. We confirmed that the results from the two different reflectometers were in good agreement. Relativity high reflectivity (over 30%) was achieved for the Mg-based multilayer mirrors. The reflectivity of the SiC/Mg multilayer mirror reached a practical level (40%) over a wide bandwidth. The results indicate that, when the IAP is focused without aberrations, a maximum intensity of over 10^{16} W/cm^2 can be expected at the focus of a two-bounce objective consisting of SiC/Mg multilayer mirrors.

42.4 Summary

Multilayer mirrors were developed for a focusing objective in the 40-nm wavelength region. For high reflectivity, we applied a selection criterion based on wave optics theory and chose six material pairs. Trial mirrors were then fabricated and tested. We successfully confirmed a practically high reflectivity (40%) and wide bandwidth for the SiC/Mg multilayer mirror. Group delay is also important for ultrafast optics because it modulates pulse duration on the focus. This aspect will be examined in future work.

Acknowledgements This work was performed under the Cooperative Research Program of the Network Joint Research Center for Materials and Devices, IMRAM, Tohoku University. This work was partially supported by the Research Foundation for Opto-Science and Technology and JSPS KAKENHI (Grant Nos. 14522729 and 16779300).

References

1. Takahashi, E.J., Lan, P., Mücke, O.D., Nabekawa, Y., Midorikawa, K.: Attosecond nonlinear optics using gigawatt-scale isolated attosecond pulses. Nat. Commun. **4**, 2691 (2013)
2. Toyoda, M., Yamasoe, K., Hatano, T., Yanagihara, M., Tokimasa, A., Harada, T., Watanabe, T., Kinoshita, H.: At-wavelength extreme ultraviolet lithography mask observation using a high-magnification objective with three multilayer mirrors. Appl. Phys. Express **5**(11), 112501 (2012)
3. Toyoda, M.: Flat-field anastigmatic mirror objective for high-magnification extreme ultraviolet microscopy. Adv. Opt. Technol. **4**(4), 339–346 (2015)
4. Yamamoto, M., Namioka, T.: Layer-by-layer design method for soft-x-ray multilayers. Appl. Opt. **31**(10), 1622–1630 (1992)
5. Windt, D.L.: Imd—software for modeling the optical properties of multilayer films. Comput. Phys. **12**(4), 360–370 (1998)

Chapter 43
Manufacture of High Precision, Multilayer Based Polarimeter Designed for Wide Energy Range from EUV to Soft X-Ray

H. Takenaka, N. Kuwabara, N. Kamachi, S.-Y. Liu, K. Endo, T. Ohchi, S. Ichimaru, H. Kimura, J. Laksman, F. Hennies, W. Grizolli and R. Sankari

Abstract The development of a high precision five rotation axes polarimeter using transmission multilayers as polarizers and reflection multilayers as analyzers is presented. In order to cover a wide energy range from EUV to soft X-ray, a set of Mo/Si, Cr/C, Sc/Cr, and W/B_4C multilayers for transmission and reflection have been designed and fabricated. A hexapod support allows to align the polarimeter easily relative to the optical axis, and the instrument is designed to be moved conveniently between different beamlines.

43.1 Introduction

Circularly polarized synchrotron radiation has attracted growing interest since the introduction of elliptically polarizing undulators [1, 2]. Tunable photon energy combined with full control of the polarization can be applied for the study of a variety of phenomena in biology, chemistry, physics, and material science. In order to perform a complete polarization analysis of light in a wide energy range, a multilayer polarimeter was designed and manufactured.

H. Takenaka (✉) · N. Kuwabara · N. Kamachi · S.-Y. Liu · K. Endo
Toyama Co., Ltd, 3816-1, Kishi, Yamakita, Kanagawa 258-0112, Japan
e-mail: takenaka@toyama-jp.com

T. Ohchi · S. Ichimaru
NTT Advanced Technology Co., 3-1, Morinosato Wakamiya, Atsugi, Kanagawa 243-0124, Japan

H. Kimura
JASRI/SPring-8, 1-1-1, Kouto, Sayo, Hyogo 679-5198, Japan

J. Laksman · F. Hennies · W. Grizolli · R. Sankari
MAX IV Laboratory, Lund University, 22100 Lund, Sweden

T. Kawachi et al. (eds.), *X-Ray Lasers 2016*, Springer Proceedings in Physics 202, https://doi.org/10.1007/978-3-319-73025-7_43

(a)

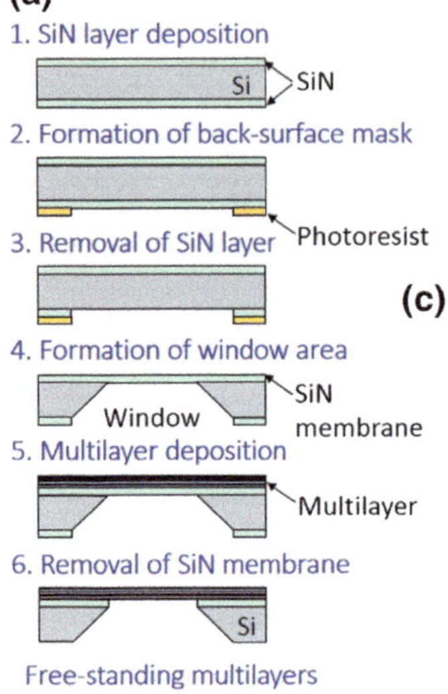

(b)

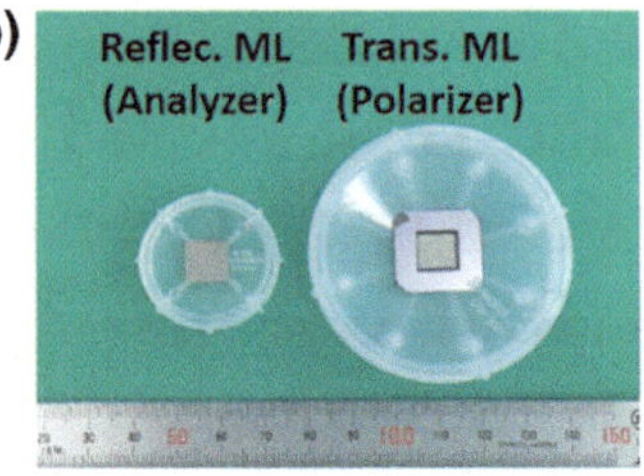

(c)

Materials	d[a] (nm)	N[b]	Γ[c]	ρ[d] (nm)
Mo/Si	9.40	100	0.33	< 0.4
Cr/C	3.13	280	0.33	< 0.4
Cr/Sc	2.57	400	0.5	< 0.4
W/B_4C	1.73	350	0.24	< 0.4

[a] d: period length.
[b] N: the number of layer pairs.
[c] Γ: the ratio of layer thickness of the 1st element to d.
[d] ρ: Roughness of the interfaces and surfaces.

Fig. 43.1 **a** Step-by-step fabrication process for multilayers. **b** Photograph of Mo/Si reflection and transmission multilayers (ML). **c** Characteristics of multilayers

43.2 Fabrication of the Multilayers

The fabrication procedure of the multilayers is shown in Fig. 43.1a and described in detail elsewhere [3, 4]. Figure 43.1b is a photograph of fabricated Mo/Si reflection and transmission multilayers. The transmission multilayers have up to 8 × 8 mm window area that can accept a beam diameter of 2 mm and down to 21° grazing incidence angle without cutting the transmitted beam.

Four pairs of multilayers (Mo/Si, Cr/C, Sc/Cr, and W/B_4C) are designed for a broad energy range from 100 to 1200 eV. The polarizers are completely free-standing multilayers and the characteristics of these multilayers are listed in Fig. 43.1c. A sample exchange unit can hold up to five polarizers and five analyzers; the multilayers can be exchanged in-vacuum during the operation via two transfer wobble sticks mounted on the vacuum chamber.

43.3 Mechanical Design of the Polarimeter

The polarimeter chamber contains the optical table components which are mounted on a hexapod, providing six degrees of freedom via motorized motions to allow a high accuracy, automatized alignment of the instrument (see Fig. 43.2a). This makes it possible to place the polarimeter accurately with high reproducibility. The optical table is decoupled from the vacuum chamber to avoid deformation under

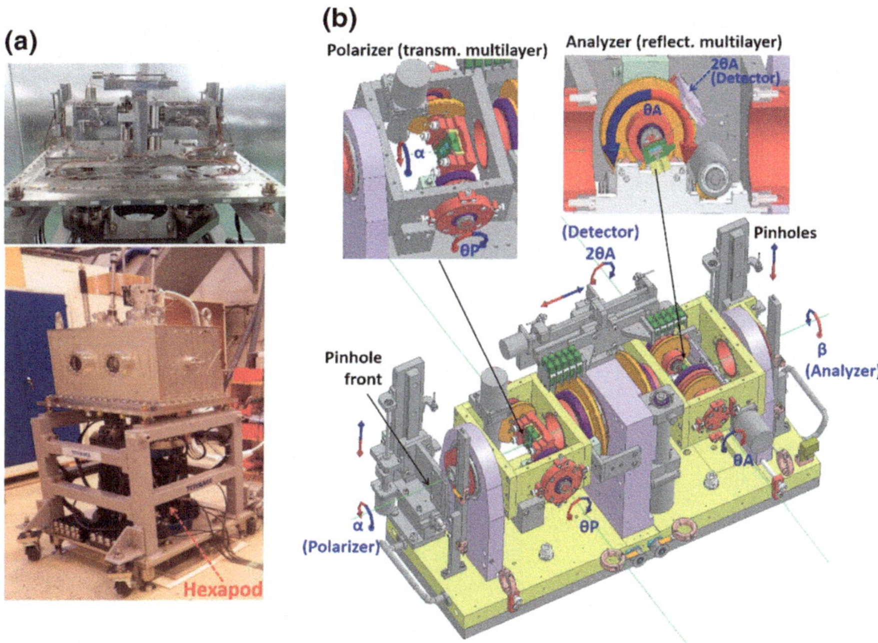

Fig. 43.2 **a** Photograph of the in-vacuum assembly of the polarimeter and the vacuum chamber with the hexapod system. **b** Schematic setup of the polarimeter showing all the motorized motions and their assignment

vacuum. The whole instrument is supported on a rigid frame with four jacking castors so that it can be transferred easily between different beamlines.

The schematic setup of the polarimeter is shown in Fig. 43.2b. The polarimeter contains the transmission multilayer polarizer (acts as phase shifter) and reflection multilayer analyzer, both of which can rotate independently about the optical axis of the beam. These two optical elements are mounted on two azimuthally rotating stages (α for polarizer and β for analyzer). These two rotation stages are identical except that the analyzer has an additional stage for housing an independently controlled 2θ detector arm.

In order to make accurate measurements, simple data analysis, and easy operation, two independent rotation axes for the upstream polarizer and the downstream analyzer must coincide with each other and stay stable. For this purpose, we proposed that increased accuracy in manufacturing and assembling allows an instrument which is supported not only at the middle but also at both ends. The same principle of relying on high accuracy manufacturing has been found very effective in the whole MAX IV Laboratory design. A set of pinholes are used for aligning the instrument relative to the photon beam [5]. The set of upstream pinholes allows light to pass, another set of downstream pinholes has a large-area photodiode behind it that allows the beam current to be measured. The detailed motions of the

Table 43.1 Ranges and accuracy of movement together with concentricity and parallelism of all the axis pairs

Description	Symbol	Range	Resolution
Polarizer rotation	α	−5° to 365°	0.0125°
Analyzer rotation	β	−5° to 365°	0.0125°
Incidence angle, polarizer	θ_P	−5° to 185°	0.0010°
Incidence angle, analyzer	θ_A	−5° to 95°	0.0010°
Detector angle	$2\theta_A$	−5° to 185°	0.0010°
Front pinhole, vertical		64 mm	0.0025 mm
Front pinhole, horizontal		−5 to 5 mm	0.0100 mm
Rear pinhole, vertical		64 mm	0.0025 mm
Rear pinhole, horizontal		−5 to 5 mm	0.0100 mm
Filter/I_0-detector, vertical		110 mm	0.0100 mm
Sample exchange unit		−71 to 71 mm	0.0100 mm
Axis pair alignment		Angular	Lateral
α, β		0.0075°//	7.6 μm
α, θ_P		0.0016° ⊥	33 μm
β, θ_A		0.0010° ⊥	48 μm
β, $2\theta_A$		0.0044° ⊥	43 μm
θ, $2\theta_A$		0.0084° ⊥	5 μm
Multilayer polarizer, θ_P		0.0006°//	0.2 μm
Multilayer analyzer, θ_A		0.0004°//	4.1 μm

polarimeter system are listed in Table 43.1. These values are obtained by a combination of autocollimator, laser tracker, and linear height gauge measurements performed at Toyama Co. Ltd.

We presented a high precision, multilayer-based polarimeter for soft X-ray synchrotron radiation. This versatile polarimeter was commissioned at beamline I1011 at MAX-II synchrotron in Lund [6].

References

1. Sasaki, S., et al.: Design of a new type of planar undulator for generating variably polarized radiation. Nucl. Instrum. Methods Phys. Res. Sect. A **331**, 763–767 (1993)
2. Onuki, H.: Polarizing undulators and wigglers. In: Undulators, Wigglers and Their Applications, pp. 216–236. CRC Press (2002)
3. Haga, T., et al: Superflat high-reflectivity semitransparent polarizer for soft X-ray. Proc. SPIE 2873 (1996)
4. Takenaka, H., et al.: EUV beam splitter for use in the wavelength region around 6 nm. J. Electron Spectrosc. Relat. Phenom. **144**, 1043–1045 (2005)

5. Nahon, L., et al.: SU5: a calibrated variable-polarization synchrotron radiation beam line in the vacuum-ultraviolet range. Appl. Opt. **43**, 1024–1037 (2004)
6. Grizolli, W., et al.: Multilayer based soft-X-ray polarimeter at MAX IV Laboratory. Rev. Sci. Instrum. **87**, 025102 (2016)

Chapter 44
Proposal of Hypereutectic Al–Si-Based Multilayer Mirrors for Wavelength Between 20 nm and 25 nm

M. Hatayama, S. Ichimaru, T. Ohchi and S. Oku

Abstract An extreme ultraviolet (XUV) multilayer mirror (MLM) for the wavelength region of 20–25 nm has been proposed. Hypereutectic Al–Si-based MLMs can be designed to have almost 55% calculated reflectivity assuming ideal conditions in this wavelength region and are expected to have the possibility of high thermal stabilities comparing with those of the conventional pure-Al-based MLMs. In addition, a narrowband MLM has been experimentally demonstrated based on the calculations. Measured reflectivity of fabricated $Zr/Al_{0.7}Si_{0.3}$ MLM has shown good agreement with the calculated value. These results show that the MLMs are useful for variety of applications such as XUV spectroscopy, astronomy, plasma physics, and attosecond-science.

44.1 Introduction

A multilayer mirror (MLM) is one of the typical optical devices in extreme ultraviolet (XUV) region as a focusing, steering, polarization control, and spectroscopy device. It is well known that reflectivity of the XUV-MLM is limited to 70% due to the absorption of materials constructing the MLM. Therefore, absorption edge, where the extinction coefficient takes a local minimum value, of lower index materials is taken into account for designing high-reflectivity MLM, such as Si (k-edge at 12.5 nm) based MLMs for wavelength region around 13–17 nm, Al (k-edge at 15 nm) for 17–20 nm, and Mg (k-edge at 25 nm) for 27–50 nm, respectively [1]. On the other hand, there are limited MLM demonstrations around 20–25 nm [2, 3] although several demonstrations are about Al-based MLM at 17–19 nm [4–8].

In this work, we proposed Al–Si-based MLMs for wavelength region between 20 and 25 nm. Using hypereutectic Al–Si, which is well known in mechanical engineering field as a high thermal conductivity and good strength material [8], the

M. Hatayama (✉) · S. Ichimaru · T. Ohchi · S. Oku
NTT Advanced Technology Corporation, Atsugi, Kanagawa, Japan
e-mail: masatoshi.hatayama@ntt-at.co.jp

T. Kawachi et al. (eds.), *X-Ray Lasers 2016*, Springer Proceedings in Physics 202, https://doi.org/10.1007/978-3-319-73025-7_44

MLMs are expected to have several advantages comparing with pure-Al-based MLMs for not only optical properties but also thermal properties. Designed reflectivities of MLMs were almost 55% assuming ideal conditions, and broadband mirrors and narrowband mirrors were also theoretically demonstrated. Measured reflectivity of a developed $Zr/Al_{0.7}Si_{0.3}$ narrowband mirror was in good agreement with the calculated value. These results show that these new MLMs can be useful for several applications such as XUV spectroscopy, astronomy, plasma physics, and attosecond science.

44.2 Multilayer Design

For designing the MLMs, Al_xSi_{1-x} was chosen as a lower index material. Upon the thermal property, the ratio of Si should be higher than 12.6%, which is the eutectic point of Al and Si. Under hypereutectic condition, it is expected that the crystallization and interdiffusion are decreased. On the other hand, it should be less than 50% because the absorption should be inhibited. Figure 44.1a and 44.1b show optical constants of Al and Si at XUV region, respectively [9]. As shown in Fig. 44.1a, b, the extinction coefficient of Si is almost as twice as that of Al at a wavelength region of 20–25 nm, and it is not a negligible value for constructing MLMs. In addition, considering the fabrication procedures such as deposition rates and fabrication tolerances, the amount of Si was estimated to 30%.

For the higher index material, high refractive index and low-absorption material are required. In Fig. 44.2, refractive index and extinction coefficient of typical higher index material at a wavelength of 20 nm are shown. Zr, C, and Y are considered as higher index material. Figure 44.3a–44.3d is calculated MLMs reflective profiles of $Zr/Al_{0.7}Si_{0.3}$, $C/Al_{0.7}Si_{0.3}$, $Y/Al_{0.7}Si_{0.3}$, and standard Mo/Si, respectively. In these calculations, ideal conditions were assumed and the higher index-material ratios were optimized for highest reflectivity. Each material combination has a possibility to obtain the higher reflectivity than Mo/Si. From these results Zr was selected as a higher index material.

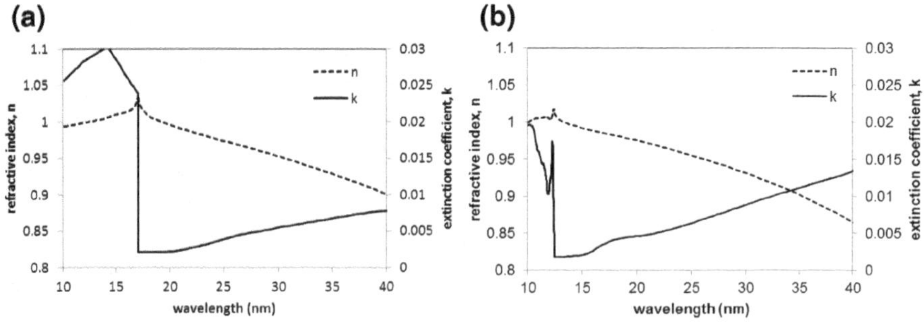

Fig. 44.1 Refractive index and extinction coefficient of **a** Al and **b** Si

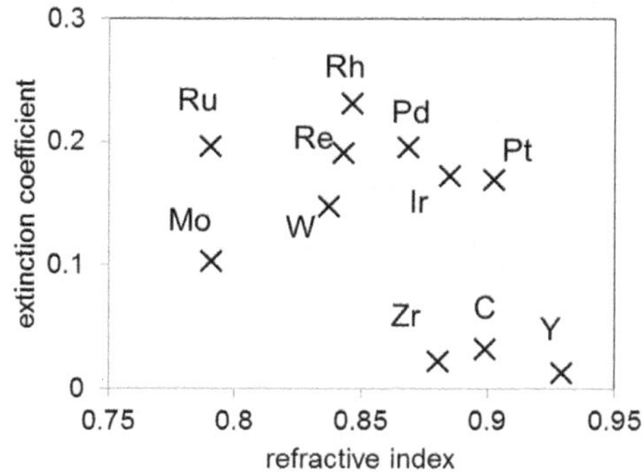

Fig. 44.2 Refractive index and extinction coefficient of higher index materials at 20 nm

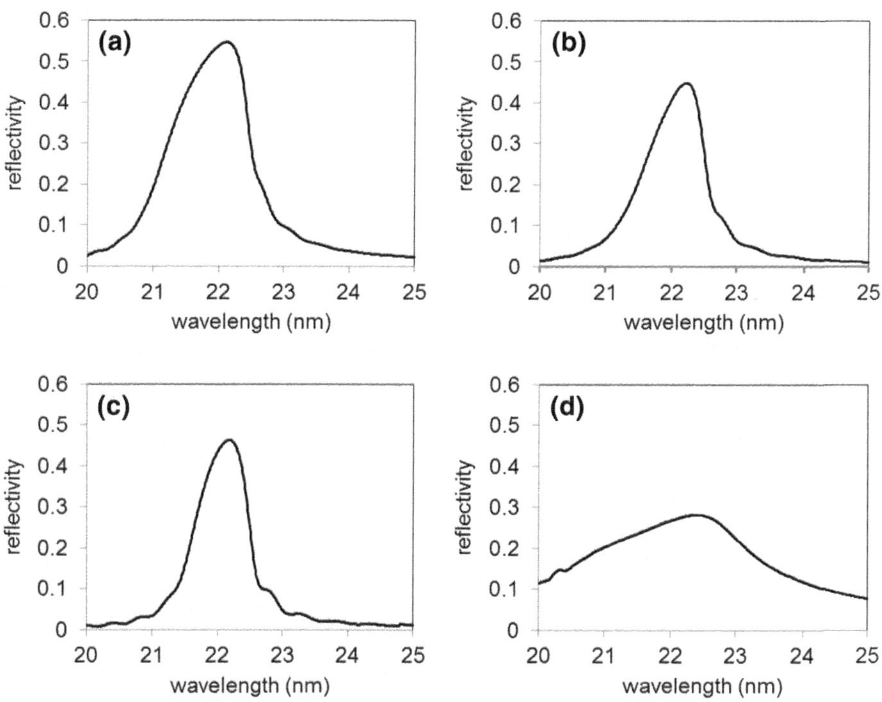

Fig. 44.3 Calculated MLM reflectivity profiles of **a** $Zr/Al_{0.7}Si_{0.3}$, **b** $C/Al_{0.7}Si_{0.3}$, **c** $Y/Al_{0.7}Si_{0.3}$, and **d** Mo/Si, respectively

For designing the multilayer structures, 3 types of mirrors will be required from the application side: (a) high-reflectivity mirrors, (b) narrowband mirrors, and (c) broadband mirrors. High-reflectivity mirrors can be useful for several XUV applications such as focusing devices and steering devices, and narrowband mirrors and broadband mirrors are used for ultrafast laser pulse applications such as

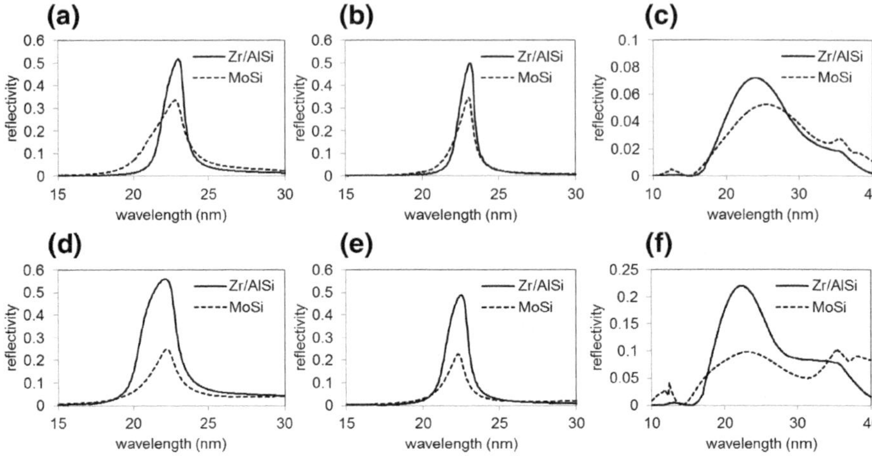

Fig. 44.4 Calculated reflectivity profiles of Zr/$Al_{0.7}Si_{0.3}$ MLMs comparing with Mo/Si MLMs. **a** High-reflectivity mirror for an incident angle of 0°, **b** narrowband mirror for 0°, **c** broadband mirror for 0°, **d** high-reflectivity mirror for 45°, **e** narrowband mirror for 45°, and **f** broadband mirror for 45°, respectively

high-order harmonics isolations and attosecond pump–probe experiments, respectively. Narrowband mirrors are also useful for plasma physics applications, and broadband mirrors also can be used for astronomy applications. Calculated profiles of reflectivity optimized, narrowband optimized, and broadband optimized Zr/$Al_{0.7}Si_{0.3}$ MLMs at wavelength region between 20 and 25 nm for an incident angle of 0° are shown in Fig. 44.4a to 44.4c, and for an incident angle of 45° are shown in Fig. 44.4d to 44.4f, respectively, comparing with Mo/Si MLMs. These figures show significant differences between Zr/$Al_{0.7}Si_{0.3}$ and Mo/Si MLMs especially mirrors for incident angle of 45°. Table 44.1 shows specifications of corresponding multilayer stuck. Note that, in calculations, ideal conditions such as roughness, surface oxidation, and material density were assumed.

44.3 Fabrication and Characterization

A 60-period multilayer stuck of Zr (1.6 nm)/$Al_{0.7}Si_{0.3}$ (14.4 nm) was deposited using DC magnetron sputtering [10] to evaluate the calculated values. The controlled deposition system can deposit each material with a thickness error less than 1%. The XUV reflectivity of this narrowband mirror was measured at Beamline 6.3.2 of the Advanced Light Source [11]. The measured reflectivity at angle of incident of 45° is shown in Fig. 44.5, comparing with a calculated reflectivity profile. Note that, a surface oxidation and a roughness were estimated 4 nm and 0.5 nm, respectively. The measured result shows good agreement with the calculated value. The peak reflectivity and full-width half maximum (EWHM)

Table 44.1 Specification of multilayer structure corresponding to Fig. 44.4a to 44.4f, where *d* is a periodic length, *Γ* is a ratio of high-index material, and *N* is a number of periods, respectively

No	$Zr/Al_{0.7}Si_{0.3}$			Mo/Si		
	d (nm)	*Γ*	*N*	*d* (nm)	*Γ*	*N*
(a)	12.0	0.3	40	12.0	0.3	40
(b)	12.0	0.2	50	12.0	0.15	50
(c)	11.5	0.45	2	11.5	0.45	2
(d)	17.0	0.3	40	17.0	0.25	40
(e)	17.0	0.15	50	17.0	0.15	50
(f)	17.0	0.45	2	16.0	0.35	2

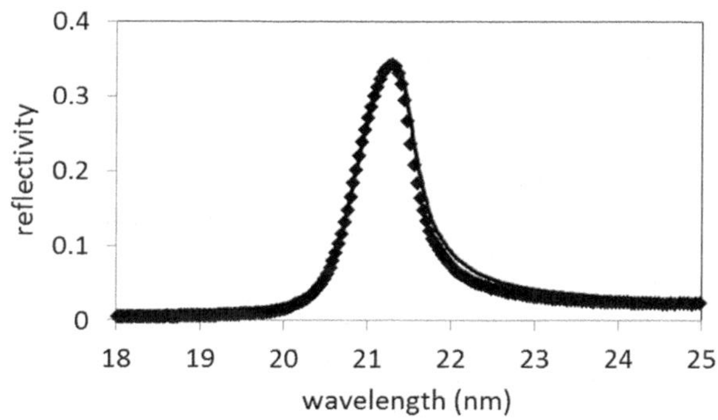

Fig. 44.5 Measured reflectivity of fabricated $Zr/Al_{0.7}Si_{0.3}$ MLM (diamond) comparing with calculated value (line)

bandwidth of this narrowband mirror were 34% and 0.8 nm, respectively. Note that the input XUV beam from ALS BL6.3.2 includes 10% p-polarized light and this should be considered for particular characterization.

44.4 Conclusion

We have designed the hypereutectic Al–Si-based MLMs at a wavelength region of 20–25 nm and experimentally demonstrated the calculations. The fabricated Zr/AlSi narrowband mirror has a peak reflectivity of 34%, and a FWHM bandwidth of 0.8 nm. These results show that the MLMs are useful for several applications in the wavelength region of 20–25 nm such as astronomy, plasma physics, attosecond-science, spectroscopy, and high-order harmonics isolation [12].

Acknowledgements The authors would like to thank Dr. Eric M. Gullikson of the Lawrence Berkeley National Laboratory for the multilayer reflectivity measurement. The Advanced Light Source is supported by the Director, Office of Science, Office of Basic Energy Sciences, of the U.S. Department of Energy under Contract No. DE-AC02-05CH11231.

References

1. X-ray Multilayer Results. http://henke.lbl.gov/multilayer/survey.html
2. Gautier, J.: Study of normal incidence three component multilayer mirrors in the range 20 nm–40 nm. Appl. Opt. **44**, 384–390 (2005)
3. Zuppella, P.: Iridium/silicon multilayers for extreme ultraviolet applications in the 20–35 nm wavelength range. Opt. Lett. **36**, 1203–1205 (2011)
4. Windt, D.L.: Performance, structure, and stability of SiC/Al multilayer films for extreme ultraviolet applications. Appl. Opt. **48**, 4932–4941 (2009)
5. Meltchakov, E.: Development of Al-based multilayer optics for EUV. Appl. Phys. A **98**, 111–117 (2010)
6. Zhong, Q.: Optical and structural performance of the Al(1%wtSi)/Zr reflection multilayers in the 17–19 nm region. Opt. Express **20**, 10692–10700 (2012)
7. Zhong, Q.: Enhancement of the reflectivity of Al/Zr multilayers by a novel structure. Opt. Express **21**, 14399–14408 (2013)
8. Ye, H.: An overview of the development of Al-Si-Alloy based material for engine applications. Eng. Perf. **12**, 288–297 (2003)
9. Henke, B.L.: A preliminary report on x-ray photoabsorption coefficients and atomic scattering factors for 92 elements in the 10–10,000 eV region. Lawrence Berkeley Laboratory Report, LBL-26259 (Lawrence Berkeley National Laboratory, 1988)
10. Takenaka, H.: Soft and hard x-ray reflectivities of multilayers fabricated by alternating-material sputter deposition. Proc. SPIE **1345**, 213–224 (1991)
11. Underwood, J.H.: Calibration and standards beamline 6.3.2 at the advanced light source. Rev. Sci. Instrum. **67**, 3372 (1996)
12. Hatayama, M.: Wide-range narrowband multilayer mirror for selecting a single-order harmonic in the photon energy range of 40–70 eV. Opt. Express **24**, 14546–14551 (2016)

Chapter 45
Irradiation Damage Test of Mo/Si, Ru/Si and Nb/Si Multilayers Using the Soft X-Ray Laser Built at QST

S. Ichimaru, M. Ishino, M. Nishikino, M. Hatayama, N. Hasegawa, T. Kawachi, T. Maruyama, K. Inokuma, M. Zenba and S. Oku

Abstract The irradiation damage for Mo/Si, Ru/Si, and Nb/Si multilayers are investigated using the soft X-ray laser system at QST. Scanning electronic microscopy observations reveal that the size of the damage on the Nb/Si multilayer is smaller than the other multilayers. This suggests that the damage fluence by the EUV irradiation of the Nb/Si multilayers is greater than the other multilayers.

45.1 Introduction

EUV (Extreme ultra-violet) multilayer mirror is the most important device for the developing of 13.5 nm next-generation lithography. Mo/Si multilayer mirror has been used with the combination of light source of laser-produced plasma (LPP) and X-ray free electron laser (XFEL) in EUV lithography system [1].

While the recently developed high power light sources are very effective for high throughput of EUVL, they also make serious problems for damages of Mo/Si multilayer mirrors [2–4]. These damages reduce the reflectivity and lifetime of Mo/Si multilayer mirror, resulting in the performance deterioration of productions and measurements using these mirrors and sources. Therefore, damage investigation is important for practical use of EUV multilayer and the future development of high resistance EUV multilayer mirror. Some previous works using LPP and XFEL estimated the threshold of Mo/Si multilayer and tried to obtain a high resistance EUV multilayer mirror, but a high resistance EUV multilayer mirror has not been developed until now [5–8].

In this paper, the high resistance EUV multilayer for high fluence EUV light using the soft X-ray laser (SXRL) system built at QST is demonstrated [9, 10].

S. Ichimaru (✉) · M. Hatayama · T. Maruyama · K. Inokuma · M. Zenba · S. Oku
NTT Advanced Technology Corporation, Atsugi, Kanagawa, Japan
e-mail: satoshi.ichimaru@ntt-at.co.jp

M. Ishino · M. Nishikino · N. Hasegawa · T. Kawachi
Kansai Photon Science Institute, National Institutes for Quantum and Radiological Science and Technology (QST), Kizugawa, Kyoto, Japan

T. Kawachi et al. (eds.), *X-Ray Lasers 2016*, Springer Proceedings in Physics 202, https://doi.org/10.1007/978-3-319-73025-7_45

45.2 Experiment Method

For the irradiation damage test of EUV multilayer, Mo/Si, Ru/Si and Nb/Si multilayer were prepared. Mo/Si multilayer is well known and usually used around the wavelength of 13.5 nm, and the others are also candidate multilayers for this wavelength region [11]. All these samples were fabricated by a magnetron sputtering method [12]. These multilayers were deposited on 4-inch silicon wafers and cut into 10 mm squares. The multilayers were designed such that the reflection peak wavelength of 13.9 nm and AOI of 6° decided by the experimental setup are shown below. The actual reflectivity of Mo/Si, Nb/Si and Ru/Si multilayer was 67%, 64% and 52% at the wavelength of 13.9 nm and AOI of 6°, respectively.

For the irradiation damage tests of EUV multilayers, the SXRL system built at QST, which was already used for the observation of irradiation damages of some materials, was used [13–16]. This SXRL has the following features: (a) the short timescale for 7 ps, (b) narrow bandwidth of $<10^{-4}$, (c) small divergence beam of <1m rad and (d) high brightness of <1 μJ/pulse [9, 10]. Using two Ag targets, 13.9 nm EUV light is produced in our experiments. Though our final target is the development of high resistance multilayer for the wavelength of 13.5 nm, the difference for the optical constants of materials for 13.5 and 13.9 nm are a little [1].

Figure 45.1 shows the schematic for the experimental setup of the irradiation damage tests [13, 17]. The SXRL from Ag targets is focused on a Mo/Si multilayer-coated concave mirror. The beam size of the sample is ~30 μm Dia. The intensity of SXRL is changed using the different thickness of Zr filters, i.e. 0, 0.1, 0.2 and 0.3 μm, that is corresponded to the transmittance of 1, 0.69, 0.48, 0.33 and 0.16, respectively. The beam intensity was estimated to be 20–30 mJ/cm^2 when the Zr filter of 0.2 μm is used [13]. Three sets of damage tests were applied to each multilayer as changing the Zr filter thickness, as shown in Figs. Figs. 45.2, 45.3 and 45.4.

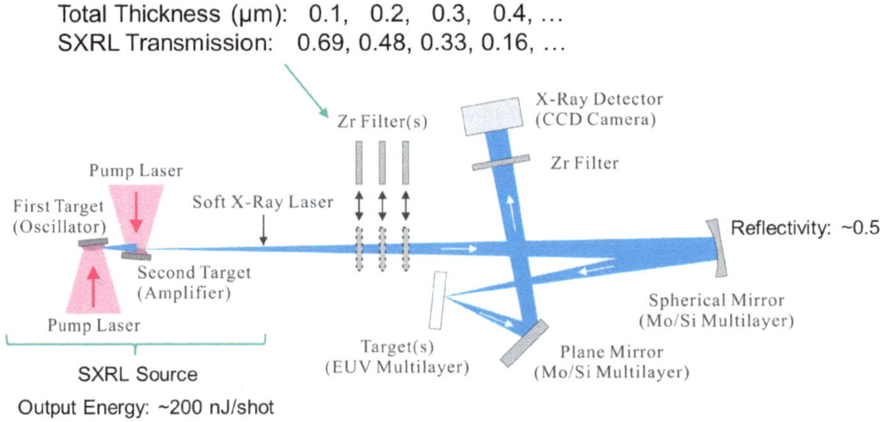

Fig. 45.1 The schematic for the experimental setup of the irradiation damage test [13, 17]

45.3 Results

The irradiation damages were observed using a scanning electron microscopy (SEM). Figures 45.2, 45.3 and 45.4 show the observation SEM images of one damage test set for Mo/Si, Nb/Si and Ru/Si multilayer, respectively. In all figures, the upper number indicates the shot number of each point. '1' means the result of 1-on-1 damage test. The left number indicates the Zr filter thickness. The area size of each SEM image is 50 μm squares. The black enclosure of SEM images means the irradiation damage was not detected visually. Clearly, the irradiation damage of Nb/Si multilayer at the Zr filter thickness of 0.2 μm is very minimal, in contrast to other multilayers that display irradiation damages. The result of the other damage test set also shows the same tendency. Therefore, the Nb/Si multilayer is not damaged at the Zr filter thickness of 0.2 μm at not only 1-shot irradiation but also 10 cycle shots, but the Mo/Si and Ru/Si multilayer are damaged at the same Zr filter thickness that corresponds to the irradiated EUV intensity of 20–30 mJ/cm^2. In addition, the irradiation damage size of Nb/Si multilayer is roughly estimated to be half the size of the other multilayers at the same condition. These results indicate that Nb/Si multilayer has a higher resistance than Mo/Si and Ru/Si multilayer for the higher fluence EUV light.

Though the SEM images of irradiation damages with the Zr filter thickness of 0.3 and 0.4 μm are not shown, irradiation damages were not found in all these SEM images, resulting in no damage under the EUV light intensity of ~15 mJ/cm^2.

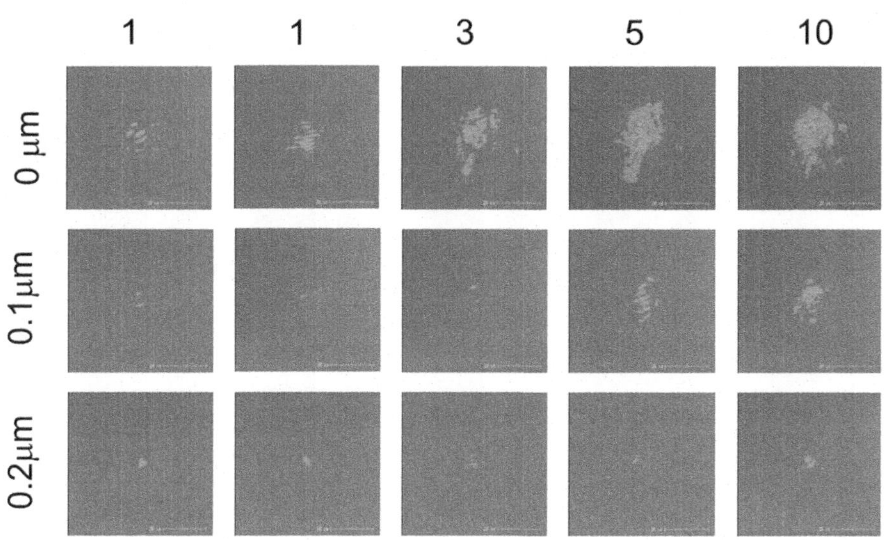

Fig. 45.2 SEM images of the irradiation damages for Mo/Si multilayer. The horizontal numbers indicate the shot number on the same position. The vertical number shows the thickness of Zr filter

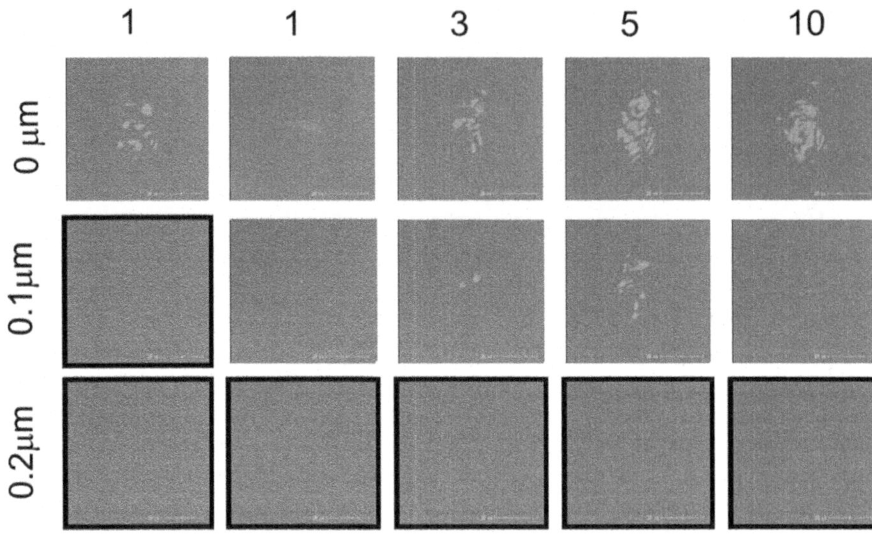

Fig. 45.3 SEM images of the irradiation damages for Nb/Si multilayer. The horizontal numbers indicate the shot number on the same position. The vertical number shows the thickness of Zr filter

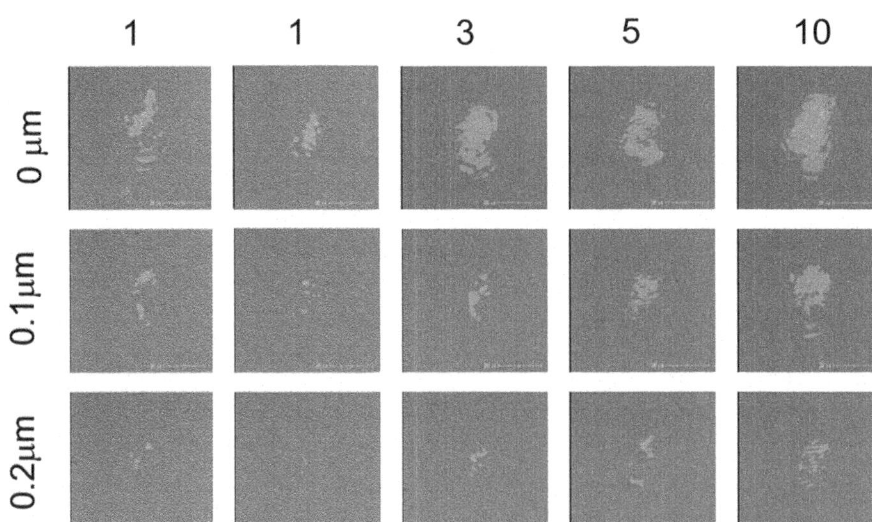

Fig. 45.4 SEM images of the irradiation damages for Ru/Si multilayer. The horizontal numbers indicate the shot number on the same position. The vertical number shows the thickness of Zr filter

45.4 Summary

In this paper, high resistance EUV multilayer for high fluence EUV light is demonstrated. For this purpose, the irradiation damage test of Mo/Si, Ru/Si and Nb/Si multilayer was carried out using the soft X-ray laser generation system built at QST. The created damages were observed under a scanning electron microscopy. These observations show that the damage size of Nb/Si multilayer is smaller than the others. In addition, the Nb/Si multilayer is not damaged by EUV lights at an intensity of 20–30 mJ/cm^2 in contrast with the other multilayers damaged at the same intensity. These results indicate the possibility that Nb/Si multilayer is superior to the other multilayers for high fluence EUV irradiation. Now, we advance to examine the quantitative evaluation of each damage. These results will be clear the difference of these damage thresholds.

References

1. Attwood, D.: Soft X-ray and Extreme Ultraviolet Radiation. Cambridge University Press (1999)
2. Kato, E., et al.: Development of superconducting accelerator with ERL for EUV-FEL. In: 2014 International Workshop on EUV and Soft X-Ray Sources, Dublin, Ireland, Nov 2014
3. RIKEN topics: http://www.riken.jp/pr/topics/2016/20160722_2/
4. Gigaphoton Inc.: News release, http://www.gigaphoton.com/en/news/4657
5. Khorsand, A.R., et al.: Single shot damage mechanism of Mo/Si multilayer optics under intense pulsed XUV-exposure. Opt. Express **18**(2), 700 (2010)
6. Sobierajski, R., et al.: Damage mechanisms of MoN/SiN multilayer optics for next-generation pulsed XUV light sources. Opt. Express **19**(1), 193 (2011)
7. Muller, M., et al.: EUV damage threshold measurements of Mo/Si multilayer mirrors. Appl. Phys. A **108**, 263 (2012)
8. Barkusky, F., et al.: Damage threshold measurements on EUV optics using focused radiation from a table-top laser produced plasma source. Opt. Express **18**(5), 4347 (2010)
9. Ochi, Y., et al.: Demonstration of submicroJoule, spatially coherent soft-X-ray laser pumped by 0.1 Hertz, 10 Joule, picosecond laser. Jpn. J. Appl. Phys. **48**, 120212 (2009)
10. Nishikino, M., et al.: Characterization of a high-brilliance soft X-ray laser at 13.9 nm by use of an oscillator–amplifier configuration. Appl. Opt. **47**, 1129 (2008)
11. Takenaka, H., et al.: Evaluation of large-area Mo/Si multilayer soft X-ray mirrors fabricated by RF magnetron sputtering. Jpn. J. Appl. Phys. **34**(1), 9A (1995)
12. Takenaka, H., et al.: Soft and hard X-ray reflectivities of multilayers fabricated by alternating-material sputter deposition. Proc. SPIE **1345**, 213 (1991)
13. Ishino, M., et al.: Observations of surface modifications induced by the multiple pulse irradiation using a soft picosecond X-ray laser beam. Appl. Phys. A **110**, 179 (2013)
14. Starikov, S.V., et al.: Soft picosecond X-ray laser nanomodification of gold and aluminum surfaces. Appl. Phys. B **116**, 1005 (2014)

15. Ishino, M., et al.: Nanoscale surface modifications and formation of conical structures at aluminum surface induced by single shot exposure of soft X-ray laser pulse. J. Appl. Phys. **109**, 013504 (2011)
16. Ishino, M., et al.: Very low electron temperature in warm dense matter formed by focused picosecond soft X-ray laser pulses. J. Appl. Phys. **116**, 183302 (2014)
17. Ishino, M. et al.: in this proceedings

Chapter 46
Mo/Si Multilayer-Coated Photodiode Detector for Monitoring Soft X-Ray Laser Intensity

T. Imazono

Abstract For the purpose of monitoring the beam intensity of laser-driven plasma soft X-ray laser (XRL), a Mo/Si multilayer reflector was coated on an X-ray photodiode detector. With synchrotron radiation at a wavelength of 13.9 nm, the s-polarized reflectance of the multilayer-coated photodiode, MP, was measured to be 52.5% at an angle of incidence of 45°, and the transmittance was 8.96%, which is defined as the ratio of the photodiode current measured at 45° to that at normal incidence. Moreover, the correlation coefficient between the reflectance and transmittance measured with XRL was 0.965. These results indicate that the MP serves as an excellent beam intensity monitor as well as provides an intense beam for XRL applications.

46.1 Introduction

The laser-driven plasma soft X-ray laser (XRL) system operated at the Kansai Photon Science Institute (KPSI) of National Institutes for Quantum and Radiological Science and Technology (QST) provides pulsed soft X-rays at 13.9 nm wavelength [1] and has been used for a variety of applications. As an XRL pulse forms nanostructures on a material surface, intensive studies on surface processing by a single or multiple exposures are recently in progress [2]. An XRL output is quite sensitive to the gain of the lasing medium. Thus, a slight gain fluctuation leads to a large variation of the XRL beam intensity. It is quite important to measure quantitatively the number of photons per pulse in an incoming beam every shot. In a 13.9 nm range a free-standing multilayer beam splitter or membrane-supported one is often used for monitoring the beam intensity, but such a beam splitter has not been installed in the beamline at KPSI because of some issues such as rigidity

T. Imazono (✉)
Kansai Photon Science Institute, National Institutes for Quantum and Radiological Science and Technology, Kyoto, Japan
e-mail: imazono.takashi@qst.go.jp

T. Kawachi et al. (eds.), *X-Ray Lasers 2016*, Springer Proceedings in Physics 202, https://doi.org/10.1007/978-3-319-73025-7_46

reduction, reflection loss, absorption increase by membrane, and roughness increase due to ion etching [3].

Silicon p–n junction photodiodes for the detection of X-rays have been developed and commercially available, e.g., AXUV Series [4]. The active area is fabricated from a polished wafer [5]. Most of them are windowless devices, but some are filtered with a bandpass filter. A well-designed multilayer film on the active area shows high reflectance [6]. A Mo/Si multilayer is the most suitable coating for XRL. When an X-ray beam is irradiated onto a Mo/Si multilayer-coated photodiode, MP, the MP generates a photodiode current as well as reflects the incident beam. If there is a strong positive correlation between the reflectance and the photodiode current, the incident beam intensity can be calibrated from the measured current. It means that the MP works as a beam splitter with an X-ray sensor as well as avoids the above-described problems on multilayer beam splitters.

This paper describes the performance of MP measured with synchrotron radiation (SR) and XRL. It is revealed that MP works as an excellent beam intensity monitor as well as provides an intense beam for XRL applications.

46.2 Sample Preparation

A Mo/Si multilayer was designed as follows: multilayer period of 10.4 nm; Mo-layer thickness/period of 0.35; the number of layers of 47; and the topmost layer of Mo. It was deposited on the 100-mm-square active area of an X-ray photodiode detector (AXUV100G) [4] as well as on a silicon wafer as a reference sample by ion beam sputtering at KPSI. The photodiode except for the active area was masked with polyimide tape during deposition so as to prevent adverse effect on the photodiode. X-ray reflectivity measurements of the reference sample were performed at 0.154 nm, and the period and thickness ratio were determined to be 10.36 nm and 0.34, respectively. Furthermore, another ordinary uncoated photodiode (AXUV100G) was prepared to be used for reflectance measurements of MP with SR and XRL.

46.3 Results and Discussion

The characterization of the MP was carried out by means of a soft X-ray reflectometer installed in the BL-11 beamline at the SR Center of Ritsumeikan University, Japan [7]. This beamline was operated by the Japan Atomic Energy Agency (JAEA) until 2012. Both the entrance and exit slits of the monochromator were set at 100 μm width and 3.6 mm height. A silicon filter of 0.5 μm thickness to reduce unwanted light was installed between the exit slit and a post-focusing mirror. The reflectance, R, of MP was defined as the ratio of the reflected light intensity from MP at the angle of incidence of 45° to the direct beam intensity, where both the

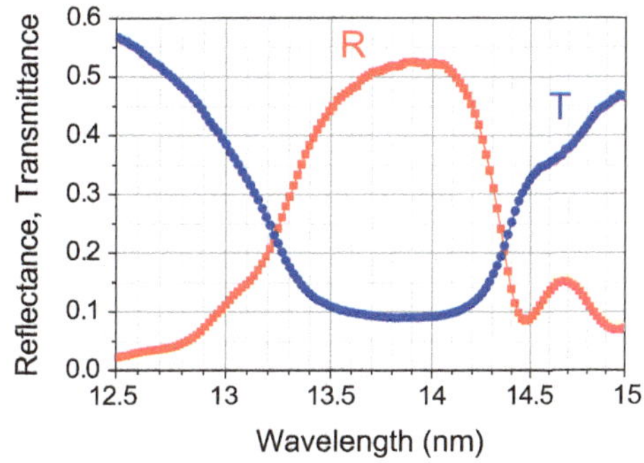

Fig. 46.1 Measured R and T of MP at 45° as a function of incident wavelength

intensities were measured with the ordinary photodiode, OP. On the other hand, the transmittance, T, of MP was defined as the ratio of the photodiode current generated by MP at 45° to that at normal incidence. It is noted that this definition is different from the literature [6]. Furthermore, the sensitivity correction factor, α, between the two independent photodiodes was defined as the ratio of the OP current to the MP current measured at normal incidence. Figure 46.1 shows the measured R and T of MP as a function of incident wavelength. We obtained $R = 52.5\%$, $T = 8.96\%$, and $\alpha = 2.91$ at 13.9 nm. Next, both the entrance and exit slit widths of the beamline monochromator were varied from 20 to 140 μm at 13.9 nm wavelength. The OP current, I_{OP}, versus the MP current, I_{MP}, is plotted in Fig. 46.2. It is found that there is a strong correlation between the two currents. This result means that I_{OP} can be calibrated using R, T, α, and I_{MP} as

$$I_{OP} = \alpha(R/T)I_{MP}. \tag{46.1}$$

As indicated by solid line in Fig. 46.2, the calibration curve obtained using (46.1) is in excellent agreement with the measured data.

Reflectance and transmittance measurements with XRL were performed by using a six-axis soft X-ray ellipsometer [8]. A single-target XRL beam from the second target having a beam divergence of ~10 mrad was focused by a Mo/Si

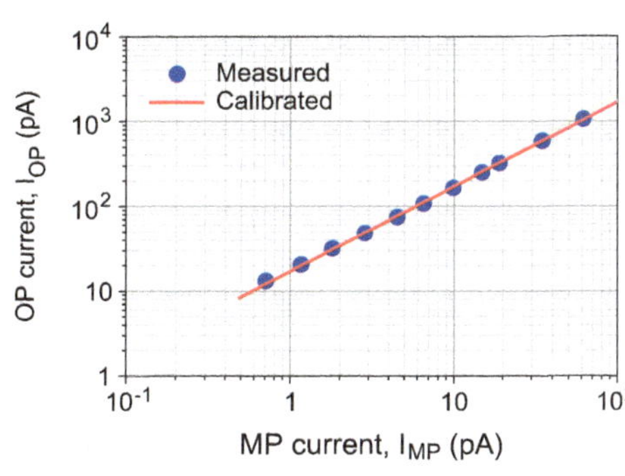

Fig. 46.2 Plot of the OP current, I_{OP}, versus MP current, I_{MP}

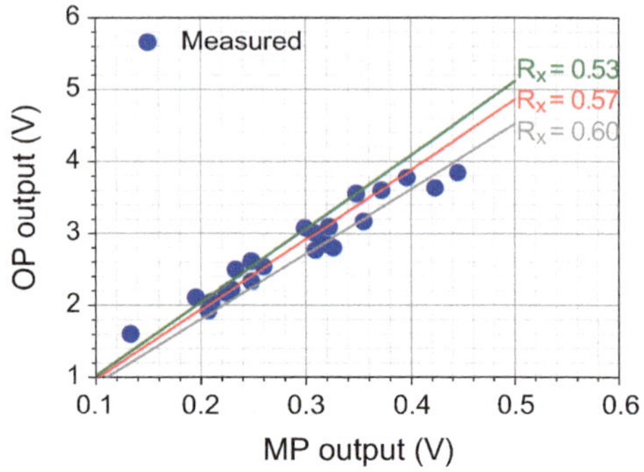

Fig. 46.3 Plot of OP output versus MP output. Solid lines are calculated by multiplying (46.1) by R_x

multilayer mirror having a 2-m radius of curvature placed at a distance of 2 m from the source point and was irradiated onto the active area of MP at 45°, on which the beam size was estimated to be 1.3 (H) × 1 (V) mm^2.

The reflected light from MP reached OP via another Mo/Si multilayer plane mirror with a reflectance of R_x which has not been characterized with SR but would be comparable to or higher than R. The photodiode output signals of MP and OP for 23 successive shots were measured by an oscilloscope (DPO4104B, Tektronix), of which the maximum values are plotted in Fig. 46.3. The correlation coefficient is calculated to be 0.965. Assuming $R_x = 0.57$, the calibration curve is successfully fitted to the measured data. In addition, the fact that the XRL beam intensity has a large variation, of which the ratio of the standard deviation to the mean is ~20% in Fig. 46.3, clearly shows that intensity monitoring is quite necessary for quantitative measurements with XRL.

46.4 Conclusions

It has been shown experimentally that the Mo/Si multilayer-coated photodiode, MP, has high reflectance at 13.9 nm, and there is a strong correlation between the reflectance and transmittance. In addition, it has been revealed that an XRL beam intensity can be derived from the measured MP output which corresponds to the transmittance. It is an evidence that MP can be used as an excellent beam intensity monitor as well as providing an intense beam for XRL applications. The validity has been confirmed in another experiment [8].

The author thanks QST staff for experimental support and fruitful discussion. This work was partially supported by JSPS KAKENHI (23760040 and 15K04685) and the JAEA President Grant (2010).

References

1. Tanaka, M., Nishikino, M., Kawachi, T., Hasegawa, N., Kado, M., Kishimoto, M., Nagashima, K., Kato, Y.: X-ray laser beam with diffraction-limited divergence generated with two gain media. Opt. Lett. **28**, 1680–1682 (2003)
2. Norman, G., Starikov, S., Stegailov, V., Fortov, V., Skobelev, I., Pikuz, T., Faenov, A., Tamotsu, S., Kato, Y., Ishino, M., Tanaka, M., Hasegawa, N., Nishikino, M., Ohba, T., Kaihori, T., Ochi, Y., Imazono, T., Fukuda, Y., Kando, M., Kawachi, T.: Nanomodification of gold surface by picosecond soft X-ray laser pulse. J. Appl. Phys. **112**, 013104 (2012)
3. Sobierajski, R., Loch, R.A., van de Kruijs, R.W.E., Louis, E., von Blanckenhagen, G., Gullikson, E.M., Siewert, F., Wawro, A., Bijkerk, F.: Mo/Si multilayer-coated amplitude-division beam splitters for XUV radiation sources. J. Synchrotron Radiat. **20**, 249–257 (2013)
4. http://optodiode.com/
5. Korde, R., Geist, J.: Quantum efficiency stability of silicon photodiodes. Appl. Opt. **26**, 5284–5290 (1987)
6. Kjornrattanawanich, B., Bajt, S., Seely, J.F.: Mo/B_4C/Si multilayer-coated photodiode with polarization sensitivity at an extreme-ultraviolet wavelength of 13.5 nm. Appl. Opt. **43**, 1082–1090 (2004)
7. Koike, M., Sano, K., Yoda, O., Harada, Y., Ishino, M., Moriya, N., Sasai, H., Takenaka, H., Gullikson, E.M., Mrowka, S., Jinno, M., Ueno, Y., Underwood, J.H., Namioka, T.: New evaluation beamline for soft X-ray optical elements. Rev. Sci. Instrum. **73**, 1541–1544 (2002)
8. Imazono, T., Koike, M.: Development of a compact polarization analysis apparatus for plasma soft X-ray laser. Thin Solid Films **571**, 513–516 (2014)

Chapter 47
Analysis of Reflection Signal from EUV Multilayer Mirror for Irradiation-Induced Damage Study

M. Ishino, S. Ichimaru, M. Hatayama, N. Hasegawa, S. Oku and M. Nishikino

Abstract To utilize intense X-ray pulse sources for scientific researches and industrial applications, it is important to study the damage mechanism caused by X-ray irradiation. The soft X-ray laser (SXRL) pulses can make damages on an extreme ultraviolet (EUV) multilayer mirror. When the SXRL pulse irradiates onto the EUV multilayer mirror, a part of input energy is reflected. The SXRL beam irradiated to a damage part, in this case, the reflected intensity is modified. If the pulse width is longer than process time of damage growth, we can observe the modulated intensity, which has been affected by the damage structure formed by the former part of duration. Then, it will be possible to confirm the start time of damage formation in the multilayer structure by use of X-ray streak camera technique.

47.1 Introduction

The soft X-ray laser (SXRL) pulses can make the nanometer scale ablation and/or modification structures [1–3]. The mechanisms of surface modifications induced by focused SXRL pulses were also investigated theoretically [1, 4]. The atomistic model reveals that the tensile stress created in materials by SXRL pulses can produce spallative ablation of the surface. The theoretical model also predicted that the damage formation in irradiated surface started in a few picosecond. The time-scale of this damage process is comparable to the pulse width of SXRL beam.

Recently, we examined the durability of extreme ultraviolet (EUV) multilayer mirrors against exposure of focused SXRL pulses. The SXRL pulse made damage structures on the multilayer surface. The observed image of SXRL beam reflected from damaged surface had diffraction patterns. We noticed that the image of the

M. Ishino (✉) · N. Hasegawa · M. Nishikino
National Institutes for Quantum and Radiological Science and Technology, Kizugawa, Kyoto, Japan
e-mail: ishino.masahiko@qst.go.jp

S. Ichimaru · M. Hatayama · S. Oku
NTT Advanced Technology Corporation, Atsugi, Kanagawa, Japan

T. Kawachi et al. (eds.), *X-Ray Lasers 2016*, Springer Proceedings in Physics 202, https://doi.org/10.1007/978-3-319-73025-7_47

reflect SXRL pulse, which was irradiated onto the fresh surface (i.e., non-damaged area), also had the diffraction patterns. This phenomenon means that if the pulse width is sufficiently longer than damage process, we can estimate the process time of degradation of the multilayer structure caused by the picosecond laser ablation.

In this chapter, we will discuss the possibility of the signal analysis of SXRL pulse, which reflects from EUV multilayer mirror, for the study of the laser-induced damage process.

47.2 Experimental Setup

Figure 47.1 shows a schematic diagram of the experimental setup. The SXRL pulse had a wavelength of 13.9 nm and a pulse width of 7 ps. The SXRL pulse was focused on the target surface using a spherical mirror, which was optimized for soft X-ray of 13.9 nm at a normal incidence. Total irradiation energy of SXRL pulse could be reduced by the insertion of Zr filters with various thicknesses, which were placed in front of the spherical mirror. The transmittance of 0.2 μm-thick Zr filter for 13.9 nm was approximately 48%. When a 0.2 μm-thick Zr filter was adopted, the total energy of the SXRL beam reached on the sample surface was approximately 48 nJ/pulse. In this case, the typical fluences on target surfaces were estimated to be 20 mJ/cm^2 [1, 2].

EUV multilayer mirrors, which were fabricated to reflect the soft X-ray of 13.9 nm near the normal incident angle, were used as irradiation targets. The part of the irradiated beam was reflected by the mirror, and the reflected beam was acquired by the X-ray CCD camera. In front of the CCD camera, Zr filter having the thickness of 0.5 μm was settled to reduce the beam intensity.

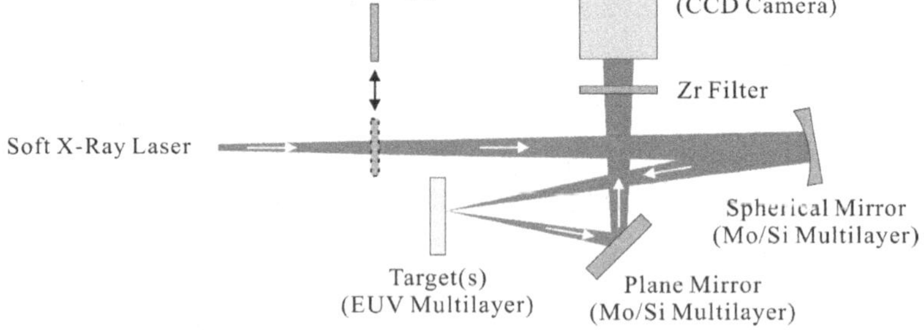

Fig. 47.1 Schematic diagram of the experimental setup. SXRL beam was focused on the multilayer surface by the spherical mirror, and the reflected beam was observed by the X-ray CCD camera

47.3 Results and Discussion

Figure 47.2 (left) shows damage structures induced by accumulation of SXRL pulses on the Mo/Si multilayer surface. In this irradiation series, Zr filter in front of the spherical mirror was not used. The numbers are indicated the total shots of pulses on each position. After the 20 shots of SXRL pulses, the damage structure with approximately of 50 μm diameter was formed. Figure 47.2 (right) shows the SXRL pattern captured by the CCD camera. This SXRL was reflected from the 50 μm diameter damage area. In the observed image, we can confirm the circular diffraction patterns, clearly. This result can be explained by the Babinet's principle [5]. We calculated $I = 0$ positions of Airy disks made at the CCD camera position, which formed by a 50 μm diameter hole settled at the target position, and we drew the calculated patterns on the images. The observed patterns and the calculated patterns have a good agreement each other.

Figure 47.3a shows the growth of reflection patterns from the Mo/Si multilayer mirror. In this irradiation process, ten pulses were accumulated on the same position. The circular diffraction patterns in the images became clear with increasing the accumulation number of SXRL pulses. Figure 47.3b shows the magnified image of the first shot of SXRL beam. Circular pattern can be seen even in the first signal. When we assumed that this pattern caused by the reflected intensity from the damage area, we could estimate the hole size to be about 17 μm from the simple calculation.

It was considered that this damage structure might be made by the first pulse itself. The former part of the beam pulse made a damage structure, and then the following later part of the pulse formed diffraction patterns on the CCD camera. If so, the reflection signal from the multilayer mirror contains a temporal information of the damage formation process, and then, it is possible to estimate the structural degradation time. This phenomenon means that if the pulse width is sufficiently

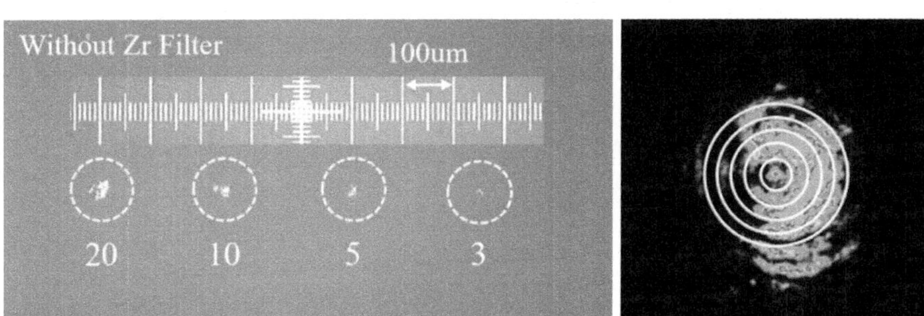

Fig. 47.2 (Left) Damages on the Mo/Si multilayer induced by accumulation of focused SXRL pulses. The indicated number of each damage is accumulation number of pulses. (Right) Reflection intensity from the damage surface, where was formed by 20 shots of pulses. Rings show the calculated $I = 0$ positions of Airy disks

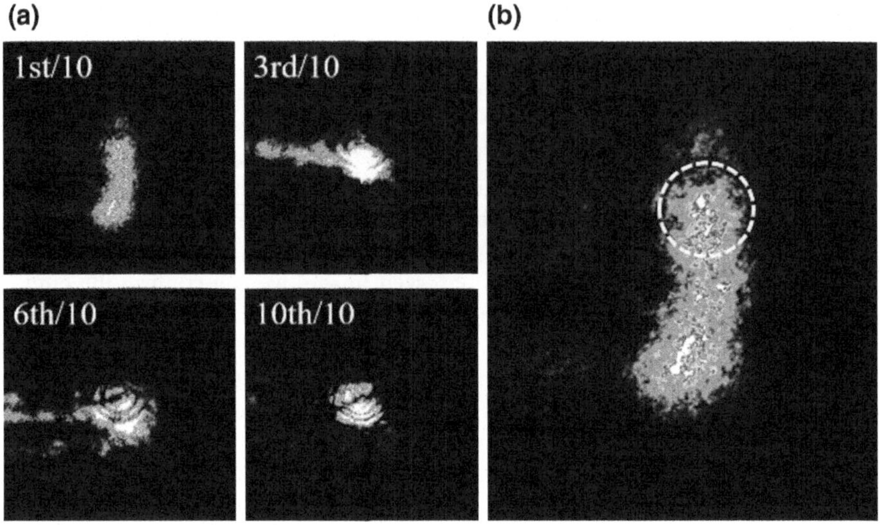

Fig. 47.3 **a** Reflection patterns of the SXRL beam (first, third, sixth, and tenth shots) from the same irradiation position on the Mo/Si multilayer. **b** Magnified the beam pattern of the first pulse

longer than damage process, we can observe the process time of degradation of the multilayer structure by use of the X-ray streak camera technique.

47.4 Summary

We observed SXRL beam patterns reflected from the Mo/Si multilayer mirror. In the observed image, we confirmed the diffraction patterns caused by the damage structure. Even in the first (single) shot of the focused SXRL beam pulse, we also confirmed the Airy disk like diffraction pattern. It is considered that this diffraction pattern was made by the damage structure induced by the SXRL pulse itself. This means that the start time of structural degradation of the multilayer mirror generated by the picosecond SXRL pulse can be confirmed by use of the X-ray streak camera technique.

References

1. Ishino, M., et al.: Nanoscale surface modifications and formation of conical structures at aluminum surface induced by single shot exposure of soft X-ray laser pulse. J. Appl. Phys. **109**, 013504 (2011)
2. Ishino, M., et al.: Observations of surface modifications induced by the multiple pulse irradiation using a soft picosecond X-ray laser beam. Appl. Phys. A **110**, 179–188 (2013)

3. Ishino, M., et al.: Very low electron temperature in warm dense matter formed by focused picosecond soft X-ray laser pulses. J. Appl. Phys. **116**, 183302 (2014)
4. Norman, G., et al.: Nanomodification of gold surface by picosecond soft X-ray laser pulse. J. Appl. Phys. **112**, 013104 (2012)
5. Hecht, E.: Optics, 3rd edn. Addison-Wesley (1998)

Chapter 48
Investigation of Surface Excitation Effect for Ablation of 4H-SiC Substrate Using Double-Pulse Beam

K. Matsunaga, T. Hayashi, S. Kurokawa, H. Yokoo, N. Hasegawa, M. Nishikino, T. Kumada, T. Otobe, Y. Matsukawa and Y. Takaya

Abstract The authors investigate a low-fluence laser processing system with a femtosecond double-pulse beam and surface excitation of a power semiconductor wafer. The double-pulse laser processing method enables a semiconductor surface to be processed at a lower fluence and prevents penetration damage of the processed surface. The first pulse of the double beam is considered to have a role in exciting the semiconductor surface to increase the efficiency of light energy absorption. In this report, to verify the feasibility of low-fluence processing, we measure the damage threshold in the low-fluence ablation process.

48.1 Introduction

Power semiconductor (i.e., SiC, C, and GaN) substrates exhibit characteristics such as high electric breakdown fields, high thermal conductivity, and stable operation at high temperatures [1, 2]. These characteristics are suitable for application to high-voltage and low-loss power devices [3–5].

However, they also cause the process of grinding and chemical mechanical polishing to be time-consuming in the semiconductor engineering field. There is a strong demand for developing the process, to achieve a high efficiency, high accuracy, and high processing efficiency of semiconductor substrates [6].

K. Matsunaga · T. Hayashi (✉) · S. Kurokawa · H. Yokoo · Y. Matsukawa
Department of Mechanical Engineering, Kyushu University, 744 Motooka, Nishi-ku, Fukuoka-shi, Fukuoka, Japan
e-mail: thayashi@mech.kyushu-u.ac.jp

N. Hasegawa · M. Nishikino · T. Kumada · T. Otobe
Quantum Beam Science Directorate, Kansai Photon Science Institute, Japan Atomic Energy Agency, Umemidai, Kizugawa, Kyoto, Japan

Y. Takaya
Department of Mechanical Engineering, Osaka University, 2-1 Yamadaoka, Suita, Osaka, Japan

T. Kawachi et al. (eds.), *X-Ray Lasers 2016*, Springer Proceedings in Physics 202, https://doi.org/10.1007/978-3-319-73025-7_48

The authors suggest a novel laser processing method for power semiconductor substrates by using low-fluence femtosecond laser pulses to prevent surface damage by penetration during laser ablation. In this study, the double-pulse laser processing method is applied to process a SiC substrate. The characteristics of this method are described below. The double-pulse beam has two peak pulses within several hundred picoseconds of each other. The main purpose of the first pulse is to excite the sample surface before the second pulse is irradiated, whereas. That of the second pulse is to process the photo-excited surface at low peak fluence.

If the photon energy is larger than the semiconductor band gap, the electrons in the valence band are excited to the conduction band, and then relax by Auger and radiative recombination within several picoseconds [7–9]. If the interval of the double pulse is short enough, it is possible for the second pulse to be transmitted to the sample surface before the decay of the photo-excitation effect.

In this report, we investigate the effect of processing the SiC photo-excited surface by using a low-fluence beam. In order to compare the ablation threshold between the photo-excited and non-excited surfaces, the SiC substrate is processed using the double-pulse beam with an increasing pulse interval between the first and second pulses. The peak fluence of the double pulse is maintained under the damage threshold of SiC using a single-pulse beam. These experiments enable to confirm the lowering effects of the peak fluence from processing the photo-excited surface. The ablation using the low-fluence beam is effective in reducing the penetration of the damage from the surface to the interior, and it is considered to improve the precision of the SiC substrate surface processing.

48.2 Experimental Setup

By using the double-pulse processing method, the second pulse of the double-pulse beam is irradiated before the decay of the photo-excitation effect due to the first pulse.

Figure 48.1 illustrates the setup of the double-pulse laser processing system. The light source is divided into two light paths. The divided paths have a role in photo-excitation (first pulse) and laser processing with the low peak fluence beam (second pulse). The pulse interval is controlled by adjusting the spatial position of the mirror (M2). Given the optical path difference of Δl, the mirror can maintain the pulse interval between the double-pulse at $\Delta t = 2\Delta l/c$, where c is the speed of light. The pulses are superimposed on the target surface after focusing by a lens (f = 250 mm). The pulse duration of the light source is 64 fs, the central wavelength is 790 nm and the repetition rate is 10 Hz.

The peak fluence of the beam can be controlled by using the half-wave plate before the polarizer (P1). The Gaussian laser beam is focused onto the front side of the sample to a spot diameter of 80 μm. The target is the Si polished surface (0001) of 4H-SiC, whose band gap is 3.26 eV, and the polarization direction of the laser is set to the (11-20) Si surface.

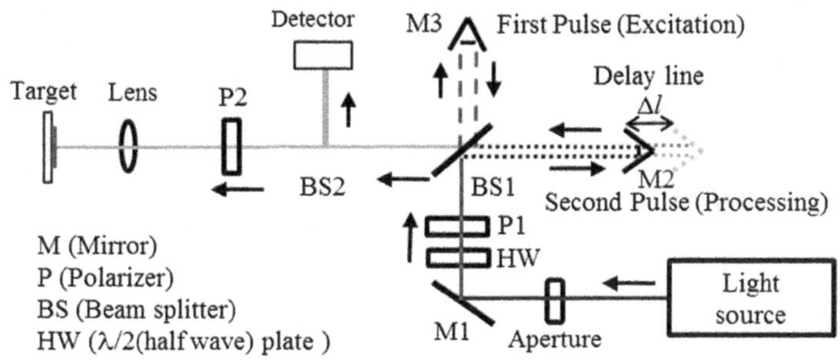

Fig. 48.1 Double-pulse laser processing system

48.3 Experimental Result

In order to compare the difference in the processing of the non-excited and photo-excited surfaces, two types of processing experiments are performed. One is single-pulse processing, where a single pulse is transmitted to the non-excited surface. The other is double-pulse processing, where the second pulse of the double-pulse beam is transmitted to the surface while maintaining the photo-excitation effect after the emission of the first pulse. As compared to the condition of the pulse energy in the single-pulse processing, the pulse energy of the double-pulse beam is equal to that of the single-pulse beam. The peak fluence of the first and second pulse is half the fluence of the single-pulse beam.

In this chapter, the damage threshold is defined as the minimum fluence of the ablation. It can be measured by analyzing the atomic force microscopy (AFM) image of the ablated crater.

First, the diameter of the focused laser beam is measured by the beam profile at the position of the sample surface, and it determines the spatial distribution of the fluence of the irradiated laser based on their Gaussian beam profile and pulse energy. Next, the cross-sectional profile of the processed sample is measured by AFM and determines the boundary of the ablated crater. Finally, the centers of both the spatial fluence and cross-sectional profiles of the ablated crater are combined, which enables us to determine the minimum fluence of the ablation at the boundary of the ablated crater.

48.3.1 Single-Pulse Processing

Figure 48.2 shows the damage threshold of single-pulse processing on the non-excited surface. The peak fluence of a Gaussian beam profile is adjusted from 1050 to 1600 mJ/cm^2 by changing the pulse energy from 25 to 36 μJ. The average damage threshold is determined as 990 mJ/cm^2 for the single-pulse processing.

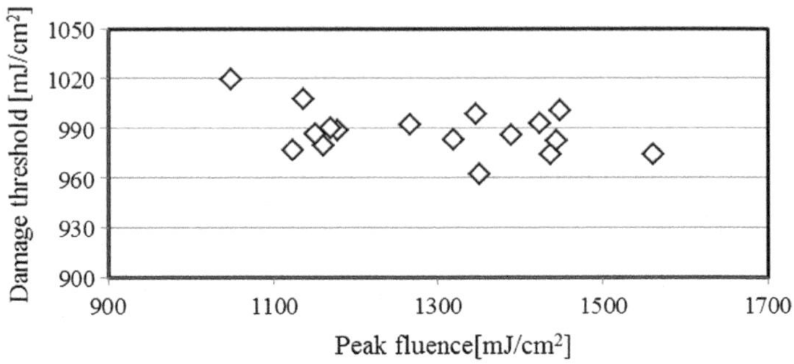

Fig. 48.2 SiC damage threshold of single-pulse processing

48.3.2 Double-Pulse Processing

In the double-pulse processing, the pulse interval is varied from 100 fs to 100 ps to investigate the dynamics of the photo-excitation effects. When the pulse interval is shorter than 100 fs, the interference between the first and second pulses makes the fluence destabilizing. The total pulse energy of the double pulse is in the range of 30–40 μJ. In this case, the SiC target cannot be processed by the first pulse itself because of the lower peak fluence on the damage threshold of SiC.

Figure 48.3 shows the damage threshold of the double-pulse processing (F_{th}-double: rhombus) with an increasing pulse interval. The plot presents the average damage threshold and the error bars represent their standard deviations. The damage threshold is measured from the cross-sectional profile of the ablated crater. Three samples, which are irradiated at the same fluence, are selected, and four spatial profiles of an ablated crater are selected on each sample. The total fluence of the double-pulse beam is higher than the damage threshold of the single-pulse beam (F_{th}-single: dot line).

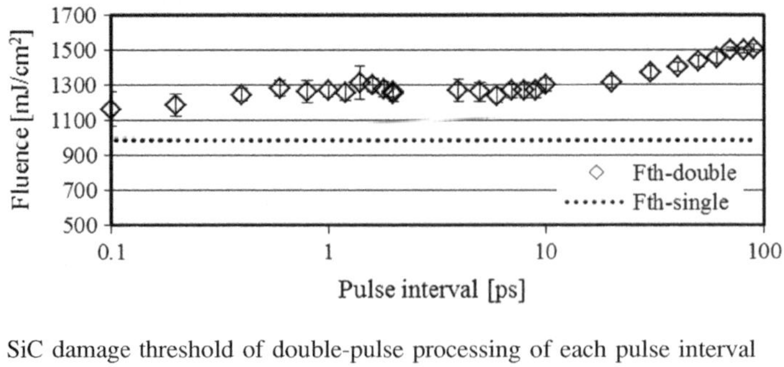

Fig. 48.3 SiC damage threshold of double-pulse processing of each pulse interval

Nevertheless, the peak fluence of the second pulse, which is used to process the photo-excited surface, is lower than the damage threshold of the SiC target. It is possible to process the photo-excited surface by using the low-fluence beam before the extinction of the photo-excitation effect.

48.4 Discussion

Figure 48.3 shows that the damage threshold gradually increases from 10 to 100 ps. A small increase in the damage threshold between at 0.1 and 1 ps can also be observed. The authors consider that it exhibits the extinction dynamics of photo-excitation, and thermal effect during double-pulse processing.

The first pulse is assumed to induce surface excitation with multi-photo ionization and transfer the carrier electrons in the conduction band, which can be the seed electrons for avalanche ionization [10, 11]. The energy of the second pulse is absorbed via both the multi-photon ionization and avalanche ionization process. The energy transfer process enables us to process the SiC at lower peak fluence.

48.5 Conclusion

To reduce the damage of laser processing for wide-band gap semiconductors, the target is processed using a low-fluence double-pulse beam.

In the double-pulse processing method, SiC can be processed with low-fluence beam. The photo-excitation effects can be maintained within the 10 ps of light emission.

The double-pulse processing method is applicable for processing a wide-band gap semiconductor with low-fluence beam. By controlling the area for photo-excitation, it becomes easy to change the processed area and depth.

It is considered that the double-pulse processing method can process the material without penetrating damage to deep inside from the ablated surface.

References

1. Morkoc, H., et al.: J. Appl. Phys. **76**(3), 1363–1398 (1994)
2. Matsunami, H., et al.: Mater. Sci. Eng. R: Rep. **20**(3), 125–166 (1997)
3. Cooper, J.A., et al.: IEEE Trans. Electron Devices **49**(4), 658–664 (2002)
4. Inoue, S., et al.: EEE Trans. Power Electron. **22**(2), 535–542 (2007)
5. Chung, G.Y., et al.: IEEE Electron Device Lett. **22**(4), 535–542 (2001)
6. Adelmann, B., Hellmann, R.: Appl. Phys. A **122**(7), 176–178 (2016)
7. Sundaram, S.K., Mazur, E.: Nat. Mater. **1**(4), 217–224 (2002)

8. Kumada, T., et al.: Appl. Phys. Lett. **108**(1), 011102 (2016)
9. Otobe, T., et al.: Phys. Rev. B **77**(16), 165104.8 (2008)
10. Vogel, A., et al.: Chem. Rev. **103**(2), 577–644 (2003)
11. Yu, X., et al.: Appl. Phys. Lett. **102**(10), 101111 (2013)

Chapter 49
Ablation of LiF and CsI by EUV Nanosecond Laser Pulse

O. Frolov, K. Kolacek, J. Schmidt, J. Straus, A. Choukourov and P. Pira

Abstract In this chapter, we present results of study an interaction of nanosecond EUV laser pulses at a wavelength of 46.9 nm with lithium fluoride (LiF), caesium iodide (CsI) and golden (Au) samples. The laser beam is focused with a spherical Si/Sc multilayer-coated mirror on samples. Samples were irradiated with 1, 2, 5, 10 and 20 laser shots with various energies. At the same time, laser ablation plumes were observed. Ablation depth of craters was analysed by atomic force microscope (AFM).

49.1 Introduction

The removal of material from solid surfaces by the technique of laser ablation has become increasingly important in a variety of applications such as advanced micromachining, surgery, X-ray laser generation, pulsed laser deposition (PLD), mass spectrometry of biomolecules, art cleaning/restoration and fundamental physics studies. Ablation with EUV radiation is much more effective than can occur with UV-Vis-NIR radiation of the same pulse energy and focal spot diameter.

Conventional laser with long wavelength was frequently used for preparation CsI thin films by PLD technics [1]. There are several strong motivations for deposition CsI on substrate. A thin layer of CsI could be used not only as a photocathode [2], but also as a traditional material for scintillation detectors [3].

O. Frolov (✉) · K. Kolacek · J. Schmidt · J. Straus
Pulse Plasma Systems Department, Institute of Plasma Physics of the Czech Academy of Sciences, Prague, Czech Republic
e-mail: frolov@ipp.cas.cz

A. Choukourov
Department of Macromolecular Physics, Faculty of Mathematics and Physics, Charles University in Prague, Prague, Czech Republic

P. Pira
Department of Spectroscopy, J. Heyrovsky Institute of Physical Chemistry of the Czech Academy of Sciences, Prague, Czech Republic

T. Kawachi et al. (eds.), *X-Ray Lasers 2016*, Springer Proceedings in Physics 202, https://doi.org/10.1007/978-3-319-73025-7_49

Recently, femtosecond visible laser was used for the non-thermal ablation of CsI [4]. Thin LiF coating could be interesting for several reasons. The LiF layer on metal electrodes [5] reduces their working function and causes more efficient electron injection and emission of field. Performance of organic electroluminescence devices has been improved due to better electron injection on cathodes with LiF coating [6]. Ablation of transparent dielectrics by ultrashort picosecond and femtosecond pulses [7] has been more successful in comparison with excimer lasers (laser radiation is absorbed nonlinear at high intensities). EUV and soft X-ray lasers are more efficient for ablation of such type of dielectrics due to better linear absorption of radiation by these crystals. Recently, LiF [8, 9] and CsI [10] samples were ablated by EUV capillary discharge laser (with a wavelength of 46.9 nm).

In the present chapter, we study not only ablation of Au, CsI and LiF samples by nanosecond laser pulses with a wavelength of 46.9 nm but also PLD of these materials on MgO substrate. Samples were irradiated with 1, 2, 5, 10 and 20 laser pulses in the range of energies from 60 to 150 μJ (all energies were estimated in a primary beam).

49.2 Experimental Setup

Capillary discharge driver CAPEX consists of a Marx generator, a coupling section, a coaxial pulse forming line, a self-breakdown spark gap and a ceramic capillary (Fig. 49.1). At present, amplitude and duration of pre-ionisation current are 100 A and 20 μs and the main capillary current is typically 15–20 kA. For experiments with ablation and PLD the apparatus was extended to vacuum chamber with a spherical multilayer mirror (with reflectivity of about 13%), interaction tube with samples and for registered detectors. A detailed description of our capillary discharge driver CAPEX can be found in the early-published papers [11–13]. Laser pulse energy has been defined from measurements of intensity of EUV radiation with Al filters of different thickness [14, 15].

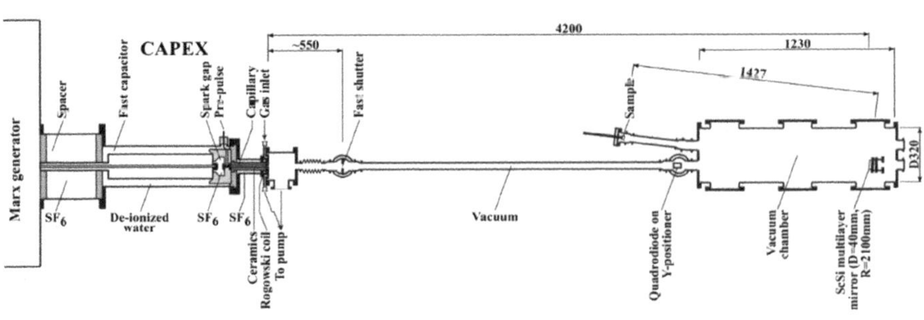

Fig. 49.1 Experimental apparatus CAPEX with extension for ablation experiments

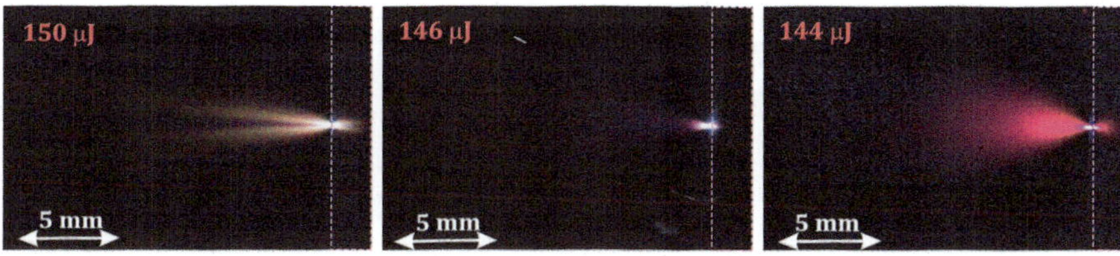

Fig. 49.2 Ablation plumes from layer of Au (left), CsI (middle) and LiF (right) for series with "high" energy

49.3 Results

In this work, visible ablation plumes from layer of Au (thickness of 50 nm) on PMMA, CsI and LiF samples were studied for PLD purpose (Fig. 49.2). All materials were irradiated in two series with 'low' (about 90 μJ) and 'high' (about 145 μJ) energy. In the case of thin layer of Au all material was ablated during the first laser shot. Ablated depth of craters from the LiF sample estimated from AFM images is shown on Fig. 49.3. Despite that ablated plume from CsI was visible, the AFM analysis of this sample was not successful and depth of ablated craters wasn't measured yet. Nevertheless, it found that all these materials are suitable for PLD. MgO substrate was placed parallel to the samples (distance about 1 mm). The angles between laser beam and substrate were shown in Fig. 49.4. In the case of Au, the substrate was deposited by single pulse and 10 pulses. LiF and CsI samples were irradiated by 1, 2, 5, 10, 20, 40 and 80 laser pulses. The deposited area is clearly visible optically in the case of Au and CsI.

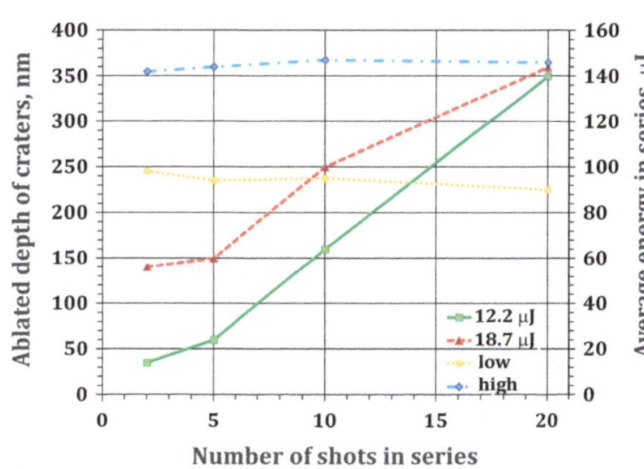

Fig. 49.3 Depth of ablated craters on LiF sample

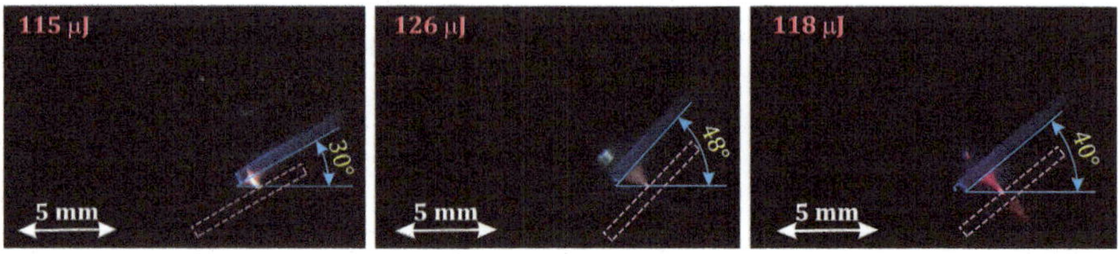

Fig. 49.4 PLD from Au (left), CsI (middle) and LiF (right) samples on MgO substrate

49.4 Conclusions

In this chapter we investigated the potential use of EUV ablation of Au, LiF and CsI (which is difficult to ablate by UV–Vis–NIR lasers) for PLD. It turned out that ablated plumes from all these materials are clearly visible. Deposition area from Au and CsI are visible even with the naked eye. In more detailed analysis of deposited substrates, it will be realised in the near future.

Acknowledgements This work supported by the Czech Science Foundation under Contract 14-29772S and by the Grant Agency of the Ministry of Education, Youth and Sports of the Czech Republic under Contract LG15013.

References

1. Fernandez-Lima, F., Collado, V.M., Ponciano, C.R., Farenzena, L.S., Pedrero, E., Da Silveira, E.F.: Laser ablation of CsI analyzed by delayed extraction. Appl. Surf. Sci. **217**(1–4), 202–209 (2003)
2. Buzulutskov, A., Breskin, A., Chehik, R., Va'vra, J.: Study of photocathode protection with thin dielectric films. Nucl. Instrum. Method A **371**, 147–150 (1996)
3. Nikl, M.: Scintillation detectors for X-rays. Meas. Sci. Technol. **17**, R37–R54 (2006)
4. Hada, M., Zhang, D.F., Pichugin, K., Hirscht, J., Kochman, M.A., Hayes, S.A., Manz, S., Gengler, R.Y.N., Wann, D.A., Seki, T., Moriena, G., Morrison, C.A., Matsuo, J., Sciaini, G., Miller, R.J.D.: Cold ablation driven by localized forces in alkali halides. Nat. Commun. **5** (3863), 1–8 (2014)
5. Schlaf, R., Parkinson, B.A., Lee, P.A., Nebesny, K.W., Jabbour, G., Kippelen, B., Peyghambarian, N., Armstrong, N.R.: Photoemission spectroscopy of LiF coated Al and Pt electrodes. J. Appl. Phys. **84**, 6729–6736 (1998)
6. Hung, L.S., Tang, C.W., Mason, M.G.: Enhanced electron injection in organic electroluminescence devices using an Al/LiF electrode. Appl. Phys. Lett. **70**, 152–154 (1997)
7. Henley, S.J., Fuge, G.M., Ashfold, M.N.R.: Comparing the short and ultrashort pulsed laser ablation of LiF. J. Appl. Phys. **97**, 023304-1–023304-9 (2005)
8. Ritucci, A., Tomassetti, G., Reale, A., Arrizza, L., Zuppella, P., Reale, L., Palladino, L., Flora, F., Bonfigli, F., Faenov, A., Pikuz, T., Kaiser, J., Nilsen, J., Jankowski, A.F.: Damage and ablation of large bandgap dielectrics induced by a 46.9 nm laser beam. Opt. Lett. **31**, 68–70 (2006)

9. Pira, P., Burian, T., Vysin, L., Chalupsky, J., Lancok, J., Wild, J., Strizik, M., Zelinger, Z., Rocca, J.J., et al.: Ablation of ionic crystals induced by capillary-discharge XUV laser. In: Proceedings of SPIE, vol. 8077, pp. 807719-1–807719-7 (2011)
10. Pira, P., Zelinger, Z., Burian, T., Vysin, L., Juha, L., Lancok, J., Danis, S., Nverly, V., Kudrna, P., Tichy, M., Grisham, M.E., Heinbuch, S., Rocca, J.J., Wild, J.: Ablation of CsI by XUV capillary discharge laser. Bull. APS **60**, 40 (2015)
11. Kolacek, K., Straus, J., Schmidt, J., Frolov, O., Prukner, V., Shukurov, A., Holy, V., Sobota, J., Fort, T.: Nano-structuring of solid surface by extreme ultraviolet Ar^{8+} laser. Laser Part. Beams **30**, 57–63 (2012)
12. Kolacek, K., Schmidt, J., Straus, J., Frolov, O., Prukner, V., Melich, R., Choukourov, A.: A new method of determination of ablation threshold contour in the spot of focused XUV laser beam of nanosecond duration. In: Proceedings of SPIE, vol. 8777, pp. 87770 N–87770 N (2013)
13. Frolov, O., Kolacek, K., Straus, J., Schmidt, J., Prukner, V., Choukourov, A.: Application of EUV optics to surface modification of materials. In: Proceedings of SPIE, vol. 8777, pp. 877707–877707 (2013)
14. Frolov, O., Kolacek, K., Schmidt, J., Straus, J.: Ablation plume induced by laser EUV radiation. Springer Proc. Phys. **169**, 397–403 (2016)
15. Kolacek, K., Schmidt, J., Straus, J., Frolov, O.: Calibration of windowless photodiode for extreme ultraviolet pulse energy measurement. Appl. Opt. **54**(35), 10454–10459 (2015)

Chapter 50
Laser-Induced Damage on Silica Glasses by Irradiation of Soft X-Ray Laser Pulse

K. Mikami, M. Ishino, T.-H. Dinh, Y. Yamamoto, N. Hasegawa, M. Nishikino, T. Kawachi, S. Motokoshi and T. Jitsuno

Abstract Laser-induced damage thresholds (LIDTs) on silica glass was evaluated with soft X-ray laser (SXRL) pulses. Silica glass is one of the most useful materials for near-infrared and visible laser systems. The high-photon energy of the SXRL would have great potential to study the laser-induced damage mechanisms. The LIDTs on silica glass measured by 13.9-nm, 7-ps SXRL pulses were several dozen of mJ/cm^2 as the experimental results. The damage morphologies are estimated by a simple 1D calculation.

50.1 Introduction

Laser-induced damage (LID) of optical components occurred often by irradiation of laser pulses and the output energy of high-power laser systems is limited by the laser-induced damage thresholds (LIDTs) of their optics. Laser-induced damage mechanism (LIDM) has been studied over 40 years to improve the LIDTs. Initial free electrons in optics are generated by laser pulse irradiation and then the electrons are multiplied by electron avalanche. When a number density of the multiplied electrons reaches a critical density, LID is occurred. The generation processes of initial free electrons by infrared- and visible-lasers are complex because of the photon energies less than the band-gap energies of the optics. Therefore, the LIDMs are modelled with many hypothesizes for the initial processes [1, 2]. Soft X-ray laser (SXRL) has a great potential to quantify the initial free electrons because of

K. Mikami (✉) · M. Ishino · T.-H. Dinh · Y. Yamamoto · N. Hasegawa · M. Nishikino
T. Kawachi
Kansai Photon Science Institute, National Institutes for Quantum and Radiological Science and Technology, Kyoto, Japan
e-mail: mikami.katsuhiro@qst.go.jp

S. Motokoshi
Institute for Laser Technology, Osaka, Japan

T. Jitsuno
Insititute of Laser Engineering, Osaka University, Osaka, Japan

T. Kawachi et al. (eds.), *X-Ray Lasers 2016*, Springer Proceedings in Physics 202, https://doi.org/10.1007/978-3-319-73025-7_50

the high-photon energy. The quantified initial free electrons will accelerate studies for the electron avalanche, dramatically.

In this study, we measured LIDTs on silica glass by 13.9-nm, 7-ps SXRL pulses and the LID morphologies were observed by a differential interference contrast microscope (DICM) and an atomic force microscope (AFM). The measured LIDTs and the LID morphologies were compared with a simple 1D calculation and the LIDM was extrapolated.

50.2 Experimental Setup and Results

A fully spatial coherent SXRL generated by TOPAZ (Twin OPtical Amplifiers using Zigzag slab) laser system [3] was used as damage testing laser and the TOPAZ was operated with 0.1 Hz pumped by two 10-J lasers. A wavelength and a pulse width of the SXRL were 13.9 nm and 7 ps, respectively. Irradiation energy of the SXRL was adjusted by the thickness of Zr filters. The SXRL was focused on a silica glass sample by Mo/Si concave mirror (curvature radius $R = 1000$ mm) with 1.5° incident angle and a diameter of the focusing spot was about 30 μm. The experimental sample was polished and the surface roughness was under 1 nm (RMS).

The LIDTs were measured with (a) 1-on-1, (b) 5-on-1, and (c) 10-on-1 testing scheme as shown in Fig. 50.1. About the testing scheme, for example ten-SXRL pulses irradiated on one evaluating the site and a presence of the laser-induced damage judged with DICM in the case of 10-on-1 testing. The LIDTs were defined as the maximum fluence without LIDs. The LIDTs were about 19 mJ/cm^2 regardless of the testing schemes and the LID probabilities were different clearly by the irradiation pulse numbers. Figure 50.2 shows the LID morphologies by

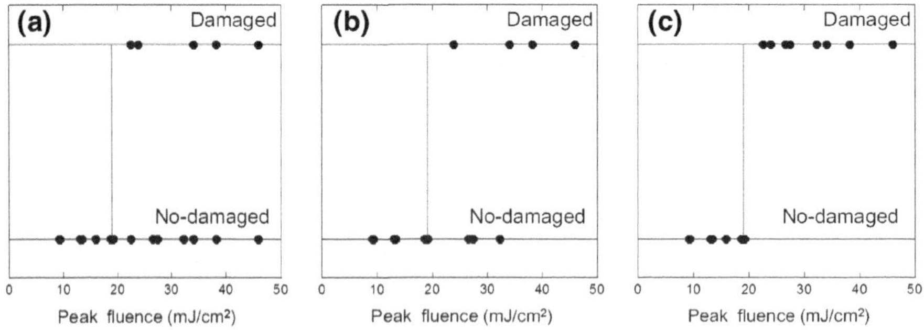

Fig. 50.1 Experimental results of damage testing with **a** 1-on-1, **b** 5-on-1 and **c** 10-on-1 testing schemes. Red lines indicate the LIDTs

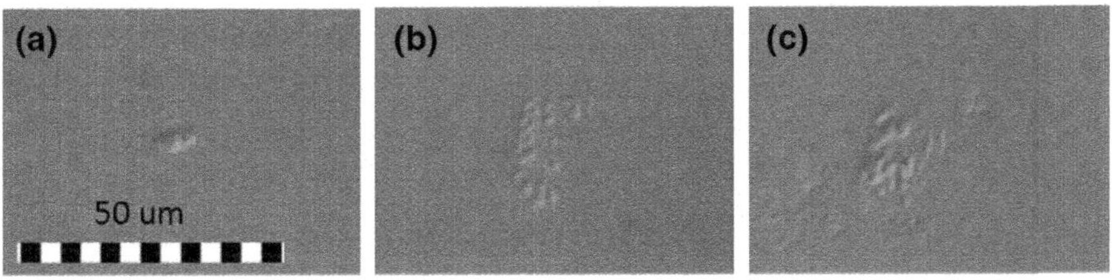

Fig. 50.2 LID morphologies observed by a differential interference contrast microscope with 38-mJ/cm^2 SXRL-pulse irradiation at **a** 1-on-1, **b** 5-on-1 and **c** 10-on-1 testing, respectively

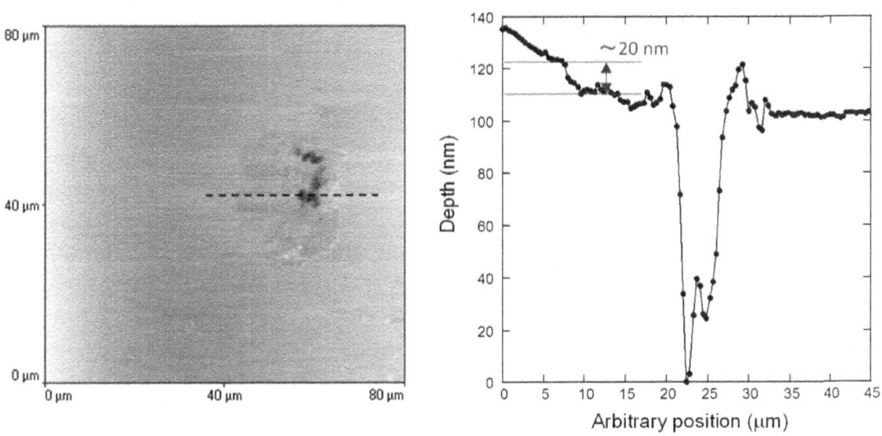

Fig. 50.3 LID morphology observed by an AFM with 34-mJ/cm^2 ten-SXRL-pulses irradiation (left image) and the depth profile as indicated dashed line in left image (right graph)

38.2-mJ/cm^2 SXRL-pulse irradiation at (a) 1-on-1, (b) 5-on-1 and (c) 10-on-1 testing schemes, and these were observed by the DICM. The LID morphologies were punctate depressions and the area of LID was increased with increasing irradiated pulse number. The left image in Fig. 50.3 shows AFM image of damage by 34-mJ/cm^2 and 10-pulses irradiation and a wide crater which cannot be observed with DICM was confirmed.

The maximum depths of the depression were about 100 nm and the depth of the wide crater was about 20 nm. The depths were decreased with decreasing the pulse number of pulses of the irradiation. In the case of 5-on-1 and 1-on-1 testing schemes, the wide crater depths were about 10 nm and 5 nm, respectively. Diameters of the craters were about 25 µm regardless of the irradiation pulse number.

50.3 Modelling and 1D Calculation

LIDT was defined with a conditional equation (50.1),

$$n_{mol} \leq n_{elec} = \beta \frac{N}{\pi r^2 L_{att}}, \quad (50.1)$$

where, n_{mol} is number density of SiO_2 molecule (~2.2×10^{22} cm^{-3}), n_{elec} is number density of free electrons generated by SXRL-pulse irradiation, β is electron multiple coefficient, N is number of free electrons generated by single photon absorption, r is radius of the irradiation spot size, and $L_{att} = 4.95$ nm is attenuation length of the SXRL for SiO_2. LIDs occurred when (50.1) is satisfied and the LIDTs were calculated by a number of generated free electron N as a function of irradiation laser fluence. The electron multiple coefficient β was varied for supposed LIDMs and the β was assumed with three cases, such as single photon absorption, energy distribution and multiple ionization. The photon energy of the SXRL-pulse has 89.3 eV (13.9-nm wavelength) and the band-gap energy of the silica glass is about 9 eV. In the case of single photon absorption, 1 of the SXRL photon would pump only 1 electron with 9 eV ($\beta = 1$). One photon energy of the SXRL would pump the maximum 9 electrons ($\beta = 9$) with energy balance between the photon energy and the band-gap energy under consideration with the linear optical phenomenon and the short attenuation length (i.e. high absorption coefficient). Finally, in the case

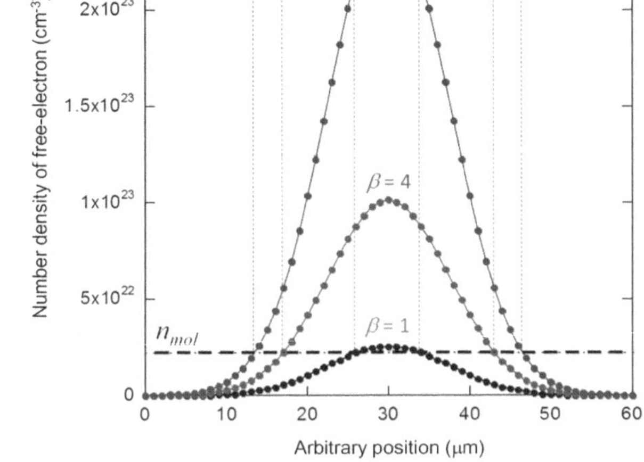

Fig. 50.4 1D calculation results for the distribution of the free electron number densities generated by SXRL-pulse irradiation with different electron multiple coefficient β

of the multiple ionization, 1 of the SXRL photon would generate four free electrons with 63.7 eV ($\beta = 4$), the summation of the first to third ionization energies of Si element and the first ionization energy of O element. All LIDMs were assumed that the surplus energies were deactivated without influence for LIDMs. The 1D calculation result of 34-mJ/cm^2 SXRL-pulse irradiation was shown in Fig. 50.4. The irradiation spot was assumed Gaussian profile having 30 μm in diameter (1/e^2). As a hypothesis, LID was occurred in 14 ps (2 times of the SXRL pulse width) and the electron dispersion was neglected. In the case of calculated result for a diameter over the n_{mol} with multiple ionization ($\beta = 4$, $d = 27$ μm), indicated a good agreement with the actual diameter of LID ($d = 25$ μm) occurred by irradiation of SXRL-pulse as shown in Fig. 50.3.

50.4 Conclusions

LIDTs of a silica glass were evaluated by 13.9-nm, 7-ps SXRL pulses. The LIDTs were the same regardless of the number of irradiation pulses. The wide crater in LID morphologies confirmed by an AFM. The damage area calculated by simple 1D modelling showed a good agreement with the experimental results and multiple ionization as a generation process of free electrons by a SXRL-pulse irradiation might be concluded. Therefore, a number density of the free electron in dielectric materials will be able to predict by the irradiation of the SXRL pulse and the quantifiable number density will be able to accelerate studies of LIDMs.

References

1. Sparks, M., et al.: Theory of electron-avalanche breakdown in solids. Phys. Rev. B **24**, 3519 (1981)
2. Mikami, K., et al.: Temperature dependence of laser-induced damage threshold of optical coatings at different pulse widths. Opt. Express **21**, 28719 (2013)
3. Nishikino, M., et al.: Characterization of a high-brilliance soft X-ray laser at 13.9 nm by use of an oscillator-amplifier configuration. Appl. Opt. **47**, 1129 (2008)

Part VII
EUV Lithography

Chapter 51
Performance of over 100 W HVM LPP-EUV Light Source

S. Okazaki, H. Nakarai, T. Abe, K. M. Nowak, Y. Kawasuji, H. Tanaka, Y. Watanabe, T. Hori, T. Kodama, Y. Shiraishi, T. Yanagida, G. Soumagrne, T. Yamada, T. Yamazaki, T. Saitou and H. Mizoguchi

Abstract Laser Produced Plasma (LPP) light source is widely accepted to be the most promising solution for the 13.5 nm high power light source for high volume manufacturing (HVM) EUV lithography. We have been developing LPP systems using CO_2 laser and tin (Sn) target for extreme ultraviolet (EUV) light source since 2003. In the development of LPP light source, various kinds of original technologies, such as the combination of pulsed CO_2 laser and Sn droplets, dual wavelength laser pulse shooting, and debris mitigation with magnetic field have been developed in Gigaphoton Inc.. By the use of Proto #2 device, 108 W EUV power (clean power at intermediate focus (I/F) in burst mode) under 80 kHz, 24 h operation was demonstrated. Currently, we are constructing the first practical source for HVM "GL200E Pilot #1" aiming for 250 W EUV output power by 27 kW pulsed CO_2 laser system.

51.1 Introduction

The development of ultra large-scale integration (ULSI) devices is promoted by the miniaturization of the minimum feature size of devices on a Si chip. Currently, ArF immersion exposure is used for the most advanced device fabrication with multiple patterning technologies. The purpose of the miniaturization is obtaining higher integration level, higher performance of the device, and lower cost. However, the adoption of the multiple patterning requires larger number of processing steps and longer processing time. The cost of the multiple patterning becomes very high

S. Okazaki (✉) · H. Mizoguchi
Gigaphoton Inc, 400 Yokokurashinden, Oyama Tochigi 323-8558, Japan
e-mail: shinjiokazaki@mbr.nifty.com

H. Nakarai · T. Abe · K. M. Nowak · Y. Kawasuji · H. Tanaka · Y. Watanabe · T. Hori
T. Kodama · Y. Shiraishi · T. Yanagida · G. Soumagrne · T. Yamada · T. Yamazaki
T. Saitou
Gigaphoton Inc, 3-25-1, Shinomiya Hiratsuka Kanagawa 254-8567, Japan

T. Kawachi et al. (eds.), *X-Ray Lasers 2016*, Springer Proceedings
in Physics 202, https://doi.org/10.1007/978-3-319-73025-7_51

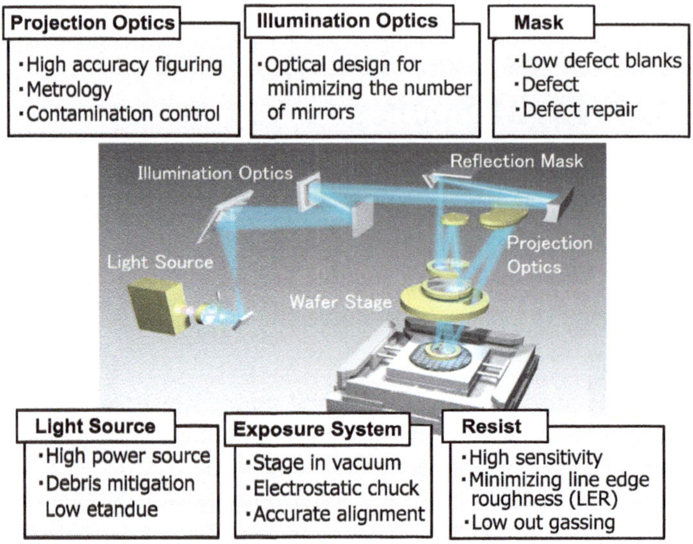

Fig. 51.1 Technical issues of EUVL

comparing to the conventional single exposure scheme [1]. According to such a situation, the introduction of extreme ultraviolet lithography (EUVL) is strongly required in the semiconductor industry.

Figure 51.1 shows the issues of EUVL. There are several technical issues to be developed for the industrialization. Every year, we have an international symposium on EUVL and discuss the focus areas. In these discussions, the output power is always thought to be the most critical.

Figure 51.2 shows the history of light source power development. As shown in this figure, before 2012, the output power of EUV light source was very low. Because, the ionization technology was not well controlled and the power of CO_2 driver laser was also low. By optimizing pre-pulse technology and developing higher driver laser output, the EUV output power dramatically improved after 2013.

EUV light source has been being developed together with exposure tool suppliers. At the first stage, the output power level was lower than 10 W. ASML shipped "alpha demo tool" in 2007 [2] and Nikon shipped "EUV-1" in 2008 [3]. ASML had developed NXE 3100 at the beginning of 2011 with higher output power light source [4, 5]. The actual power level was less than 20 W. ASML developed "NXE 3300" for the development of preproduction tool for HVM [6]. Around ten NXE 3300 series systems are working with 40–80 W in the world [7, 8].

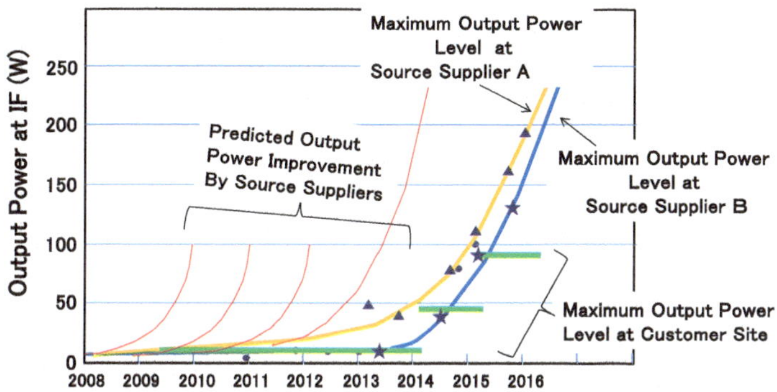

Fig. 51.2 History of output power of EUV light source

51.2 Development of Laser Produced Plasma EUV Light Source

51.2.1 Sn Debris Mitigation with Magnetic Field

Figure 51.3 shows the basic concept of our HVM EUV light source. Various kinds of theoretical studies and original technologies are developed [9, 10]. The theoretical [11] and experimental [12] results have clearly shown the combination of CO_2 laser and Sn plasma gives high conversion efficiency (CE) from the driver CO_2 laser to EUV light.

The Sn debris mitigation procedure with the magnetic field is a unique procedure in our system. Figure 51.4 shows the mitigation procedure. First, a Sn droplet is irradiated with a pre-pulse laser. By the irradiation of a pre-pulse laser, a Sn droplet is broken into submicron mist. The mist expands and forms a Sn cloud. Second, the

Fig. 51.3 Basic concept of gigaphoton's LPP source

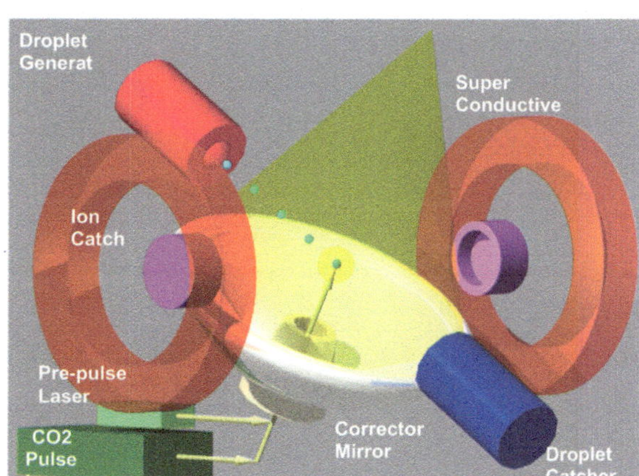

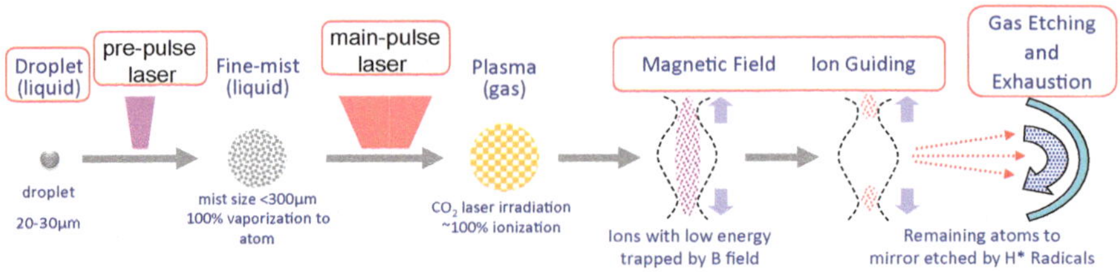

Fig. 51.4 Sn Debris mitigation scheme with magnetic field

cloud is irradiated by a pulsed CO_2 laser beam after a certain delay time. The Sn cloud becomes a high-temperature plasma.

Sn ions have several ionization states and emit EUV light during recombination process. With the magnetic field, most of the Sn ions are trapped and swept away from the chamber. Even the magnetic mitigation procedure, there are residual Sn atoms and fragments which can deposit on the collector mirrors. This residual debris can be etched away in the photochemical reaction of the environmental H_2 gas with EUV light.

51.2.2 Pre-pulse Technology [13, 14]

Conversion efficiency (CE) of 4.7% with 20 μm in diameter droplet has been demonstrated in the experimental device by optimizing the pre-pulse laser conditions as shown in Fig. 51.5. The effects of pre-pulse length are also examined and found that a pulse length of 10 ps gives higher CE than that of 10 ns.

Fig. 51.5 Conversion efficiency optimization with pre-pulse condition

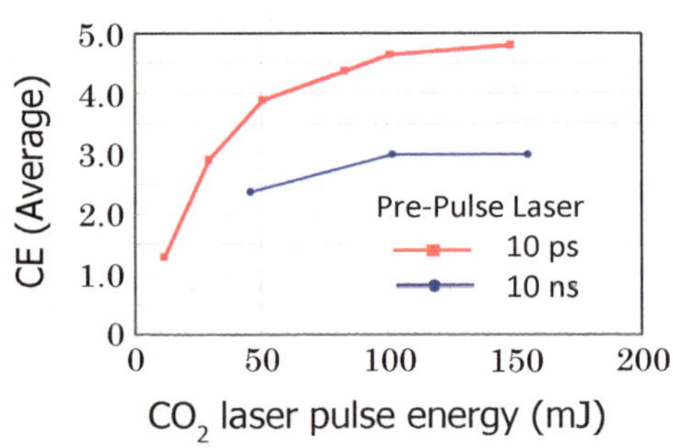

51.2.3 IR Reduction Technology on Collector Mirror

EUV light is collected by multilayer covered collector mirror. However, EUV plasma emits not only EUV light but also ultraviolet (UV) light, visible light, and infrared (IR) light, respectively. These light components are called out of band light. Within these out of band lights, IR light mainly heat up the collector mirror. So IR light must be eliminated before the intermediate focus (I/F) point. To eliminate IR light, we put diffraction grating structures fitted for IR light on the collector mirror surface. By the interference effects, we can selectively diffract IR light and eliminate it at the intermediate focus point as shown in Fig. 51.6 [15].

51.2.4 Experimental Results of Prototype System

Currently, we are operating two prototype devices and constructing a pilot device. Major differences between these three devices are the CO_2 laser power and the output angle. The detailed specifications are listed in Table 51.1.

Proto #1 device is mainly used for the development of debris mitigation procedure. The CO_2 laser power in this device is around 5 kW and the output angle of the EUV emission is horizontal.

Proto #2 device is mainly used for obtaining higher EUV output power. The CO_2 laser power of this device is around 20 kW and the output angle of EUV emission is 62 degrees upward matched to the NXE EUV exposure system. Figure 51.7 shows the picture of proto #2 EUV chamber. The new CO_2 laser amplifier has been developed in cooperation with Mitsubishi Electric supported by NEDO [16, 17]. The preamplifier of the CO_2 laser system in Proto #2 is the system developed by Mitsubishi Electric [18].

Recently we demonstrated 108 W (2% in the band, clean power) with a feedback loop, 50% duty cycle and 24 h operation by proto #2 EUV source as shown in Fig. 51.8. During the operation, the typical dose stability was <0.5% within

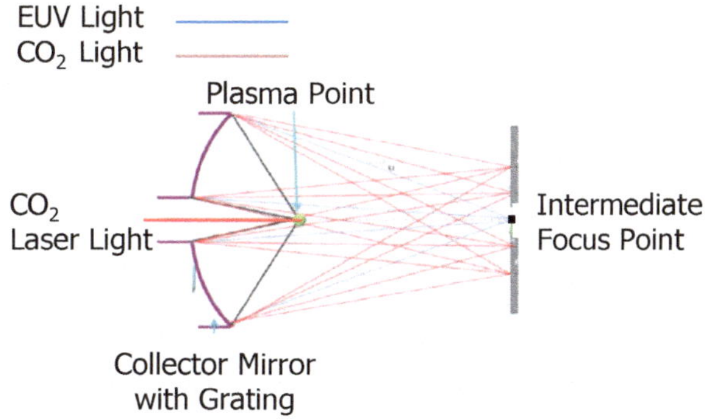

Fig. 51.6 Light rays from plasma to intermediate focus (IF)

Table 51.1 Specifications of gigaphoton EUV light sources

	Specification	Proto #1	Proto #2	Pilot #1
Target performance	EUV power (W)	25	>100	250
	CE (%)	3	3.5	4
	Output angle	Horizontal	62° upper	62° upper
	Availability	1 week operation	1 week operation	>75%
Technology	Droplet generator (μm)	20–25	20	<20
	CO_2 laser (kW)	5	20	27
	Pre-pulse laser	Pico-second	Pico-second	Pico-second
	Debris mitigation	Magnetic field	10 days	>3 month

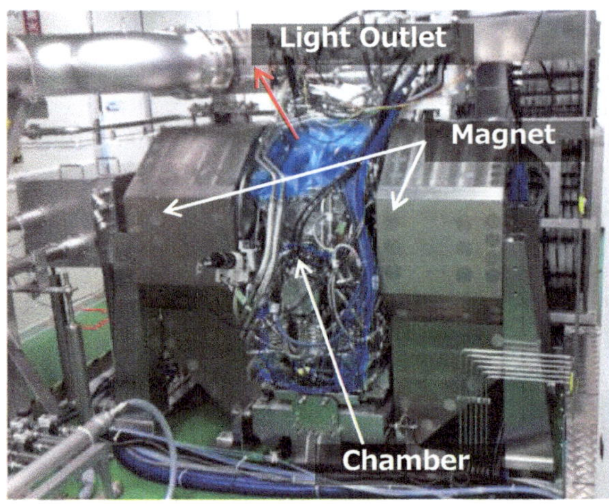

Fig. 51.7 Photograph of Proto #2 system

3 sigma, and the 20 kW CO_2 laser input power has been controlled between 40% and 60% of the full power.

51.3 New High Power HVM LPP-EUV Source

A new high power EUV light source named Pilot #1 device has been just constructed. The illustration of the device is shown in Fig. 51.9. The target specifications are listed in Table 51.1 along with the specification of Proto #1 and Proto #2 devices for comparison.

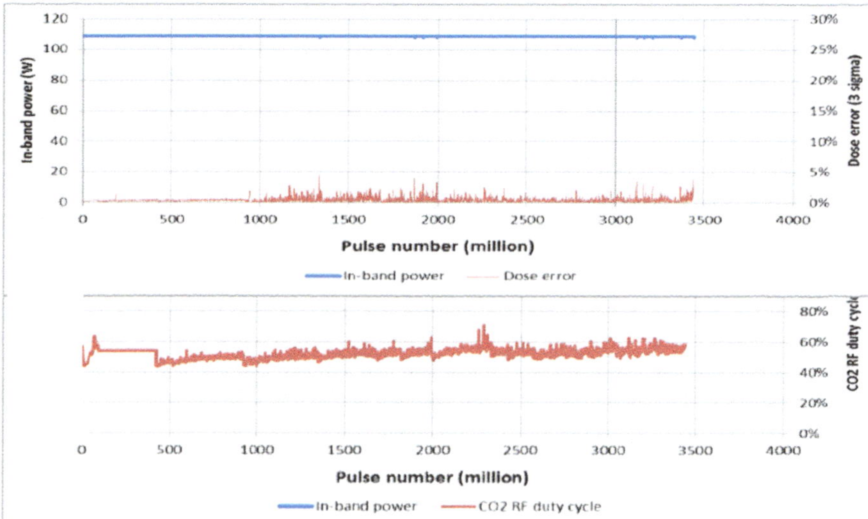

Fig. 51.8 24 h EUV source operation with 108 W in burst mode

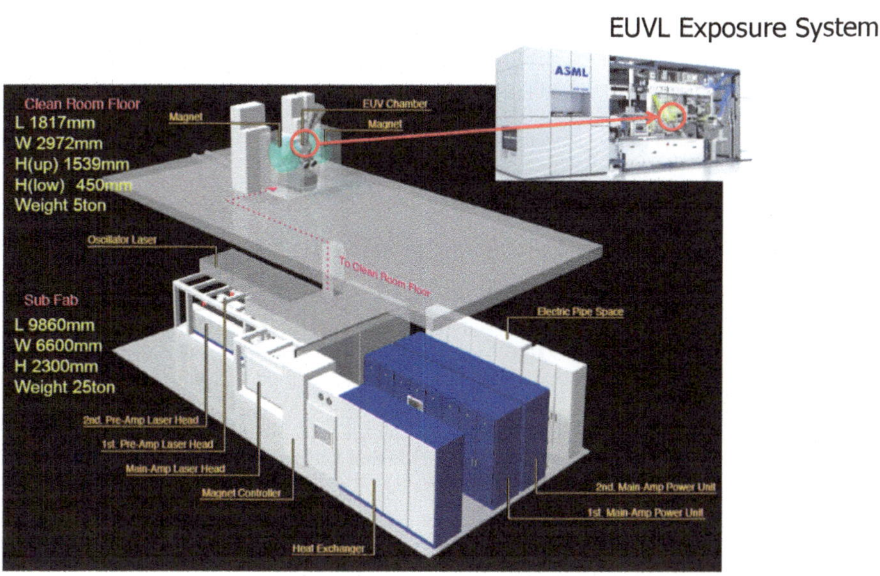

Fig. 51.9 Illustration of Pilot #1 device

51.4 Summary

We have developed Proto #1 device, Proto #2 device. Over 100 W operation was demonstrated with Proto #2 device. We obtained the output power of 108 W, 24 h at 80 kHz stable operation. The device is using a 20 kW CO_2 laser amplifier system consisting of a preamplifier by Mitsubishi Electric and three main amplifiers by TRUMPF. A new device named Pilot #1 has just been constructed. The target performance of pilot #1 device is 250 W EUV output power at I/F with stable operation.

Acknowledgements The work was partially supported by the New Energy and Industrial Technology Development Organization (NEDO) Japan. The authors are also grateful to Dr. J. Sunahara, Dr. K. Tomita, Prof. K. Uchino, and Dr. A. Endo.

References

1. Okazaki, S.: High resolution optical lithography or high throughput electron beam lithography: the technical struggle from the micro to the nano-fabrication evolution. Microelectron. Eng. **133**, 23 (2015)
2. Harned, N., et al.: EUV Lithography with the Alpha Demo Tools: status and challenges. In: Proceedings of the SPIE, 6517, 6517-06 (2007)
3. Miura, T., et al.: Nikon EUV1, development progress updates. In: Proceedings of SPIE No. 6921, pp. 6921-0 M (2008)
4. Wagner, C., et al.: EUV into production with ASML's NXE platform. In: Proceedings of the SPIE No. 7636, 7636-1H (2010)
5. Wagner, C., et al.: Performance validation of ASML's NXE3100. In: Proceedings of the SPIE No. 7969, 7969-49 (2011)
6. Peters, R., et al.: ASML's NXE platform performance and volume introduction. In: Proceedings of SPIE No. 8679, 8679-50 (2013)
7. J. J.H. Chen.: Progress on enabling EUV lithography for high volume manufacturing. In: 2015 EUVL Symposium (2015)
8. Philips, M.: EUVL readiness for 7 nm. In: 2015 EUVL Symposium (2015)
9. Endo, A., et al.: Laser produced EUV light source development for HVM. In: Proceedings of the SPIE No.6517, 65170-O (2007)
10. Mizoguchi, H., et al.: 100 W 1st generation laser-produced plasma light source system for HVM. In: Proceedings of the SPIE No.7969, 7969-08 (2011)
11. Nishihara, K., et al.: EUV Sources for Lithography, Chap. 11. In: Bakshi, V. (ed.) (2005)
12. Tanaka, H., et al.: Comparative study on emission characteristics of extreme ultraviolet radiation from CO_2, and Nd:YAG laser-produced tin plasmas. Appl. Phys. Lett. **87**, 041503 (2005)
13. Yanagida, T., et al.: Characterization and optimization of tin particle mitigation and EUV conversion efficiency in a laser produced plasma EUV light source. In: Proceedings of the SPIE No.7969, 7969-100 (2011)
14. Fujimoto, J., et al.: Development of the reliable 20 kW class pulsed carbon dioxide laser system for LPP EUV light source. In: Proceedings of the SPIE No. 7969, 7969-99 (2011)
15. RIGAKU Technical Display: IR rejection collector optic manufacturing process. In: 2013 EUVL Symposium (2013)

16. Tanino, Y., et al.: A Proposal for an EUV light source using transverse flow CO_2 lasers. In: 2012 EUVL Symposium (2012)
17. Novak, K.M., et al.: EUV driver CO2 laser system using multi-line nano second pulse high stability master oscillator for Gigaphoton's EUV LPP system. In: 2013 EUVL Symposium (2013)
18. Mizoguchi, H., et al.: Update of one hundred watt HVM LPP-EUV source performance. In: 2015 EUVL Symposium

Chapter 52
EUV Free-Electron Laser Requirements for Semiconductor Manufacturing

Erik R. Hosler and Obert R. Wood II

Abstract Laser-produced plasma extreme-ultraviolet (EUV) sources currently power EUV lithography tools, supporting advanced semiconductor manufacturing research and development. However, a source with sufficient power to support high-volume manufacturing has yet to be realized, but the sheer number of installed tools indicates that the transition to EUV lithography is no longer a question of "if" but "when". After the initial insertion of EUV lithography into manufacturing, the prerequisite dose scaling with technology node will be steep and therefore drive the EUV cost-per layer. Free-electron lasers may offer a low-cost, high-power alternative and facilitate further expansion of lithographic capabilities, including the development of high-NA and high-throughput NA 0.33 scanners. Adaptation and development of existing scientific light source knowledge, components, and infrastructure for development of an industrial light source requires collaboration between the academic and manufacturing communities. Considerations for an integrated free-electron laser lithography light source are discussed, focusing on the specific needs of the semiconductor industry.

52.1 Introduction

Extreme-ultraviolet lithography (EUVL) is widely viewed as the next iteration in wavelength scaling for the semiconductor industry. Over the past decade, device scaling has largely been supported by 193 nm immersion lithography, but at the most advanced nodes the necessity for complicated integration and overlay/alignment schemes is driving EUVL insertion to high-volume manufacturing. Immersion lithography reached its physical direct patterning limit at 28 nm technology node dimensions, requiring a historic increase in process complexity through multi- and self-aligned patterning processes. As a result, both process complexity and per layer patterning cost continue to increase with each progressive

E. R. Hosler (✉) · O. R. Wood II
GLOBALFOUNDRIES, 400 Stone Break Road Extension, Malta, NY 12020, USA
e-mail: erik.hosler@globalfoundries.com

T. Kawachi et al. (eds.), *X-Ray Lasers 2016*, Springer Proceedings in Physics 202, https://doi.org/10.1007/978-3-319-73025-7_52

technology node. However, successive delays in EUVL insertion, since its initial proposed technology interception at the 65 nm node in 2005, may have limited EUVL single patterning as a direct means to scale to only a few technology nodes (assuming insertion at the 7 nm node). Successful and widespread utilization of EUVL therefore requires hardware/implementation alternatives to extend its applicability. Free-electron lasers (FELs) as an alternative light source offer the potential to further reduce EUV exposure cost, allowing expansion of its use beyond the most complex/critical layers and facilitating additional area cost scaling [1].

Examining these critical levels, typically in the middle-of-line of semiconductor manufacturing, illustrates the various multi-patterning techniques employed. At the most basic level, there are two types of multi-patterning: (1) multicolor and (2) self-aligned patterning. Figure 52.1 illustrates how these multi-patterning techniques could relate to the desired pattern. While multicolor techniques (e.g., repetition of the litho-etch, $(LE)^n$, process on a single layer) may closely replicate the designed layout, self-aligned techniques (e.g., self-aligned **d**ouble/**q**uadruple/**o**ctuple/... patterning (SA**x**P)) require different integration and design layouts to replicate the performance of a drawn structure/device. However, depending on the drawn dimensions, EUVL may be able to print the desired pattern in a single exposure.

Though the simplicity of EUVL patterning is appealing for semiconductor manufacturing flows, the exposure cost as well as current infrastructure issues (low-defectivity lithography masks, high-resolution photoresist materials, actinic inspection tools, etc.) creates a barrier to entry. Moreover, the reliability and productivity of existing EUV lithography scanners are somewhat unpredictable, though significant advances are expected in the future [2]. Figure 52.2 shows the approximate EUVL lithography costs versus a subset of mature multi-patterning techniques as a function of scanner productivity (based on photoresist dose) and EUV source power. Given the current 250 W EUV lithography source target for high-volume manufacturing (HVM) tools, cost parity can be expected with LE^3 assuming a photoresist dose between 30 and 35 mJ/cm^2. Concerns regarding

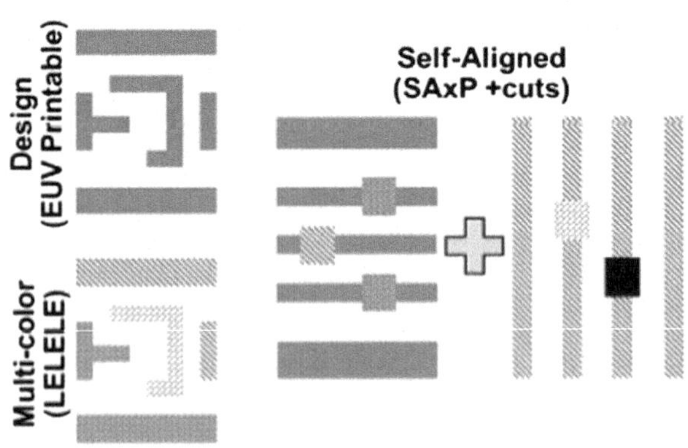

Fig. 52.1 Device layout illustration depicting how EUV lithography may simplify device patterning as compared to multicolor litho-etch or self-aligned patterning with cut masks techniques

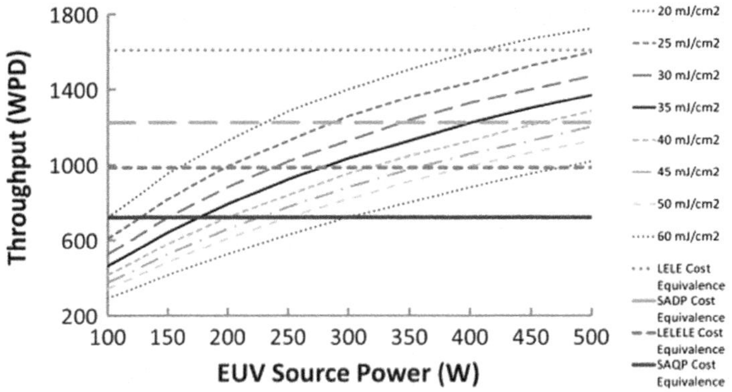

Fig. 52.2 EUV lithography scanner average wafer throughput per day (WPD) as a function of source power at select photoresist doses (mJ/cm^2) and benchmarked to various patterning schemes

productivity are further compounded by the tight photoresist patterning performance requirements (critical dimension, line-edge roughness, and etch selectivity) for upcoming technology nodes. Furthermore, to achieve the required patterning performance, it is expected that the photoresist dose will increase especially as the desired critical dimension continues to shrink [3].

In order to meet a technology node's roughness and resolution requirements by photoresist dose scaling and provide increased scanner productivity, the prerequisite EUV source power is expected to likewise increase [4]. Developing laser-produced plasma (LPP) sources [5] beyond 250 W has been theoretically explored and is feasible [6]; however, the cost of such light sources may not match the economic needs of the semiconductor industry. FELs offer a high-power, low-cost alternative to LPP scaling [1], having separately demonstrated both EUV emission [7] and high-power operation [8]. Should FELs be successful in lithography operations, they must be adapted from their current role as scientific light sources to provide for the specific needs of the semiconductor community, including robust, 24/7 operation, capability to power multiple end stations simultaneously, and adaptation to existing manufacturing sites.

52.2 EUV Free-Electron Laser Adaptation

Converting FELs from their current form, single-user scientific facilities, into tools for HVM will require extensive engagement between academia and industry on many issues. Here, only a few examples are discussed. First and foremost, integration of a large light source facility into semiconductor fab, where space is typically at a premium, must be determined for all end users. Second, FEL usability must be expanded from a single to a many end station facility. In a semiconductor fab, the end stations may constitute a variety of instruments, including lithography scanners, metrology/inspection tools, or failure analysis equipment. Here, multiple end station support is critical since primarily many scanners must be supported by a single FEL light source facility to justify the increased capital and operational costs versus LPP sources. Finally, the capabilities of an FEL light source must be fully reviewed by both communities to ensure full entitlement capture for an FEL/accelerator facility integrated with a manufacturing plant.

52.2.1 Fab Integration

Integration of an FEL light source into a semiconductor fab requires pre-emptive planning prior to any construction, let alone tool installation. Due to the sheer size of a 13.5 nm FEL (approx. 100 m × 20 m, operating at 800 MeV), placement of the facility must be external to the fab. Figure 52.3 shows an example integration of an FEL light source. Of particular importance is the alignment of the incoming EUV beam relative to the orientation of the lithography scanners (or other toolsets). Due to the relatively low reflectivity of EUV optics (~70% for multilayer polysilicon mirrors), minimization of the optical train is preferred and therefore places a restriction on the orientation of each element relative to the beam (grazing incidence or near normal). Moreover, the entry angle of the EUV beam into the scanner must match the current LPP source angle to ensure equivalent integration. Similar restriction applies to any existing toolset requiring EUV light.

In addition to mechanical integration of an EUV FEL into a fab, light source reliability must also be considered. To ensure 100% availability of the EUV beam to the fab lithography scanners, two FELs must operate simultaneously but independently to provide an indistinguishable source to all applications. Figure 52.3 shows an example of a redundant FEL light source integrated with a fab using a common mirror apparatus at near normal incidence angle. The caveat of such a configuration is that the distance between the FEL, defocusing grazing incidence optics, and collimating normal incidence optic is correlated with the FEL output beam power density and therefore dictates the size of the overall facility.

52.2.2 EUV/Electron Beam Distribution

Beyond integration of the FEL light source facility with a semiconductor fab, a means of supplying EUV light to each targeted end station must also be developed. This may be accomplished by either splitting the photon beam or the electron beam. Each splitting method has its advantages and disadvantages.

Splitting the photon beam simplifies scanner integration and reduces total facility cost, but the lack of high reflectivity 13.5 nm optics makes the entire system intrinsically lossy, increasing the FEL output and optic/longevity requirements. However, splitting the electron beam can be accomplished with high efficiency but

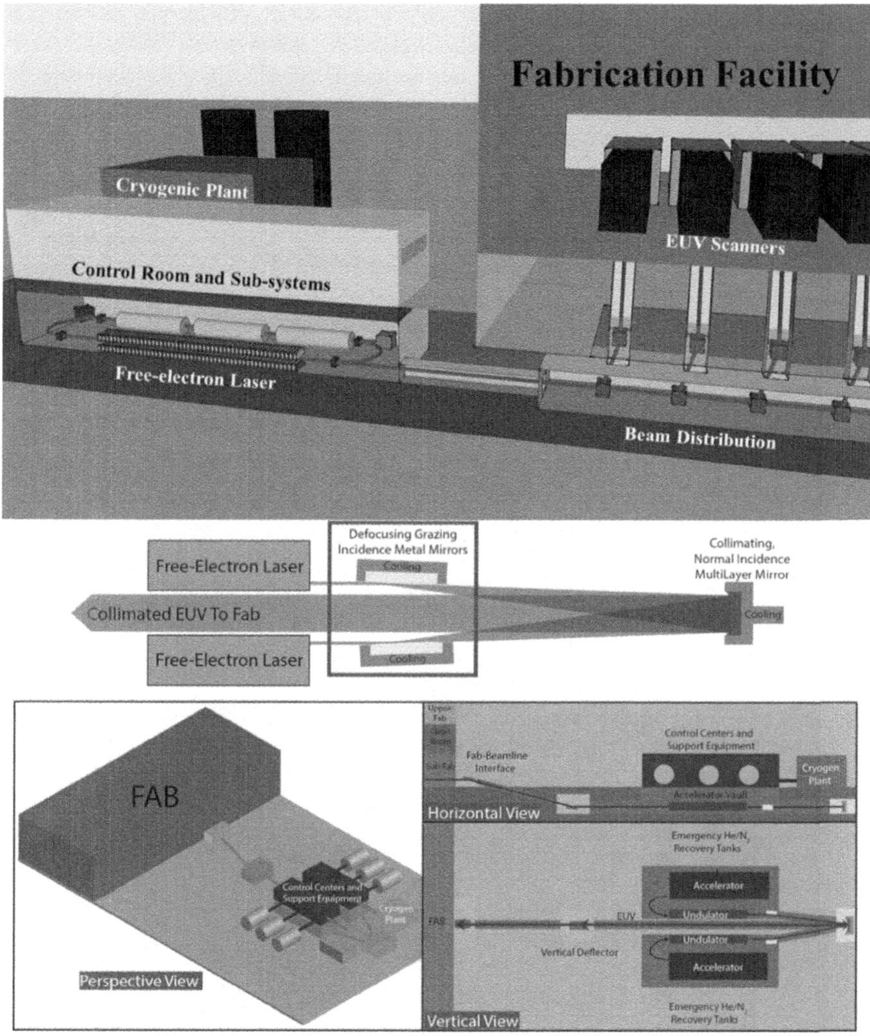

Fig. 52.3 (Top) Schematic diagram of a potential free-electron laser facility integrated with a semiconductor manufacturing facility [9]. (Bottom) Schematic layout of a redundant FEL light source facility and EUV common mirror apparatus (Middle) for delivering an indistinguishable EUV source to all end stations [10]

may require an equivalent number of undulator radiators to the number of electron beams, proportionally increasing the facility cost. Alternatively, an undulator elongated along the transverse axis of the electron beam propagation vector could facilitate multi-beam radiating from a single apparatus. Such a device would require precision machining but is not outside the capability of existing capabilities. Proposals for each beam distribution system are presented in Fig. 52.4 [11, 12].

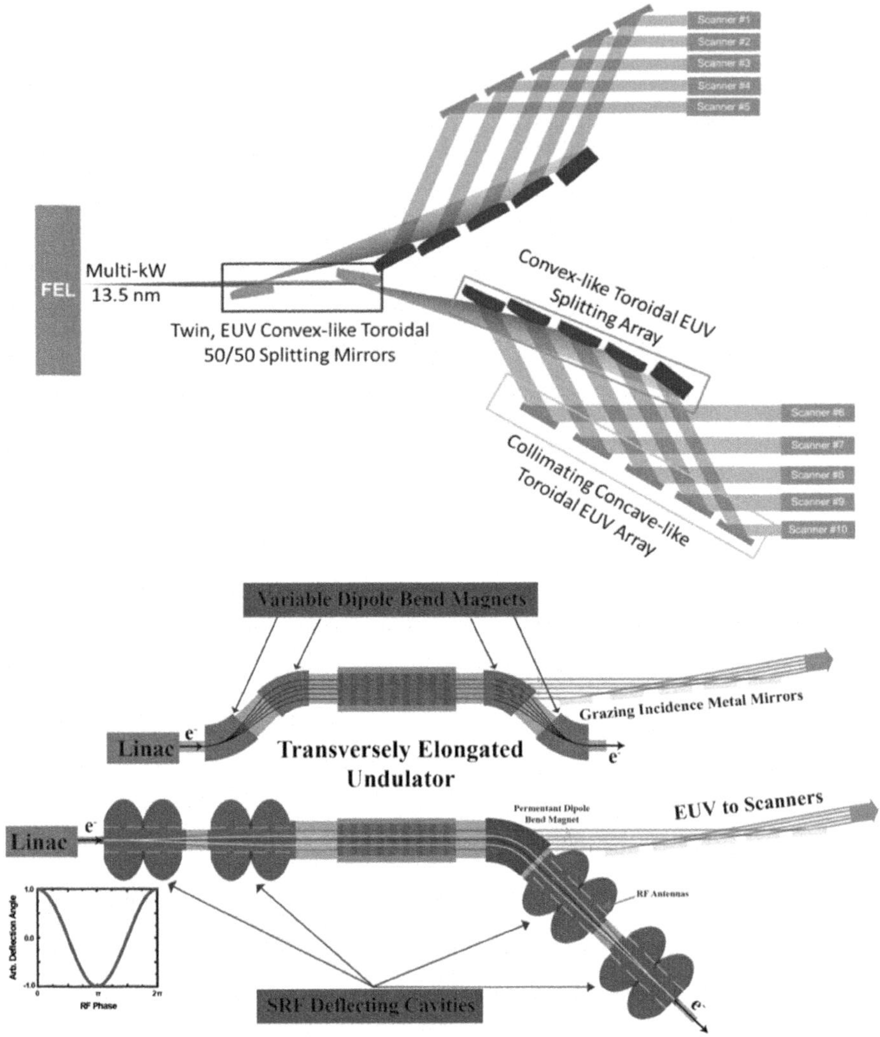

Fig. 52.4 (Top) Schematic diagram of a proposed EUV beam distribution apparatus for multiple end stations [11]. (Middle/Bottom) Schematic diagram of two proposed electron beam splitting apparatus around an undulator. (Middle) Electron beam splitting is accomplished by varying the magnetic field of a dipole magnet. (Bottom) Electron beam splitting is accomplished by a varying electric field in a radiofrequency cavity [12]

52.2.3 Potential Metrology Applications

Besides the capability of powering multiple applications simultaneously, an onsite accelerator facility offers many potential advantages to a semiconductor manufacturing operation, including a metrology source. A primary concern of EUVL

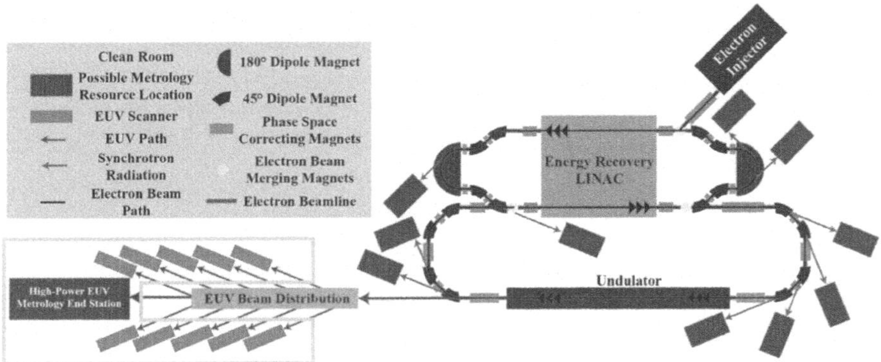

Fig. 52.5 Schematic diagram of an EUV free-electron laser facility incorporating multiple metrology end stations which utilize the synchrotron radiation from the recirculating bends of the accelerator as well as the EUV beam [14]

adoptees is the ability to inspect masks at the exposure wavelength. Current metrology sources lack the necessary brightness to meet the demands of a production environment. Rapid inspection requires a high-brightness, potentially coherent source which is intrinsically provided by an FEL given the emission scaling physics [13]. For lithography, the coherence of the FEL must be partially disrupted to prevent interference at the wafer plane in the photoresist; however, the direct output of the FEL can be used for imaging, elemental analysis, and various investigative applications. Furthermore, should an FEL's accelerator design involve an electron beam recirculation loop, the resulting synchrotron radiation may also be used for metrology or failure analysis instrumentation. An illustrated layout outlining potential metrology/inspection end station locations is shown in Fig. 52.5 for the case of an EUV energy recovery linear accelerator (linac) free-electron laser (EUV ERL FEL) [14]. Here, the FEL light source provides an economic means for EUVL as well as the inspection and control capabilities necessary to support an EUVL program.

52.3 Cost of Ownership

Central to semiconductor technology, scaling is cost reduction per unit area, node-on-node. As the number of lithography levels transferred from optical to EUV lithography increases, the productivity of EUV scanners must likewise increase. Also, as the device critical dimension of each new technology scale, both the NA of the scanner and the photoresist dose may also increase. These variables necessitate an increase in EUV source power. Though LPP sources may be scalable to higher

Tab. 52.1 Preliminary cost comparison between laser-produced plasma (LPP) and EUV energy recovery linear accelerator free-electron laser (EUV ERL FEL) lithography semiconductor manufacturing scenarios [1]

($M)		10 × 250 W LPP sources	EUV ERL FEL (10x scanners powered @ 1 kW)
Cost to power 10 scanners	OpEx	85	23
	CapEX	256	240
	Total cost First year	**>314**	**263**
	FEL CapEX savings		***16***
	FEL yearly OpEX savings		***62***
Uptime per source		Target 90%	∼100%
Average exposures per day (10 scanners, dose: 25 mJ/cm^2, 120 fields)		13,280	29,700

powers, there is concern regarding both the economics of power scaling and plasma/source stability. Alternatively, FELs have demonstrated performance at high power and wavelength, albeit separately [7, 8]. Should a high-power EUV FEL be demonstrated, the scaling economics are more favorable verses LPP. By design, FELs are intrinsically efficient, particularly for superconducting RF accelerators with cavity quality factors approaching greater than 10^{10} [15].

A preliminary cost comparison between LPP and EUV ERL FEL lithography programs is presented in Table 52.1 [1]. While the initial capital investment (CapEx) is comparable between the two programs, the annual operational expense (OpEx) savings for the FEL program is substantial. Calculating the per wafer cost bases between the two programs, the FEL-based program offers a clear advantage; however, the technology must first be proven feasible.

52.4 Conclusion

Semiconductor cost scaling has traditionally been accomplished by increasing pattern density through reducing the device critical dimension. As the fundamental atomic size limit is approached, cost saving will instead need to be accomplished through operation and equipment innovations. Driving EUV source economics and performance toward parity with conventional lithography tooling will be an essential achievement for enabling economic execution of the most advanced technologies.

References

1. Hosler, E.R., Wood, O.R., Barletta, W.A., Mangat, P., Preil, M.E.: Considerations for a free-electron laser based extreme-ultraviolet lithography program. In: Proceedings of the SPIE 9422, Extreme Ultraviolet (EUV) Lithography VI, 94220D (2015)
2. Schafgans, A.A., et al.: Performance optimization of MOPA pre-pulse LPP light source. In: Proceedings of the SPIE 9422, Extreme Ultraviolet (EUV) Lithography VI, 94220B (2015)
3. Brainard, R., Trefonas, P., Lammers, J., Cutler, C., Mackevich, J., Trefonas, A., Robertson, S.: Shot noise, LER, and quantum efficiency of EUV photoresists. In: Proceedings of the SPIE 5374, Emerging Lithographic Technologies VIII, 74 (2004)
4. van Schoot, J., et al.: EUV lithography scanner for sub-8 nm resolution. In: Proceedings of the SPIE 9422, Extreme Ultraviolet (EUV) Lithography VI, 94221F (2015)
5. Purvis, M.A., et al.: Advances in predictive plasma formation modeling. In: Proceedings of the SPIE 9776, Extreme Ultraviolet (EUV) Lithography VII, 97760K (2016)
6. Endo, A.: Extendibility evaluation of industrial EUV source technologies for kW average power and 6.x nm wavelength operation. J. Mod. Phys. **5**, 285 (2014)
7. Schneidmiller, E.A., et al.: Potential of the FLASH free electron laser technology for the construction of a kW-scale light source for next-generation lithography. J. Micro/Nanolithog. MEMS, MOEMS **11**, 021122 (2012)
8. Benson, S., et al.: High power operation of the JLab IR FEL driver accelerator. In: Proceedings of the 2007 Particle Accelerator Conference, p. 79 (2007)
9. Hosler, E.R. Wood, O.R., Preil, M.E: Extending extreme-UV lithography technology. SPIE Newsroom (2016)
10. Hosler, E.R.: Method, apparatus and system for using free-electron laser compatible EUV beam for semiconductor wafer processing. Patent number 9,392,679 (2016)
11. Hosler, E.R.: Method and device for splitting a high-power light beam to provide simultaneous sub-beams to photolithography scanners. Patent number 9,541,839 (2017)
12. Hosler, E.R., Preil, M.E.: Method, apparatus and system for providing multiple EUV beams for semiconductor processing. Patent application submitted (2015)
13. Saldin, E.L., Schneidmiller, E.A., Yurkov, M.V.: The Physics of Free Electron Lasers. Springer, Berlin, New York (2000)
14. Hosler, E.R., Mangat, P.: Method, apparatus and system for using free-electron laser compatible EUV beam for semiconductor wafer metrology. Patent application submitted (2015)
15. Drury, M., et al.: Performance of the first C100 cryomodules for the CEBAF 12 GeV upgrade project. In: Proceedings of the LINAC 2012, MOPB030, vol. 237 (2012)

Chapter 53
Coherent Lithography with Tabletop Soft X-Ray Lasers: Latest Achievements and Prospects

M. C. Marconi, W. Li, C. Kyaw, W. Rockward, C. S. Menoni and J. J. Rocca

Abstract We present the latest results in the development of a Talbot lithography technique utilizing tabletop SXR lasers. The main characteristics of the method are reviewed and a hybrid scheme that combines interferometric lithography with Talbot lithography is presented. This approach that we named Talbot Interference Lithography allows overcoming resolution limitations typical of the classical Talbot imaging.

53.1 Introduction

Photolithography schemes that require the utilization of coherent illumination are convenient ways to fabricate periodic nanostructures as compared to other methods like mask projection lithography or electron/ion beam lithography. The most direct approach to reduce the size of the smallest feature that can be printed is by reducing the wavelength of the illumination. In the extreme ultraviolet (EUV) and soft X-ray (SXR) spectral region, significant work has been done utilizing large photon facilities like synchrotron light sources [1–3]. However, these large photon facilities have the drawback of limited access, impairing a fast development of nanofabrication capabilities in small companies and university research groups.

The development of compact lasers capable to emit in the soft X-ray spectral region opened the possibility of implementing techniques that thus far were limited to synchrotrons. The highly coherent emission of tabletop SXR lasers makes them particularly well suited to implement coherent lithography schemes like interference lithography and Talbot lithography. Additionally, the high photon flux also enables

M. C. Marconi (✉) · W. Li · C. Kyaw · C. S. Menoni · J. J. Rocca
Engineering Research Center for EUV Science and Technology and Electrical and Computer Engineering Department, Colorado State University, Fort Collins, CO 80523, USA
e-mail: Mario.Marconi@colostate.edu

W. Rockward
Department of Physics, Morehouse College, Atlanta, GA 30314, USA

T. Kawachi et al. (eds.), *X-Ray Lasers 2016*, Springer Proceedings in Physics 202, https://doi.org/10.1007/978-3-319-73025-7_53

an efficient activation of the photoresist over large areas in very short exposure times.

Interference lithography is limited to pure periodic simple structures like lines, or dots arranged in a periodic lattice [4, 5]. An extra degree of freedom is added using the Talbot imaging effect, which enables the lithographic printing of arbitrary designs arranged in a periodic lattice [6]. The self-imaging phenomenon or Talbot imaging is illustrated in Fig. 53.1. A periodic diffraction mask is coherently illuminated. At certain distances along the beam path, the Talbot images replicate the mask structure. These distances are called Talbot distances (Z_T) and are determined by the wavelength λ and the periodicity of the mask *p* according to the expression $Z_T = N\frac{p^2}{\lambda}$ where N is an integer number.

Talbot lithography has unique characteristics that make this method a very robust approach. Because the Talbot imaging is the product of diffraction in the Fresnel approximation, there is not a one-to-one correspondence between the image and the Talbot mask. Thus, if any defect or imperfection develops in the mask, the image integrity is preserved and the defect is not present in the Talbot plane. This effect was studied and demonstrated to be a practical lithography application [7, 8].

The diffraction field after the periodic mask has a very rich structure. Besides the appearance of Talbot images at characteristic distances corresponding to the Talbot planes, an identical image also is observed at half the Talbot distance, but in this case displaced laterally by half a period. It is also observed that at distances corresponding to rational multiples of half the Talbot distance, $z = (p/q)Z_T/2$ where *p* and *q* are the co-prime numbers, replicas of the mask with increasing spatial frequency are obtained. This leads to the fractional Talbot effect, which produces images with a spatial frequency multiplication relative to the parent mask at

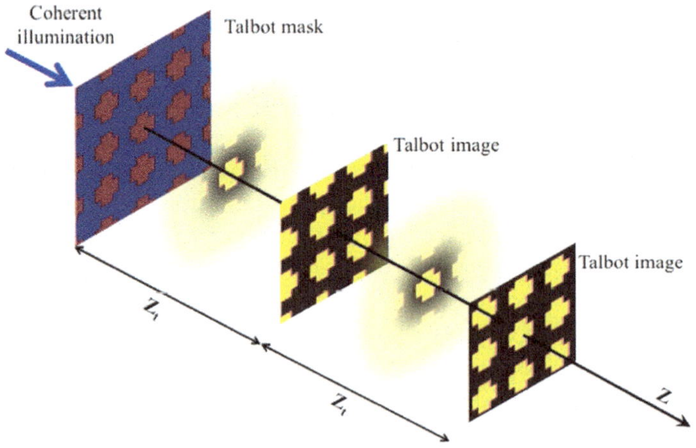

Fig. 53.1 Talbot imaging effect. A periodic diffraction mask produces replicas of itself at multiples of the Talbot distance Z_T

locations between the mask and the first Talbot plane. The fractional Talbot effect was used to obtain prints in photoresist with increased frequencies both for one- and two-dimensional Talbot masks [9, 10].

53.2 Hybrid Technique: Talbot Interference Lithography

Interference lithography has the advantage of simplicity and a relaxed depth of focus that facilitates the experimental setup. However, it is based on recording interference patterns and, consequently, it is limited to simple periodic structures, like lines and dots. On the other hand, Talbot lithography allows for printing of complex arbitrary designs arranged in a periodic lattice but has the inconvenience of a stringent depth of focus that adds an extra complication to the exposure system. A combination of both techniques can overcome these inconveniences. We have demonstrated a hybrid approach defined as Talbot Interference Lithography (TIL) by recording the interference pattern produced by two congruent Talbot images generated by the same mask under proper illumination. The scheme of the TIL setup is depicted in Fig. 53.2.

Two mutually coherent beams forming an angle θ with the optical axis illuminate a Talbot mask. Each beam produces a Talbot image in a common Talbot plane that generates interference. As a result, the final pattern obtained in the Talbot plane consists of the interference pattern created by the double beam illumination modulated by the periodic lattice generated by the Talbot image. With this approach, the minimum feature size that can be printed is determined only by the angle θ and the wavelength λ, while the larger sized periodic lattice, defined by the Talbot image, characterizes the periodic fields where the interference pattern is produced. With this method, it is possible to achieve a high resolution with a relaxed depth of focus, essentially, facilitating the experimental setup [11].

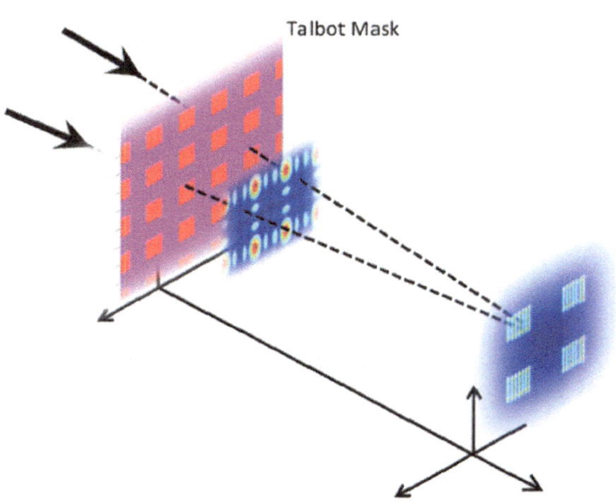

Fig. 53.2 Talbot Interference Lithography setup. A Talbot mask is coherently illuminated with two beams. The two Talbot images coincide in a Talbot plane where the interference is recorded in the photoresist

53.3 Experimental Results

For the experimental demonstration of the TIL method, we utilized a tabletop SXR capillary discharge laser. This compact source generates a highly coherent beam at $\lambda = 46.9$ nm with the sufficient photon flux to efficiently activate the photoresist in short exposure times, typically less than 1 min. The compact SXR laser produces pulses with an average energy of 400 μJ. The beam has a coherence radius of 550 μm measured at 1.5 m from the source, allowing for a full coherent illumination of the Talbot mask. Additionally, the SXR laser beam is highly temporal coherent with a narrow spectral bandwidth dominated by Doppler broadening, $\Delta\lambda/\lambda \leq 3\text{–}4 \times 10^{-5}$, corresponding to a temporal coherence length of 700 μm [12–14].

The two-beam illumination was achieved by splitting the laser beam with a grazing incidence flat mirror, in a Lloyd's configuration, as illustrated in Fig. 53.3a. The Talbot mask was a commercially available TEM grid having square holes $7.5 \times 7.5\ \mu m^2$ (Fig. 53.3b). The period of the Talbot mask was 12.5 μm, corresponding to a Talbot distance of 6.66 mm. The sample was a Si substrate spin coated with HSQ (Hydrogen silsesquioxane) photoresist located at the fifth Talbot plane. The pattern obtained at this location can be observed in Fig. 53.3c, with a zoom out inset. The square array of fields is produced by the Talbot imaging effect, while the fine lines in each one of the fields correspond to the interference of the two Talbot images.

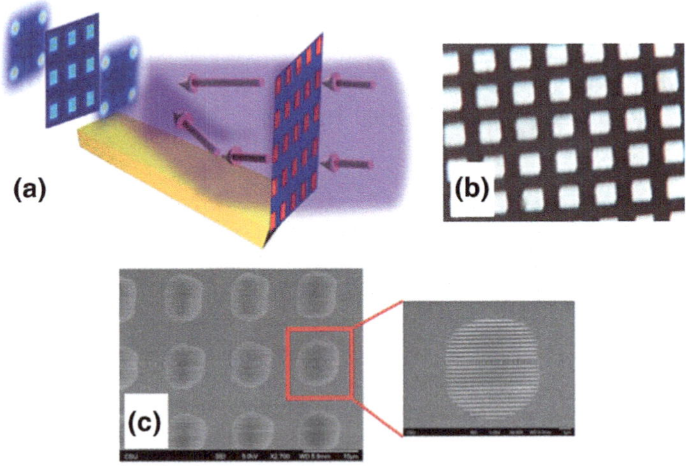

Fig. 53.3 **a** Loyd's mirror configuration to illuminate the Talbot mask with two mutually coherent beams. **b** Microscope image of the Talbot mask. **c** Print obtained at the fifth Talbot plane

53.4 Conclusions

The Talbot lithography method is a convenient lithography approach that is well adapted to the SXR spectral region, because it does not require special optics or sophisticated alignment. Features with arbitrary design arranged in a periodic lattice can, faithfully, be printed without errors and high resolution. The stringent experimental requirement consequence of the reduced depth of focus that accompanies a large numerical aperture mask is the main drawback of the method. However, combining the Talbot imaging effect with interferometric lithography, a hybrid approach—the Talbot Interference Lithography—can overcome this limitation, attaining high resolution with a much relaxed depth of focus requirement.

Acknowledgements We acknowledge the support from the National Science Foundation through award ECCS 1507907.

References

1. Langner, A., Solak, H.H., Gronheid, R., van Setten, E., Auzelyte, V., Ekinci Y., Schenau, K. V., Feenstra, K.: Measuring resist-induced contrast loss using EUV interference lithography. Extreme Ultrav. (Euv) Lithogr. **7636**, 76362x (2010)
2. Solak, H.H., David, C.: Patterning of circular structure arrays with interference lithography. J. Vac. Sci. Technol. B **21**, 2883–2887 (2003)
3. Solak, H.H., He, D., Li, W., Singh-Gasson, S., Cerrina, F., Sohn, B.H., Yang, X.M., Nealey, P.: Exposure of 38 nm period grating patterns with extreme ultraviolet interferometric lithography. Appl. Phys. Lett. **75**, 2328–2330 (1999)
4. Wachulak, P.W., Capeluto, M.G., Menoni, C.S., Rocca, J.J., Marconi, M.C.: Nanopatterning in a compact setup using table top extreme ultraviolet lasers. Opto-Electron. Rev. **16**, 444–450 (2008)
5. Wachulak, P.W., Urbanski, L., Capeluto, M.G., Hill, D., Rockward, W.S., Iemmi, C., Anderson, E.H., Menoni, C.S., Rocca, J.J., Marconi, M.C.: New opportunities in interferometric lithography using extreme ultraviolet tabletop lasers. J. Micro-Nanolithogr. MEMS MOEMS **8**, 021206 (2009)
6. Isoyan, A., Jiang, F., Cheng, Y.C., Cerrina, F., Wachulak, P., Urbanski, L., Rocca, J., Menoni, C., Marconi, M.: Talbot lithography: self-imaging of complex structures. J. Vac. Sci. Technol. B **27**, 2931–2937 (2009)
7. Li, W., Urbanski, L., Marconi, M.C.: Invited article: progress in coherent lithography using table-top extreme ultraviolet lasers. Rev. Sci. Instrum. **86**, 121301 (2015)
8. Urbanski, L., Isoyan, A., Stein, A., Rocca, J.J., Menoni, C.S., Marconi, M.C.: Defect-tolerant extreme ultraviolet nanoscale printing. Opt. Lett. **37**, 3633–3635 (2012)
9. Kim, H., Li, W., Danylyuk, S., Brocklesby, W.S., Marconi, M.C., Juschkin, L.: Optical properties of 2D fractional Talbot patterns under coherent EUV illumination. J. Phys. D-Appl. Phys. **48**, 375101 (2015)
10. Kim, H.S., Li, W., Danylyuk, S., Brocklesby, W.S., Marconi, M.C., Juschkin, L.: Fractional Talbot lithography with extreme ultraviolet light. Opt. Lett. **39**, 6969–6972 (2014)
11. Li, W., Marconi, M.C.: Extreme ultraviolet talbot interference lithography. Opt. Express (2015) (Submitted)

12. Liu, Y., Seminario, M., Tomasel, F.G., Chang, C., Rocca, J.J., Attwood, D.T.: Achievement of essentially full spatial coherence in a high-average-power soft-X-ray laser. Phys. Rev. A **63**, 033802 (2001)
13. Macchietto, C.D., Benware, B.R., Rocca, J.J.: Generation of millijoule-level soft-X-ray laser pulses at a 4-Hz repetition rate in a highly saturated tabletop capillary discharge amplifier. Opt. Lett. **24**, 1115–1117 (1999)
14. Urbanski, L., Marconi, M.C., Meng, L.M., Berrill, M., Guilbaud, O., Klisnick, A., Rocca, J.J.: Spectral linewidth of a Ne-like Ar capillary discharge soft-X-ray laser and its dependence on amplification beyond gain saturation. Phys. Rev. A **85**, 033837 (2012)

Chapter 54
A 10 Hz Short-Pulse CO_2 Laser System for Extreme Ultraviolet Source

R. Amano, T.-H. Dinh, A. Sasanuma, G. Arai, Y. Fujii, A. Takahashi, D. Nakamura, T. Okada, T. Miura and T. Higashiguchi

Abstract We produced 3–15 ns pulses at 10.6 μm wavelength to investigate efficient extreme ultraviolet and soft X-ray sources. A master oscillator power amplifier (MOPA) laser operating at a repetition rate of 10 Hz was built on transversely excited atmospheric (TEA) CO_2 mediums. A scheme of a pulse-width tunable CO_2 laser was proposed. The amplification by a 14-pass amplifier provided output energies of 150 and 60 mJ per pulse at durations of 15 ns and 3 ns, respectively. An initial observation of Gd plasmas produced by the CO_2 laser shows that the spectral efficiency around 6.7 nm was improved by short laser pulses irradiation.

54.1 Introduction

The great interest of laser-produced high-*Z* plasmas is that the overlap in adjacent ion stages could yield intense unresolved transition arrays (UTA) lying within reflectance bandwidth of a multilayer mirror. This scheme can realize a powerful microlight source in extreme ultraviolet (EUV) and soft X-ray (SXR) region [1, 2]. For efficient operation, it is important to produce a high-temperature plasma, while simultaneously reducing both self-absorption and hydrodynamic expansion loss effects. The reduction of self-absorption in low-density plasmas produced by CO_2

R. Amano · T.-H. Dinh (✉) · A. Sasanuma · G. Arai · Y. Fujii · T. Higashiguchi
Department of Innovation System Engineering & CORE, Utsunomiya University, Utsunomiya, Tochigi, Japan
e-mail: thanhhungvnjp@gmail.com

A. Takahashi
Department of Health Sciences, School of Medicine, Kyushu University, Fukuoka, Japan

D. Nakamura · T. Okada
Graduate School of Information Sciences and Electrical Engineering, Kyushu University, Fukuoka, Japan

T. Miura
HiLASE Centre, Za Radnicí 828, 252 41 Dolní Břežany, Czech Republic

T. Kawachi et al. (eds.), *X-Ray Lasers 2016*, Springer Proceedings in Physics 202, https://doi.org/10.1007/978-3-319-73025-7_54

(carbon dioxide) lasers has been investigated [3, 4]. Since the critical electron density n_{ec} depends on the laser wavelength, λ_L, i.e., $n_{ec} \propto \lambda_L^{-2}$, the critical density at a laser wavelength of $\lambda_L = 10.6$ μm for a CO_2 laser is two orders of magnitude smaller than at $\lambda_L = 1.06$ μm for the solid-state laser. However, to date, CO_2 LPPs were studied by using laser pulses longer than 15 ns. A question remains on behavior of high-Z plasmas produced by shorter CO_2 laser pulses.

In this study, we generate 3–15 ns pulses CO_2 laser at 10.6 μm, in order to investigate LPP sources for high-resolution lithography and lab-scale SXR imaging systems. We develop a pulse-width tunable CO_2 laser operating at a repetition rate of 10 Hz for fundamental experiments. Observation of CO_2 LPPs of Gd also is conducted.

54.2 Experimental Results

A master oscillator power amplifier (MOPA) laser was built on transversely excited atmospheric pressure (TEA) CO_2 gain mediums, which are commercial 1-J and 10-J lasers operated at a repetition rate of 10 Hz. The initial cavity mirrors of these laser units were replaced with ZnSe windows. The MOPA consists of an electro-optic (EO) *Q*-switch, a pulse shaper, and a 14-pass amplifier. An energy meter (Gentec-EO QE95LP-H-MB) and a photovoltaic sensor (Vigo PVM-10.6) featuring a response time below 1 ns were used to measure pulse energies and temporal profiles of the CO_2 laser output, respectively.

The master oscillator is a ring cavity, which can turn to be a regenerative amplifier in future. As shown in Fig. 54.1a, we achieved a pulse width of 15 ns with the oscillator. The pulse shaper is a long lifetime Ge semiconductor switch controlled by two Nd:YAG pulses at 1.06 μm. The control pulses were used to create a high density of free charge carriers on the Ge substrate. If carrier density exceeds the critical one, the Ge surface turns into a mirror reflecting the entering laser beam. At below critical density, the CO_2 laser beam is attenuated mostly by free carrier absorption. The initial Nd:YAG pulse was 5 mJ at durations of 150 ps, while the lateral was 40 mJ at duration of 100 ns. The 150-ps pulse with an intensity of 1.7×10^7 W/cm^2 was used to provide the mirror effect, and the 100-ns pulse at an intensity of 2.0×10^6 W/cm^2 was used to increase the shutter lifetime mainly by the absorption. The separation time between the two pulses was optimized to be 40–100 ns. Figure 54.1b shows a typical temporal profile of the CO_2 pulse sliced by a single pulse of 150 ps Nd:YAG laser. The fall time of less than 1 ns was observed, but a small part of the tail still remains due to the short lifetime of the carrier in the Ge substrate: typically few hundred picoseconds. By the double control pulses, the tail of the CO_2 laser pulse was completely blocked as shown in Fig. 54.1c. The minimum pulse width was observed to be 3 ns. This limitation in our laser system is mainly caused by the raise edge of laser pulse from the *Q*-switched oscillator.

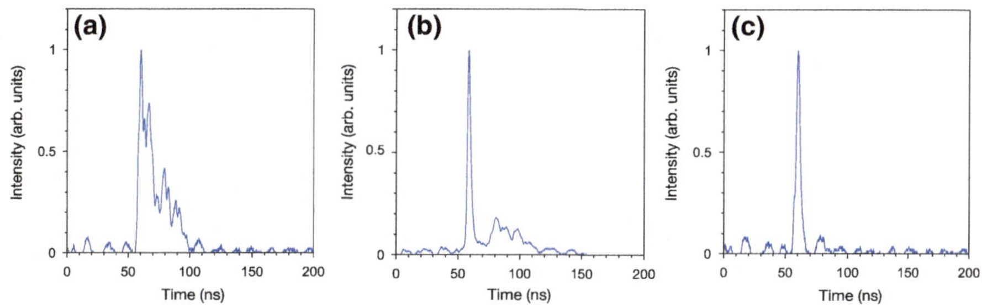

Fig. 54.1 Typical temporal profiles measured at the output of the oscillator (**a**) and the pulse shaper (**b**, **c**). **b** was produced by the 150-ps Nd:YAG pulse, while **c** was produced by both of the 150-ps and 10-ns Nd:YAG pulses

Figure 54.2a shows the output pulse energy after the 14-pass amplifier as a function of the amplifier discharge voltage. We achieved the CO_2 laser pulse of 150 and 60 mJ at the durations of 15 ns and 3 ns, respectively. The beam profile has been kept in a good quality as shown in Fig. 54.2b, c. Such CO_2 laser beam can provide laser intensities of the order of 10^{11} W/cm^2 for an efficient EUV emission at 6.*x* nm.

The developed CO_2 laser was used for EUV generation. Figure 54.3a, b shows the time-integrated spectra of the CO_2 laser-produced Gd plasmas at different pulse durations of 5 ns and 15 ns, respectively. These pulses were set at the same pulse energy of 50 mJ. The corresponded laser intensities are 2×10^{11} and 7×10^{10} W/cm^2 at pulse durations of 5 ns and 15 ns, respectively. As a result, the spectral structures are much different between the two. The spectral purity (spectral

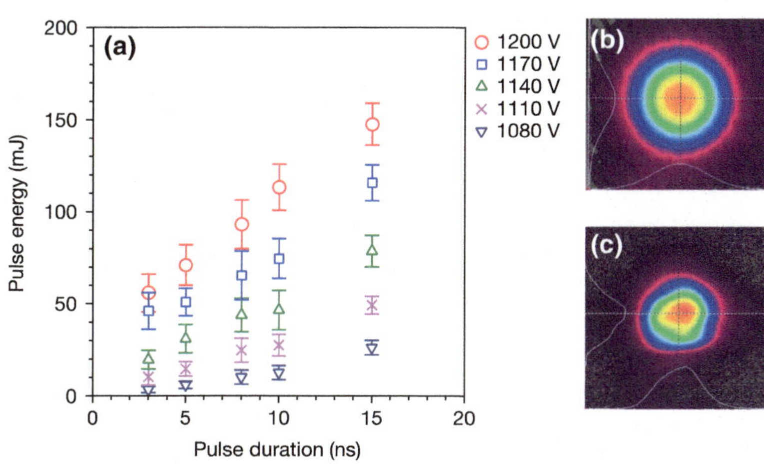

Fig. 54.2 Output pulse energy after multi-pass amplifier as a function of the amplifier discharge voltage. (**b**) and (**c**) are typical beam profiles measured at the output of the oscillator and the amplifier by a pyroelectric camera, respectively

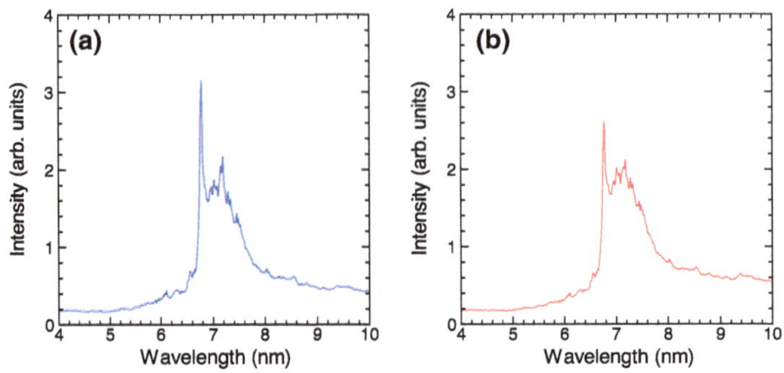

Fig. 54.3 Time-integrated spectra of Gd plasmas produced by the CO_2 laser at different pulse widths. **a** was recorded at a pulse duration of 5 ns, and an intensity of 2×10^{11} W/cm^2, while that of **b** was 15 ns, and 7×10^{10} W/cm^2, respectively

efficiency) was improved by irradiating the short pulse duration. Under the optically thin plasma conditions imposed by the CO_2 LPP, this peak, which is mainly due to the $4d^{10}\ {}^1S_0$–$4d^9 4f\ {}^1P_1$ transition of Pd-like Gd^{18+} overlapped with ${}^2F$–${}^2D$ lines of Ag-like Gd^{17+}, known to lie near 6.79 nm shows that these ions are indeed present in the plasma [5].

54.3 Summary

We reported the development of a nanosecond CO_2 laser system for investigation of an efficient UTA emission in EUV and SXR region. Pulse width of the CO_2 laser is tunable from 3 to 15 ns. By using the laser system, we observed the improvement of the spectral purity at the pulse duration of 5 ns as compared with 15 ns was found. It imposes that there is a dependence on pulse duration, ranging from 3 to 15 ns, of the UTA emission around 6.79 nm. In the next step, the detail of this dependence should be characterized. The laser output also will be improved in order to investigate the UTA emission in the water window SXR region.

Acknowledgements This work was performed under the auspices of MEXT, and "Project for Bio-Imaging and Sensing at Utsunomiya University" from MEXT. We also are grateful to the Gigaphoton Inc. for their support and providing the picosecond laser system.

References

1. Greg, T.: Optical lithography: lithography at EUV wavelengths. Nat. Photonics **4**, 809 (2010)
2. Higashiguchi, T.: Feasibility study of broadband efficient "water window" source. Appl. Phys. Lett. **100**, 014103 (2012)

3. Tanaka, H.: Comparative study on emission characteristics of extreme ultraviolet radiation from CO_2 and Nd:YAG laser-produced tin plasmas. Appl. Phys. Lett. **87**, 041503 (2005)
4. Higashiguchi, T.: Characteristics of extreme ultraviolet emission from mid-infrared laser-produced rare-earth Gd plasmas. Opt. Express **21**, 31837 (2013)
5. Li, B.: Investigation of Gd and Tb plasmas for beyond extreme ultraviolet lithography based on multilayer mirror performance. Appl. Phys. Lett. **101**, 013112 (2012)

Chapter 55
Modeling of Ablation of the Target Material for the Plasma for Coherent and Incoherent EUV Sources

A. Sasaki, A. Sunahara, K. Nishihara and T. Nishikawa

Abstract Particles are sometimes emitted during the ablation of the metal targets used for laser pumped coherent and incoherent light sources. These particles have a significant effect on the absorption efficiency of the laser, on the density and temperature distribution of the plasmas, and thus on the performance of the source. We propose a simulation model of the particle emission based on Lagrangian hydrodynamics taking the liquid-to-gas transition into account using a dynamic mesh reorganization technique.

55.1 Introduction

Laser-pumped plasma (LPP) extreme ultraviolet (EUV) light sources have been thoroughly investigated, and recently an output power of more than 100 W with 5% convergence efficiency (CE) has been achieved using the double-pulse technique [1], as shown in Fig. 55.1. First, the Sn droplet target is irradiated by a short pre-pulse laser with a temporal duration of ≈10 ps, and is broken up into small particles. Second, when the cloud of particles has expanded to a radius 10 times larger than that of the original droplet, the target is irradiated by the main CO_2 to produce a plasma with

A. Sasaki (✉)
Kansai Photon Science Institute, National Institutes for Quantum and Radiological Science and Technology, Kizugawa-shi, Kyoto 619-0215, Japan
e-mail: sasaki.akira@qst.go.jp

A. Sunahara
Institute for Laser Technology, 1-8-4 Utsubohonmachi, Nishi-ku, Osaka 550-0004, Japan

K. Nishihara
Institute for Laser Engineering, Osaka University, 2-6 Yamadaoka, Suita-shi, Osaka 565-0871, Japan

T. Nishikawa
Graduate School of Natural Science and Technology, 1-1 Tsushima-naka, Okayama 700-8530, Japan

T. Kawachi et al. (eds.), *X-Ray Lasers 2016*, Springer Proceedings in Physics 202, https://doi.org/10.1007/978-3-319-73025-7_55

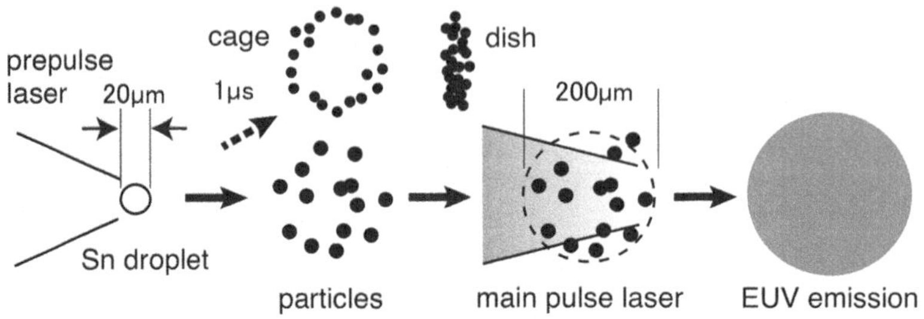

Fig. 55.1 Schematic diagram of the excitation of the LPP EUV source using the double-pulse technique

appropriate density and temperature for EUV emission [2]. We present a model of the particle emission to optimize the absorption efficiency of the laser, and thus the CE.

55.2 Model

We developed a model of the ablation of the laser-heated Sn target. The EUV emission is obtained from the heated plasma with an electron temperature $T_e \approx 50$ eV, in which Sn atoms are ionized to multiple charged states. By the irradiation of the relatively weak pre-pulse laser, initially, solid Sn target melts and evaporate at a temperature below the critical temperature, producing particles during the liquid-to-gas transition.

The present model is based on the two dimensional (2D) Lagrangian hydrodynamics on triangular grids. We take into account the properties of the target material near the solid-density and warm-temperature regions, $T_e \leq 1$ eV, on the basis of the Van der Waals equation of state,

$$P = \frac{RT}{V_m - b} - \frac{a}{{V_m}^2}, \tag{55.1}$$

where P, T, and V_m are the pressure, temperature, and specific volume per mole, respectively, and a and b are the Van del Waals constant for Sn [3]. The equations of the hydrodynamics are,

$$\rho \frac{\mathrm{d}\boldsymbol{u}}{\mathrm{dt}} = -\nabla P, \tag{55.2}$$

$$\frac{\mathrm{dU_m}}{\mathrm{dt}} = Q_m - P\frac{\mathrm{dV_m}}{\mathrm{dt}}, \tag{55.3}$$

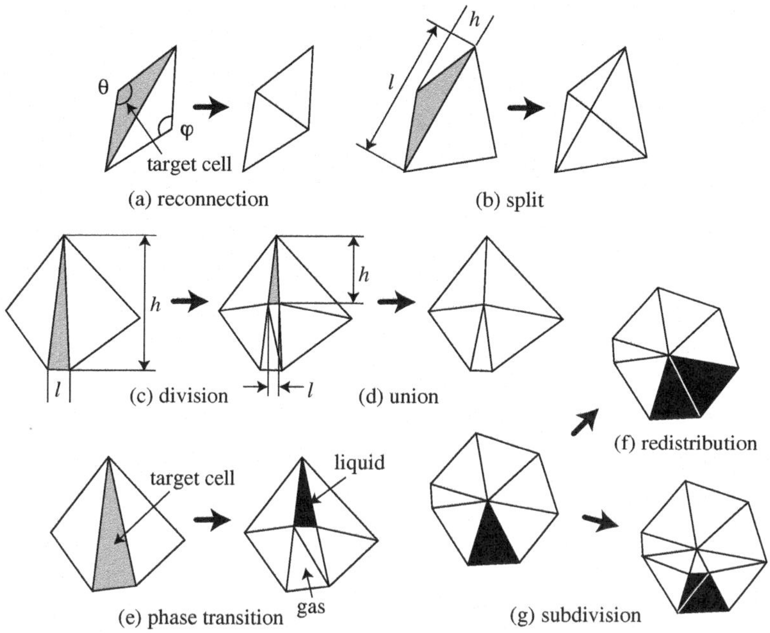

Fig. 55.2 Schematic diagram of the algorithms of mesh reorganization

where ρ, $\boldsymbol{u}$, are the density and velocity the fluid, and U_m, Q_m are the specific internal energy and heating, respectively.

The Van der Waals equation of state is the simplest model that represents the liquid-to-gas transition. The model gives a two-phase region in the density and temperature plane, where the material splits into separate liquid and gas components with a void ratio of α. In the present model, if the density and temperature of a Lagrangian cell are found to be inside the two-phase region, the cell is split into several cells to have the correct void ratio for the group of the cells.

We developed algorithms for mesh reorganization, as shown in Fig. 55.2. We used the time splitting method, in which calculation of the hydrodynamics motion and reorganization of the mesh are carried out alternately in each time step. To reduce the distortion of the mesh, we calculate the aspect ratio of each cell, and if it exceeds a certain limit, we apply one of reconnection, split, division, and union algorithms illustrated in Fig. 55.2, conserving the volume, mass, and energies of the target cells, and we place grids along the distribution of material and particles in the gas phase and bubbles in the liquid phase. As an application of these algorithms, the liquid-to-gas transition is represented by splitting a target cell into liquid and gas cells.

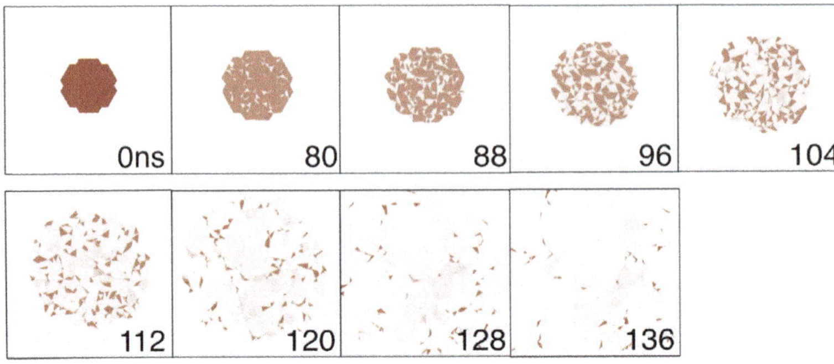

Fig. 55.3 Temporal evolution of the density distribution of a heated tin cylinder, with an initial radius of 10 μm, temperature of 2000 K, and density of 5.5×10^3 kg/m^3. The heating rate of 1.5 $\times 10^{12}$ W/mol

55.3 Result and Discussion

First, we validated the code through a comparison with a simple 1D model for the adiabatic expansion of low-density gas to obtain reasonable agreement. Second, we carried out a calculation of a heated liquid Sn cylinder. As shown in Fig. 55.3, the target is initially heated at the liquid density; then, small bubbles appear, which expand until finally the target decomposes into particles. We appreciate that the material is at rest until the gas phase becomes dominant. In future work, the present model will be applied to investigate the laser–matter interaction in the plasmas for the EUV source.

Acknowledgements This work was supported by JSPS grant No. 26610195.

References

1. Mizoguchi, H., et al.: In: Proceedings of SPIE 9776 Extreme Ultraviolet (EUV) Lithography VII, 97760J (2016)
2. Sasaki, A., et al.: Modeling of radiative properties of Sn plasmas for extreme-ultraviolet source. J. Appl. **107**, 113303 (2010)
3. Young, D.A.: Critical point of metals from the van der Waals model. Phys. Rev. A **3**, 364 (1971)

Chapter 56
Surface Layer Modification of Metal Nanoparticle Supported Polymer by Irradiation of Laser-Driven Extreme Ultraviolet Light

N. Tanaka, R. Deguchi, N. Wada, K. Yasuda, A. Yogo and H. Nishimura

Abstract Creation and control of interfacial structures between polydimethilsiloxane $(C_2H_6OSi)_n$ and nanoparticles allow us to create functional materials. Such technology is beneficial for fabrication of wearable devices. One of the concerns of modification of such materials is the effect of particle or heat loads on the sample when discharge plasma or visible to infrared light source is used for the processing. Extreme ultraviolet light could realize heat and particle load free fabrication of such materials because interaction of extreme ultraviolet radiation and materials is known for direct photoionization without thermal process. This study shows surface modification and creation of interfacial structure of AuPd nanoparticle supported polydimethylsiloxane by irradiation of high-fluence pulsed extreme ultraviolet light. It was observed that the size of nanoparticles increases as the number of EUV shots increases, and a layer of mixed Au and Si appeared after EUV irradiation.

56.1 Introduction

Recent progress of intense extreme ultraviolet (EUV) light source for photolithography has been attracting much attention from not only industry but also from research of fundamental and applied physics. EUV–matter interaction and conventional laser–matter interaction, especially in energy transfer, are expected be fundamentally different. Due to its high photon energy at ~100 eV, EUV photons directly ionize atoms by photoionization or break the chemical bonds without lattice

N. Tanaka (✉) · R. Deguchi · N. Wada · A. Yogo · H. Nishimura
Institute of Laser Engineering, Osaka University, Suita, Osaka, Japan
e-mail: tanaka-n@ile.osaka-u.ac.jp

K. Yasuda
Graduate School of Engineering, Osaka University, Suita, Osaka, Japan

T. Kawachi et al. (eds.), *X-Ray Lasers 2016*, Springer Proceedings
in Physics 202, https://doi.org/10.1007/978-3-319-73025-7_56

vibration. Thus, it is expected that the EUV radiation can change the physical and chemical properties of the material surface without or less thermal process. Polydimethylsiloxane (PDMS, $(C_2H_6OSi)_n$) is a polymer material widely used in industry. Further, it is a biocompatible elastomer and expected to be a suitable material for wearable devices with fabrication of electrodes on the surface [1, 2]. Creation of interfacial structure between PDMS and metal nanoparticles in discharge plasma has been confirmed. Although such structure allows us to create functional materials [3], thermal and particle damages are concerned. To avoid such negative effects on materials, EUV radiation would be a suitable energy source for modification of material surface. However, it has not been studied much due to difficulty of access to EUV sources with high photon flux. This study aims at showing the feasibility of EUV use for modification of surface structure and interfacial properties.

56.2 Experimental Setup

The samples were prepared by using a DC sputter device. AuPd (Au97Pd3 wt%) layer with a thickness of 10 or 20 nm was formed on the surface of PDMS with a thickness of 0.5 mm. PDMS samples without nanoparticle layer were also prepared for comparison. A pulsed laser-driven xenon plasma EUV source [4] was used for the irradiation experiment. The experimental setup is shown in Fig. 56.1a. We have utilized the wide spectrum of EUV emission from Xe plasma (Fig. 56.1b) to maximize the EUV energy. The EUV spot shape and size of the sample were controlled by using an Au-coated toroidal elliptical total reflection mirror. The EUV light was defocused to avoid sample ablation. Thus, the EUV spot on the sample was circular sector as shown in Fig. 56.1c, and the average fluence was 6.4 mJ/cm^2. The temporal pulse duration was 10 ns and the repetition rate was 10 Hz. The samples were irradiated by 4000 shots of EUV light. The samples and respective reference samples were investigated by AFM (atomic force microscopy, SEIKO instruments Nanopics 2100) to analyze surface morphology such as particle size, and XPS (X-ray photoelectron spectroscopy, XPS SYSTEM Shimadzu ESCA850) to analyze the depth profile of composition in the samples.

56.3 Results and Discussion

Figure 56.2 shows pictures of 10 nm AuPd-supported PDMS samples. A noticeable change in color in the irradiation spot was observed. Original color was transparent dark gray, and it turned into transparent wine red. The reproducibility of this change in color was confirmed. The color of bare PDMS sample did not change. Furthermore, in case of silver nanoparticles (AgPd (Ag97Pd3 wt%) layer), the color changed to yellow in another set of EUV irradiation experiment. Since it is

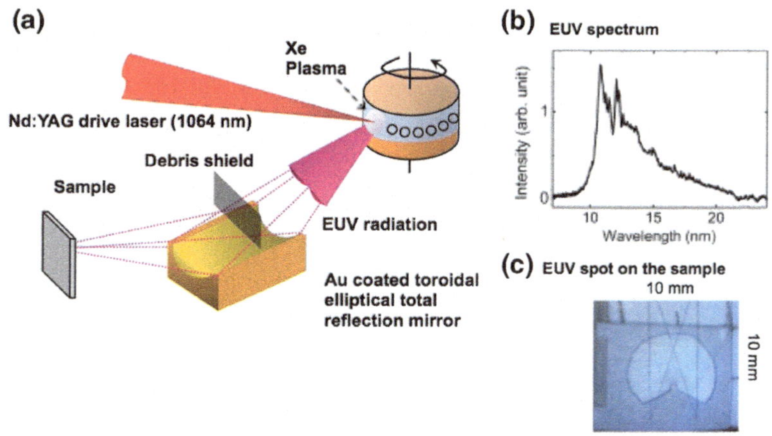

Fig. 56.1 Experimental setup. **a** The Xe EUV source, **b** EUV spectrum from the Xe plasma, and **c** defocused spot of co-aligned He–Ne laser on the sample

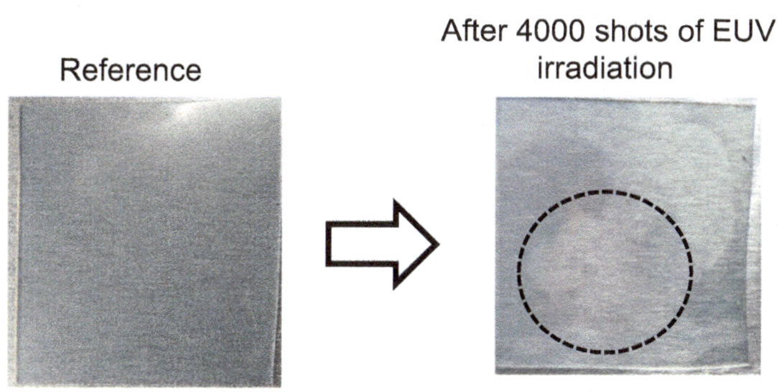

Fig. 56.2 Pictures of 10-nm AuPd-supported PDMS samples. The transparent dark gray changed to transparent wine red color after 4000 shots of EUV irradiation

well known that change in size of nanoparticles causes change in color, we have analyzed surface morphology by AFM (Fig. 56.3). The reference sample showed uniform particle distribution. Nanostructures with diameter of hundreds of nanometers were observed after 4000 shots of EUV irradiation. As increasing the shot number, the density of big structures increased. Similar trend was observed for AgPd-supported PDMS samples. Figure 56.4 shows results of XPS analysis of 20-nm AuPd-supported PDMS samples. The depth profile was obtained by Ar etching. Although the etching rates of our samples have not been determined, rough estimate calculates 100 s of etching corresponds ~10 nm. Peaks of C 1s (C–H bond), Si 2p (Si–O bond), and Au 4f were observed. The depth profile of the reference sample showed clear boundary between the nanoparticle layer and the PDMS bulk (Fig. 56.4a). It is consistent with the fact that the nanoparticles are only

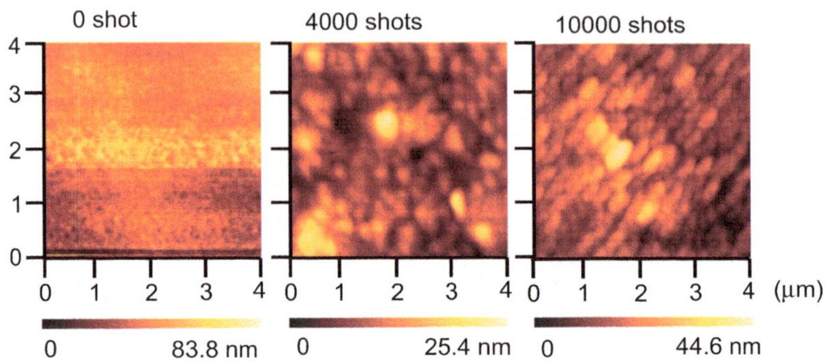

Fig. 56.3 AFM images of 10-nm AuPd-supported PDMS samples after 0 (reference), 4000, and 10000 shots of EUV irradiation

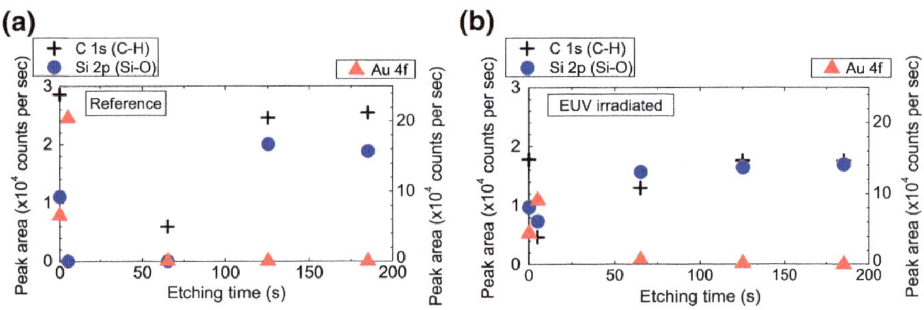

Fig. 56.4 Depth profiles of composition in the 20-nm AuPd-supported PDMS samples analyzed by XPS. **a** Reference sample and **b** after 4000 shots of EUV irradiation

supported by PDMS bulk. On the other hand, several nanometers of mixed layer of Au, Si–O, and C–H appeared after 4000 shots of EUV irradiation. It indicates that the PDMS bonds were broken by EUV, surface morphology was roughened, and the nanoparticles were repositioned in between the roughened structure. The detailed mechanism of creation of such morphology as well as EUV–sample interaction process is still in discussion.

56.4 Conclusion

Surface morphology and interfacial properties of AuPd nanoparticle supported PDMS samples irradiated by pulsed EUV light were studied to investigate feasibility of usage of pulsed EUV for material surface modification and creation of interfacial structure. The size of nanoparticles increased as the number of EUV irradiation shot increased resulting in a change in surface color. XPS analysis showed Au, Si–O, and C–H mixed structure indicating repositioning of nanoparticles on the roughed surface.

Acknowledgements This work was supported by Institute of Laser Engineering collaboration research number 2015B1-01, and partially supported by JSPS Grant-in-Aid for Young Scientists (B) number 25800303.

References

1. Kudo, H., et al.: Biosens. Bioelectron. **22**, 558 (2006)
2. Lu, N., et al.: Adv. Funct. Mater. **22**, 4044 (2012)
3. Yasuda, K.: J. Phys. Conf. Ser. **379**, 012033 (2012)
4. Masuda, M., et al.: Appl. Phys. B Lasers Opt. **119**, 421 (2015)

Chapter 57
Micrometer-Scale Photo-Direct Machining of Polydimethylsiloxane Using Laser Plasma EUV Radiations

H. Urai, T. Makimura and M. Ogawa

Abstract We have investigated photo-direct micromachining of PDMS sheets using laser plasma EUV radiations. The EUV light was radiated from laser plasma. The generated EUV light was focused using an ellipsoidal mirror and was incident on PDMS sheets through windows of a contact mask. Square micro-holes are precisely fabricated on a PDMS sheet. We have established a technique for precisely micromachining PDMS sheets in a micrometer-scale.

57.1 Introduction

Polydimethylsiloxane (PDMS) is a material used for biochips [1] and micro-total analysis systems/lab-on-chips due to its flexibility, chemical/thermodynamic stability, biocompatibility, and moldability. For further development, it is inevitable to develop a technique to fabricate precise three-dimensional structures in a micrometer-scale at high aspect ratio. So far, molding, reactive ion etching [2] and laser ablation techniques [3] have been employed to fabricate the macro-structures. It has been, however, difficult to fabricate the structures with the required precision and aspect ratios. In the present paper, we have investigated photo-direct machining of PDMS sheets using intense extreme ultraviolet (EUV) light radiated from laser plasma [4–11].

57.2 Experimental Setup

Figure 57.1 shows the experimental setup for micromachining of PDMS sheets using laser plasma EUV radiations. The EUV light was generated by irradiation of a tantalum target (T) with Nd:YAG laser light (L_1), at 532 nm, operated at 10 Hz,

H. Urai · T. Makimura (✉) · M. Ogawa
Institute of Applied Physics, University of Tsukuba, Tsukuba, Japan
e-mail: makimura@bk.tsukuba.ac.jp

T. Kawachi et al. (eds.), *X-Ray Lasers 2016*, Springer Proceedings in Physics 202, https://doi.org/10.1007/978-3-319-73025-7_57

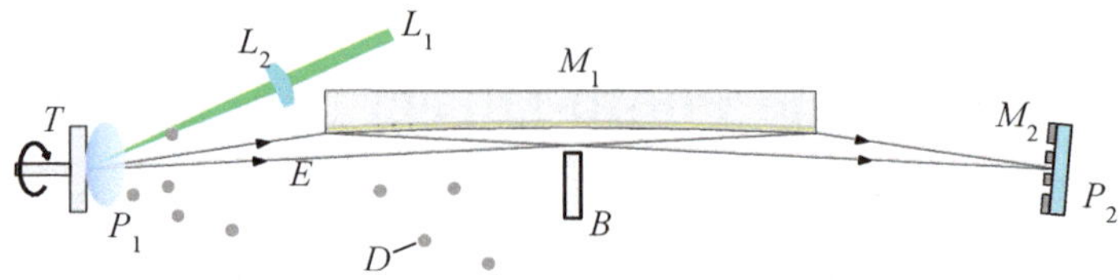

Fig. 57.1 Experimental setup for micromachining of PDMS sheets using laser plasma EUV radiations

with a pulse duration of 10 ns, at a typical pulse energy of 500 mJ/pulse. The laser-generated tantalum plasma (P_1) emits broadband EUV light (E). The EUV light was focused on PDMS sheet (P_2) using a gold-coated ellipsoidal mirror (M_1). This mirror is designed to efficiently focus EUV light near 10 nm. The PDMS sheets were irradiated through windows of a contact mask with a thickness of 4 μm (M_2) for patterning. Peak power density of EUV light on the PDMS sheets was adjusted, by placing the sheets at different distances from the focal point of the ellipsoidal mirror. The peak power density was estimated from laser-to-EUV light conversion efficiency, solid angle, and reflectivity of the ellipsoidal mirror and the beam cross section on the PDMS surfaces. A blocking plate (B) was inserted to avoid contamination of PDMS sheets with debris (D) ejected from the plasma source target.

57.3 Result and Discussion

We estimated the capability of photo-direct fabrication of high aspect structures using EUV light. For that purpose, we calculated propagation of plane light wave after passing through a circular opening with a diameter of 1 μm. Figure 57.2 shows beam profiles at given traveling distances from the opening for light waves with

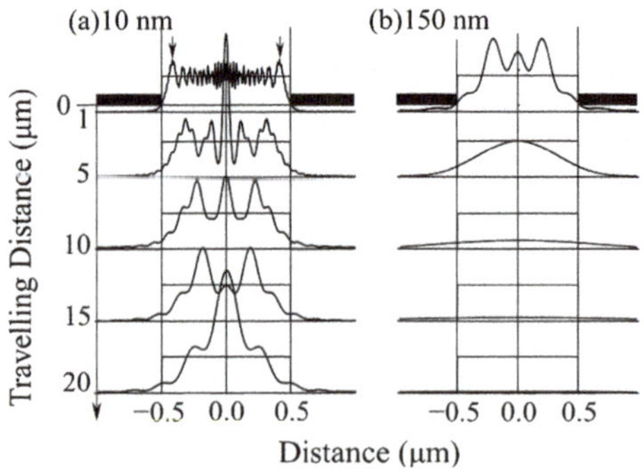

Fig. 57.2 Beam profile of light at 10 nm (**a**) and 150 nm (**b**) after passing through a circular opening with a diameter of 1 μm

Fig. 57.3 SEM images of a PDMS surface after 10 shots of irradiation with EUV light through square windows of a contact mask

wavelengths at 10 nm (a) and 150 nm (b). The intensity is normalized to that estimated by geometric optics. In the case of 10 nm light, peaks are seen at ±0.45 μm due to Fresnel diffraction at 1 μm below the opening, as indicated by the arrows. The peaks get closer as traveling distance get longer up to 20 μm below the opening. It should be noted that 79% of photons are confined in a zone expected by the geometric optics even after traveling 20 μm below the opening. In the case of 150 nm light, the Fresnel peaks meet with each other at 5 μm. Beam width gets wider as the wave travels further. Only 23% of photons are in the geometric optics zone even at 10 μm. These estimations have revealed that 10 nm light is required to fabricate through holes in 10-μm-thick sheets and that 150 nm light has a too long wavelength for that purpose.

Figure 57.3 shows a scanning electron microscope (SEM) image of a PDMS surface after 10 shots of irradiation with EUV light at a peak power density of 2×10^8 W/cm^2 through square windows of a contact mask. Square holes are precisely fabricated on the PDMS surface. The SEM observation has revealed that the surface roughness of the ablated region is as small as that of the unirradiated original region and that the walls are steep. The depth is estimated to be 3 μm by observing the fabricated structures at 45°.

Further, a PDMS sheet with a thickness of 13 μm was irradiated with EUV light through circular openings with a diameter of 3 μm. It is demonstrated that through holes with a diameter of 2 μm can be fabricated on the PDMS sheet.

57.4 Conclusion

In conclusion, we have investigated photo-direct micromachining of PDMS sheets using laser plasma EUV radiations. Precise square micro-holes are fabricated on a PDMS sheet. Further, high aspect through holes with a diameter of 2 μm is fabricated in a PDMS sheet with a thickness of 13 μm. We have established a technique for precisely micromachining PDMS sheets in a micrometer-scale.

References

1. Mosadegh, B., et al.: Uniform cell seeding and generation of overlapping gradient profiles in a multiplexed microchamber device with normally-closed valves. Lab Chip **10**, 2959–2964 (2010)
2. Chen, W., et al.: Photolithographic surface micromachining of polydimethylsiloxane (PDMS). Lab Chip **12**, 391–395 (2012)
3. Fukami, Y., et al.: Ablation of silicone rubber using UV-nanosecond and IR-femtosecond lasers. Jpn. J. Appl. Phys. **43**, 4240–4241 (2004)
4. Makimura, T., et al.: Quartz micromachining using laser plasma soft X-rays and ultraviolet laser light. Appl. Phys. Lett. **85**, 1274–1276 (2004)
5. Makimura, T., et al.: Direct micromachining of quartz glass plates using pulsed laser plasma soft X-rays. Appl. Phys. Lett. **86**, 103111 (2005)
6. Makimura, T., et al.: Silica nanomachining using laser plasma soft X-rays. Appl. Phys. Lett. **89**, 101118 (2006)
7. Torii, S., et al.: Ablation process of silica glass induced by laser plasma soft X-ray irradiation. Appl. Surf. Sci. **255**, 9840–9842 (2009)
8. Okazaki, K., et al.: Micromachining of transparent materials with Fresnel diffraction of infrared radiation. Appl. Phys. A: Mater. Sci. Process. **104**, 593–599 (2010)
9. Torii, S., et al.: Direct etching of poly (methyl methacrylate) using laser plasma soft X-rays. Appl. Phys. Express **3**, 66502 (2010)
10. Chiang, T.Y., et al.: Synchrotron-radiation-stimulated etching of polydimethylsiloxane using XeF_2 as a reaction gas. J. Synchrotron Radiat. **17**, 69–74 (2010)
11. Dyer, P.E., et al.: Diffraction modelling of laser ablation using transmission masks. Opt. Commun. **240**, 391–399 (2004)

Part VIII
Instrumentation for Advanced X-Ray Applications

Chapter 58
Broadband High-Resolution Imaging Spectrometers for the Soft X-Ray Range

A. O. Kolesnikov, A. A. Kuzin, D. V. Negrov, E. N. Ragozin, P. V. Sasorov, A. N. Shatokhin and E. A. Vishnyakov

Abstract We develop imaging (stigmatic) XUV spectrometers with the use of plane grazing-incidence varied line-space (VLS) diffraction gratings and focusing normal-incidence multilayer mirrors (MMs), including broadband aperiodic ones. A stigmatic 12–30 nm spectrometer with a resolving power of at least 500 is demonstrated.

58.1 Introduction

Grazing-incidence spectrometers with classical concave gratings for a wavelength range $\lambda \sim 20$–300 Å are astigmatic. This entails a drastic lowering in the irradiance of spectral lines and the loss of spatial resolution. Using focusing normal-incidence MMs in combination with varied line-space (VLS) diffraction gratings (DGs) makes it possible to remedy these defects.

Cornu [1] realized that monotonic variations in line spacing modified the curvature of the diffracted wavefronts and the spectral (horizontal) focal curve. Hettrick and Bowyer [2] proposed a way to obtain a stigmatic spectral image in the XUV by sending a converging homocentric beam onto a VLS DG whose local line

A. O. Kolesnikov · E. N. Ragozin (✉) · A. N. Shatokhin · E. A. Vishnyakov
P. N. Lebedev Physical Institute, Russian Academy of Sciences, 53 Leninskii prospekt, Moscow 119991, Russia
e-mail: enragozin@gmail.com

A. O. Kolesnikov · A. A. Kuzin · D. V. Negrov · E. N. Ragozin · A. N. Shatokhin
Moscow Institute of Physics and Technology (State University), 9 Institutskii per., Dolgoprudnyi, Moscow 141701, Russia

A. A. Kuzin
Institute for Spectroscopy Russian Academy of Sciences, 5 Phizicheskaya street, Troitsk, Moscow 142190, Russia

P. V. Sasorov
M. V. Keldysh Institute of Applied Mathematics, Russian Academy of Sciences, 4 Miusskaya square, Moscow 125047, Russia

T. Kawachi et al. (eds.), *X-Ray Lasers 2016*, Springer Proceedings in Physics 202, https://doi.org/10.1007/978-3-319-73025-7_58

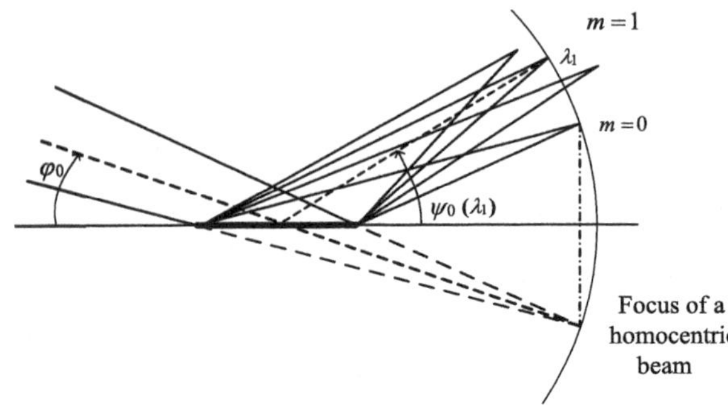

Fig. 58.1 Cancellation of astigmatism at one wavelength (λ_1) in the incidence of a converging beam on a plane VLS grating (φ_0 and ψ_0 are the grazing angles of the central rays of the incident and diffracted beams, and m is the order of diffraction)

density has to obey the local grating equation to bring the diffracted rays in the principal plane to a common focus. However, in this case, the spectral image is stigmatic at only one wavelength (Fig. 58.1).

The line density $p(w)$ of a VLS grating is conveniently expressed as

$$p(w) = p_0 + p_1 w + p_2 w^2 + p_3 w^3 + \cdots. \tag{58.1}$$

Factor p_1 modifies the spectral curve and p_2 and p_3 remove meridional coma and spherical aberration. In the approach of [2], p_0 and φ_0 are free parameters.

58.2 Spectrometer Design

Let a slightly astigmatic beam be incident on a plane VLS grating (Fig. 58.2). Let L_1 be the distance of the vertical focus of the incident beam from the grating center and L_2 be that of the horizontal one ($L_1 > L_2$). Then, it is possible to cancel astigmatism at two wavelengths, λ_1 and λ_2, at a sacrifice of one free parameter (either p_0 or φ_0). The slightly astigmatic beam incident on the grating is produced by a near normal-incidence MM, which images a point source. The sequence of steps is as follows: we define the spectral range, and then determine λ_1 and λ_2 to minimize the defocusing in the range. If we define φ_0, then $p_0 = \sqrt{(L_1/L_2 - 1)(\sin^2 \varphi_0)/\lambda_1\lambda_2}$. If, alternatively, we define p_0, then $\varphi_0 = \arcsin(mp_0\sqrt{\lambda_1\lambda_2}/\sqrt{L_1/L_2 - 1})$. The grating parameters p_i result from (58.2) after Taylor series decomposition:

$$\begin{aligned} mp(w)\lambda_{1,2} = \cos\left[\operatorname{arcctg}\left(\operatorname{ctg}\varphi_0 - \frac{w}{L_2 \sin\varphi_0}\right)\right] - \\ - \cos\left[\operatorname{arcctg}\left(\operatorname{ctg}\psi_0^{1,2} - \frac{w}{L_1 \sin\psi_0^{1,2}}\right)\right]. \end{aligned} \tag{58.2}$$

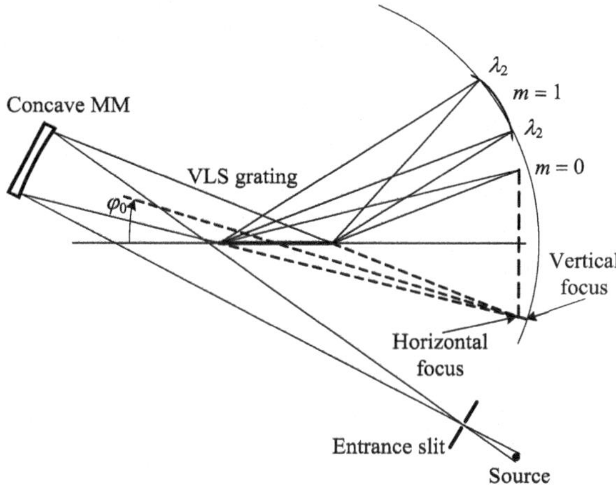

Fig. 58.2 Cancellation of astigmatism simultaneously at two wavelengths, λ_1 and λ_2

We calculated several 1-m-long spectrometer versions for the 120–300 Å range, with $\lambda_1 = 144$ Å, $\lambda_2 = 270$ Å, and $p_0 = 600$ mm^{-1}. For a grating area of 50 × 20 mm, the ray trace images of a point source are all confined to one detector pixel. The plate scales are about 5 Å/mm and, in view of the detector pixel size of 13 μm, typical practical resolution is expected at ~(0.05–0.07) Å throughout the range. The spectrometer may be equipped with a narrowband periodic multilayer mirror or a broadband aperiodic one. In the latter case, the operating range may span an octave in wavelength or more. In the instrument described below, the MM was mounted at an angle of incidence of 7.59° and the VLS grating was mounted at a grazing angle of 6.44°.

58.3 Spectrometer Implementation

The spectrometer layout is shown in Fig. 58.3. The XUV source was the plasma of a plane LiF target irradiated by a 1.06 μm, 0.5 J, 10 ns laser pulse. For a test, we took an aperiodic MM spanning a range of at least 125–250 Å [3].

The VLS grating ($p_0 = 600$ mm^{-1}, $p_1 = 2.32$ mm^{-2}) was fabricated by e-beam lithography (EBL) followed by inductively coupled plasma (ICP) etching. A 100 nm tungsten film was deposited on a glass substrate, which was next spin-coated with the positive-tone e-beam resist PMMA A4 (Microchem, 5000 rpm, 90 s). It was then exposed to EBL (beam energy 50 keV, current 15.5 nA, write field 600 × 600 μm, dwell time 0.14 ms). The resist was developed in MIBK:IPA (1:3) for 120 s and IPA for 60 s. Finally, the structure was formed with SF_6 ICP etching followed by O_2 plasma cleaning.

Due to its high light-gathering power, the spectrometer records the spectrum in one 0.5 J laser shot. The portion shown in Fig. 58.4 contains the lines of Li III and F V–F VII. The closest first-order lines resolved with a safety margin are the

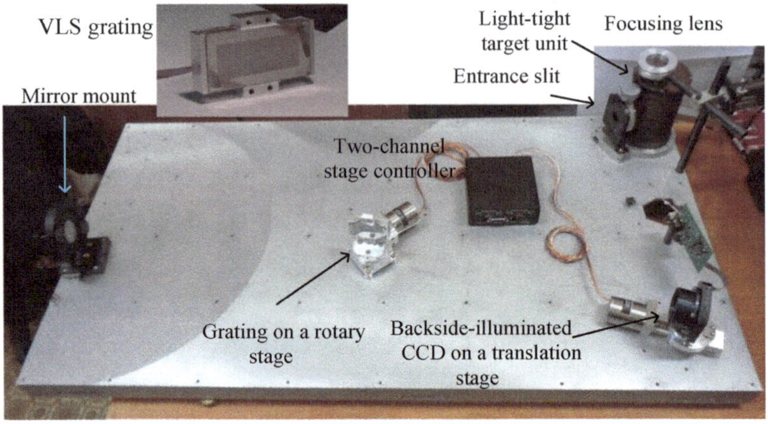

Fig. 58.3 Spectrometer elements accommodated on a 1.1-m-long base plate. VLS grating (top of drawing: tungsten VLS grating made by e-beam lithography)

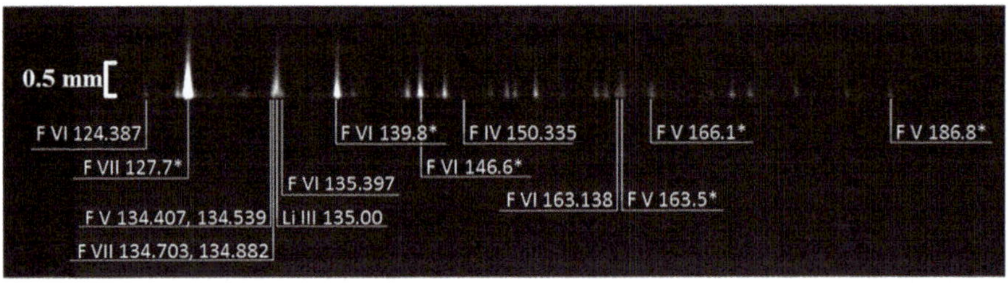

Fig. 58.4 Single-shot first-order stigmatic spectrum of an LiF target excited by a 0.5 J pulse. Asterisks indicate unresolved line arrays. Entrance slit width: 30 μm

163.138 Å line of F VI and the {163.456, 0.501, 0.558, 0.596 Å} unresolved line array of F V, yielding a conservative figure $\lambda/\delta\lambda \sim 510$. The line half-widths typically correspond to four detector pixels (52 μm). In view of the plate scale (5.5 Å/mm at 125 Å, 6.3 Å/mm at 200 Å), this corresponds to $\lambda/\delta\lambda \sim 450$ and 600, respectively. The strongest line arises from the 3d → 2p (127.653 Å, 127.796 Å) transitions in F VII, which saturates the CCD detector pixels corresponding to the near-surface portion of the space-resolved spectral image and broadens the apparent linewidth. Its second order is much weaker and is safely resolved, testifying to a resolving power of ~900.

Acknowledgements This work was supported by the Russian Science Foundation (Grant No. 14-12-00506).

References

1. Cornu, M.A.: Vérifications numériques relatives aux propriétés focales des réseaux diffringents plans. Comptes Rendus Acad. Sci. **117**, 1032–1039 (1893)
2. Hettrick, M.C., Bowyer, S.: Variable line-space gratings: new designs for use in grazing incidence spectrometers. App. Opt. **22**, 3921–3924 (1983)
3. Pirozhkov, A.S., Ragozin, E.N.: Aperiodic multilayer structures in soft X-ray optics. Phys. Usp. **58**, 1095–1105 (2015)

Chapter 59
Development of Soft X-Ray Microscope in Water Window Using Laser-Produced Plasma Light Source

T. Ejima, Y. Kondo, Y. Ono, T.-H. Dinh, T. Higashiguchi and T. Hatano

Abstract For the sake of observation of samples in solution, a contact-type soft X-ray microscope is under construction by the use of both an optical microscope for readout and a laser-produced plasma light source with a heavy metal target. Edge response is evaluated through a water layer.

59.1 Introduction

Light in the wavelength region between C and O absorption edges (λ = 2.3–4.4 nm) penetrates water and is absorbed by C and N atoms; then, it is suitable for observing organelles of bio-cells in cultural fluids [1]. As a light source in this “Water-window” wavelength region, synchrotron radiation (SR) has been used [2]. Recently, light from laser-produced plasma (LPP) with a heavy metal target, such as Au, Pb, and Bi, shows strong emission on water-window wavelength region [3]. In addition, LPP light sources are laboratory-type ones and easy to operate comparing with the SR light sources.

Nanoscopic materials in solution are fluctuated by Brownian motion; therefore, exposure time needs to be short enough to observe its shape exactly [4]. Duration times of the emissions from LPP light sources equal to those of the excitation laser; therefore, LPP light sources are considered to be suitable for the observation of nanoscopic samples in solution. For this purpose, we have developed a contact-type X-ray microscope (CXRM) with an LPP light source.

T. Ejima (✉) · T. Hatano
Institute of Multidisciplinary Research for Advanced Materials, Tohoku University, Sendai, Japan
e-mail: takeo.ejima.e7@tohoku.ac.jp

Y. Kondo · Y. Ono · T.-H. Dinh · T. Higashiguchi
Department of Electrical and Electronic Engineering, Utsunomiya University, Utsunomiya, Japan

T. Kawachi et al. (eds.), *X-Ray Lasers 2016*, Springer Proceedings in Physics 202, https://doi.org/10.1007/978-3-319-73025-7_59

59.2 Details of Contact-Type Soft X-Ray Microscope

Optical layout of CXRM is represented in Fig. 59.1. The microscope is composed of three parts: a laser plasma light source which is composed of both a heavy metal target and an excitation laser, a soft X-ray condenser optics including a scintillator plate that converts a soft X-ray image into a visible image, and an optical microscope which magnifies the visible image.

Excitation laser used in this study was an Nd:YAG laser of which the wavelength was 1064 nm, the energy was 200 mJ/pulse, and the duration time of the pulse was 150 ps. A Bi plate was used as a target, because emissions from plasmas using heavy metal targets obey quasi Moseley's law [3] and strong emission from the Bi plasma is observed in water-window wavelength region.

The condenser mirror is a toroidal mirror of which surface was coated with Au [5]. This condenser was designed to obtain a large numerical aperture; therefore, the reflectance of the mirror was smaller than those of the usual grazing incidence mirrors used in soft X-ray region [5]. The reflectance of the mirror was calculated as shown in Fig. 59.2 [6]. In order to block the visible light which is simultaneously excited from the LPP, 100-nm-thick Ti membrane was installed between the condenser mirror and a sample. Transmittance of the Ti membrane was also calculated as shown in Fig. 59.2, and a throughput of the optics was 4×10^{-3} on average in the water-window wavelength region.

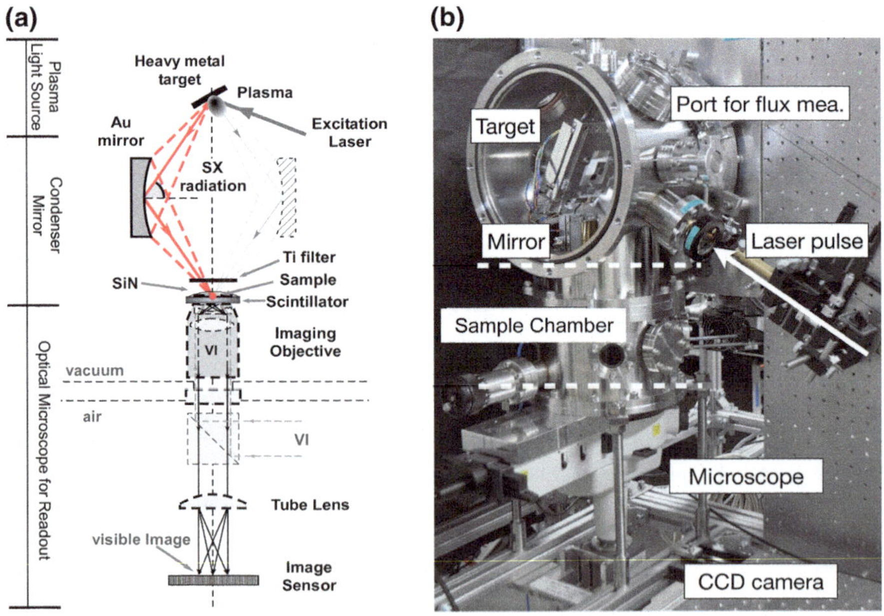

Fig. 59.1 Layout of CXRM (**a**), and CXRM under construction (**b**)

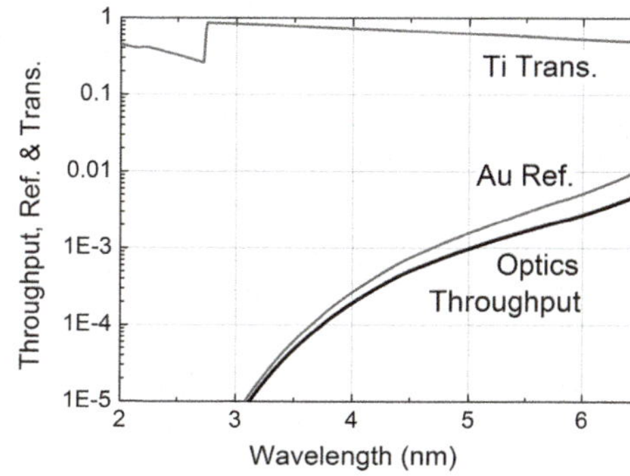

Fig. 59.2 Transmittance of 100-nm-thick Ti filter, reflectance of Au mirror (gray curves), and the throughput of the SX optics (black curve) are calculated [6]

Scintillator used in this study was Ce:LYSO (Oxide Co., Japan), because the quantum efficiency of Ce:LYSO is high in water-window wavelength region [2]. Both sides of the scintillator plate were optically polished, and the plate was placed on a sample holder.

59.3 Evaluation of CXRM

The spatial resolution of the present CXRM was evaluated by the use of an edge of an SiN membrane with a water layer that was enclosed between the SiN membrane and a scintillator plate. The edge response was taken under the conditions; exposure time was 10 s with 10 Hz repetition rate. The location of the edge used in the evaluation was indicated by a broken square in Fig. 59.3a.

The edge was separated by the polystyrene beads, which can be seen as small dots in Fig. 59.3a. When a fringe pattern is observed between parallel plates, distance d between the plates is represented as

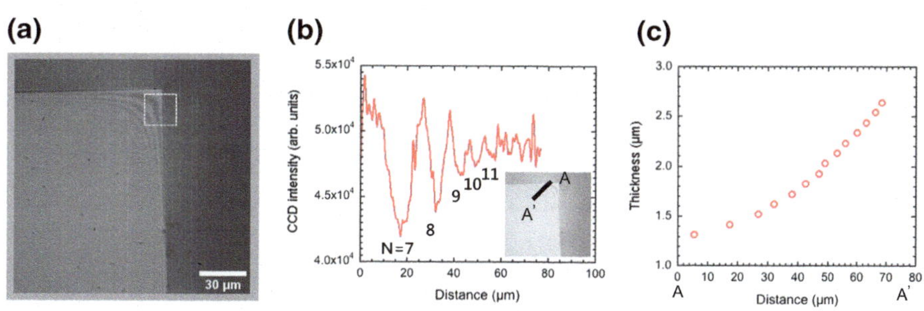

Fig. 59.3 An image of the SiN membrane edge with a water layer observed under visible epi-illumination (**a**), the interference fringe intensity at the corner edge designated from point A to point A′ in the inset (**b**), and the thickness obtained from point A to point A′ (**c**)

$$d = \mathrm{N}\lambda/(2n), \quad (59.1)$$

where λ is an observation wavelength, n is a refraction index, and N is the order of a fringe [7]. In the measurement, the wavelength of the illumination light was 545 nm, and the refraction index of water was 1.34. Figure 59.3b shows an intensity change of the fringe pattern at from point A to point A′ as represented in the inset. The order of the fringe pattern, N, was determined as indicated at each peak considering with the diameter of the polystyrene beads, 1.1 μm. The polystyrene beads were not fluctuated during the observation because the beads were pressed by both the SiN membrane and the scintillator plate. The thickness of the water layer is determined as in Fig. 59.3c. Finally, thickness of the water layer in the square region is 1.4 μm in average.

Figure 59.4 shows the primary differential values of the edge response, which is exhibited by the dots, and the fitting results, by both solid and broken curves. The response was composed of two edges with a background, which are originated from the edge structure of the SiN membrane. The HWFM values of the peak A and B were 0.83 and 0.64 μm, respectively. The main edge response is peak A and the value is more than twice as large as that is expected theoretically. The difference between the values will be caused by both the distance of the water layer and the blur through the water layer. Peak B will be also originated from the same blur. One of the origins of the blur is considered to be the incident angle of the condenser mirror to the edge.

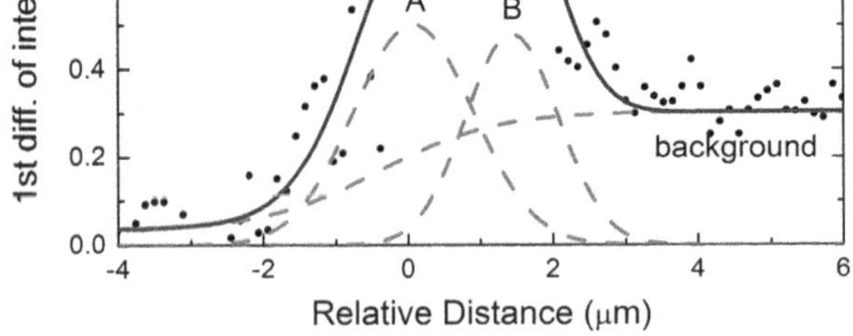

Fig. 59.4 Primary differential values of the edge response, body, and the SXR image of the edge response, inset. The response was composed of two peaks with background. Left peak represents the edge response and right one, the blur. Background is caused by SXR transmission through the water layer

59.4 Summary

A contact-type soft X-ray microscope was developed for observation of nanoscopic samples in solution. Present spatial resolution of the microscope was 0.83 μm under 10 s exposure time with a Bi target, which was determined by an edge response through a 1.4-μm-thick water layer.

References

1. Sayre, D., Kirz, J., Feder, R., Kim, D.M., Spiller, E.: Science **196**, 1339 (1977)
2. Ejima, T., Neichi, Y., Ishida, F., Yanagihara, M.: J. Phys. **463**, 012055 (2013)
3. Ohashi, H., Higashiguchi, T., Suzuki, Y., Arai, G., Otani, Y., Yatagai, T., Li, B., Dunne, P., O'Sullivan, G., Jiang, W., Endo, A., Sakaue, H.A., Kato, D., Murakami, I., Tamura, N., Sudo, S., Koike, F., Suzuki, C.: Appl. Phys. Lett. **104**, 234107 (2014)
4. Ito, A., Shinohara, K.: Cell Struc. Func. **17**, 209 (1992)
5. Ejima, T., Hatano, T., Ohno, K., Fukayama, T., Aihara, S., Yanagihara, M., Tsuru, T.: Appl. Opt. **53**, 6846 (2014)
6. Windt, D.L.: Comput. Phy. **12**, 360 (1998)
7. Born, M., Wolf, E.: Principles of Optics, Sec. 7.5.2. Cambridge University Press (1999)

Chapter 60
Development of Time-Resolved Small-Angle X-Ray Scattering System Using Soft X-Ray Laser

T. Kumada, N. Hasegawa, M. Nishikino, T. Otobe, R. Motokawa and T. Suemoto

Abstract We are developing time-resolved grazing-incidence small-angle scattering (GI-SAXS) system using a soft X-ray laser. As a first step, we carried out GI-SAXS measurements of optical grating and craters produced on gold surface by femtosecond laser ablation. We succeeded in observing the diffraction spots from the grating as expected, but the diffuse scattering from the craters was hidden in thermal noises of a soft X-ray charge-coupled device camera.

60.1 Introduction

The nonthermal effect on femtosecond laser ablation has been focused on the development of high-precision micromachining technique with small thermal effect. Since the pressure is always zero at the surface due to boundary condition, the voids are generated not at the surface but around the center depth in the photoexcited volume. The upper part of the photoexcited volume is separated as a thin layer by expanding and merging the voids [1, 2]. This ablation phenomenon is called spallation, which has been studied using time-resolved optical [3] and X-ray [4] microscopies, X-ray shadowgraph [5], and reflectivity [6–8]. Recently, we newly focus on a time-resolved grazing-incidence small-angle X-ray scattering (GI-SAXS) measurement. In principle, the GI-SAXS can measure time evolution of the morphology of the voids. The voids would be stretched from normal to the

T. Kumada (✉) · R. Motokawa
Materials Sciences Research Center, Japan Atomic Energy Agency, Ibaraki, Japan
e-mail: kumada.takayuki@jaea.go.jp

N. Hasegawa · M. Nishikino · T. Otobe
Kansai Photon Science Institute, National Institutes for Quantum and Radiological Science and Technology, Kyoto, Japan

T. Suemoto
Toyota Physical and Chemical Research Institute, Aichi, Japan

T. Kawachi et al. (eds.), *X-Ray Lasers 2016*, Springer Proceedings in Physics 202, https://doi.org/10.1007/978-3-319-73025-7_60

sample surface by the tensile stress, whereas spherical bubbles would be formed when the ablation is dominated by thermal effect. We can quantitatively evaluate the contribution of the tensile stress on the femtosecond laser ablation from the asymmetric morphology of the voids.

Lindenberg et al. [9] have reported a steep increase in intensity at low wavenumber q limit of a time-resolved GI-SAXS profile when an InSb sample was irradiated with an intense femtosecond laser pulse. They pointed out that it could be due to voids. However, their resolution using the hard X-ray (0.14 nm) was not enough to distinguish the scattering of the voids from a leakage of non-scattered direct beam. According to Bragg's law, $q = 4\pi \sin\theta/\lambda$, higher q-resolution with larger scattered angle θ can be expected at longer λ. Thus, we develop a GI-SAXS system using a soft X-ray laser and evaluate the resolution and sensitivity.

60.2 Experimental Setup

Figure 60.1 shows the experimental setup, which was slightly modified from the X-ray imaging system [4, 5]. A soft X-ray pulse with $\lambda = 13.9$ nm, dispersion $\Delta\lambda/\lambda$ of 10^{-4}, and time duration of 1 ps from the soft X-ray laser in Kansai Photon Science Institute was guided through a 0.2–0.5-μm-thick zirconium filter and a concave mirror with a curvature radius of 3 m to a sample at an incident angle of 19.2° from the surface. The reflected pulse was detected using a charge-coupled device (CCD) camera (Princeton, PI-MTE: 2048B) with the sample-to-camera distance of 124 mm. The X-ray pulse was focused to a diameter of approximately 50 μm on the CCD camera, whereas the diameters of an ellipsoidal spot at the sample surface were 400 and 100 μm in the major and minor axes, respectively. A gold-deposited optical grating for visible and infrared wavelengths (1740 lines/mm) and a gold-deposited optical mirror irradiated with 10 shots of femtosecond laser pulses with 795 nm in wavelength, 70 fs in duration, 0.6 J/cm^2 in fluence, and 50 μm in spot size are used.

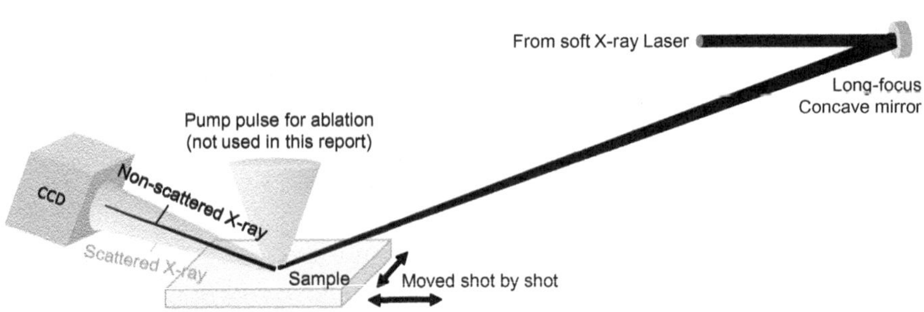

Fig. 60.1 Experimental setup of soft X-ray laser GI-SAXS diffractometer

60.3 Results and Discussion

Figure 60.2a, b shows the GI-SAXS profiles of the optical grating whose groove directions are parallel and perpendicular to the plane of incidence, respectively. Their peak positions are consistent with theoretical predictions. The full width at a half maximum of the zero-order diffraction peak (non-scattered beam) determines the q-resolution of the order of 10^{-4} nm^{-1}, which is much higher than that in [9] by 3–4 orders of magnitude.

Figure 60.2c, d compares the profiles of the gold-coated optical mirror before and after the irradiation of the femtosecond laser pulses. Any scattering from the crater was obtained even after subtraction of the scattered profiles before and after the irradiation (not shown). The crater-related scattering would be hidden in shot-to-shot fluctuation of the spot profile of the non-scattered beam and thermal

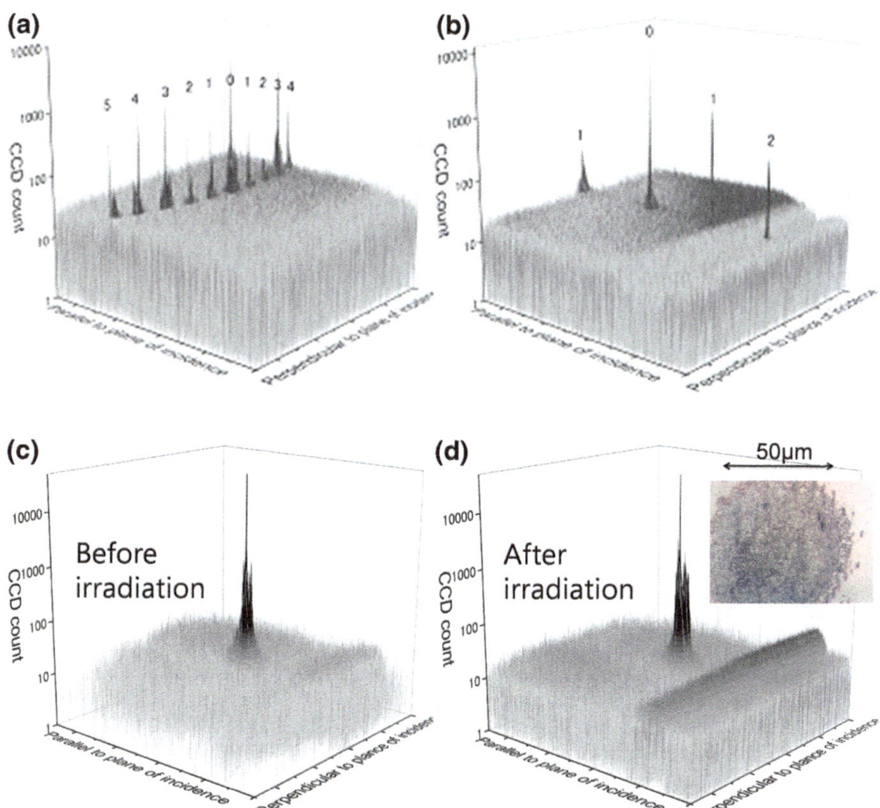

Fig. 60.2 GI-SAXS profiles of the optical grating aligned in **a** parallel and **b** perpendicular to the plane of incidence, and the gold-deposited optical mirrors **c** before and **d** after irradiation of the femtosecond laser pulses. The numbers n are attributed to the n-order diffraction peaks. The image area is 27.6 × 27.6 mm. Note that the valley in (**b**) and hill in (**d**) on the right side are due to imperfect subtraction of the background signal. The optical microscopic image of the crater, which was produced by the irradiation of the laser pulse, was shown in the right of (**d**)

noise of the CCD camera. In order to measure the GI-SAXS profiles from such disordered surfaces, the CCD camera should be replaced by a photon counting-type two-dimensional (2D) detector, and/or the optical layout should be optimized to obtain more tightly focused X-ray pulse on the 2D detector. To date, no photon counting-type 2D detector for our soft X-ray is available. The development of the detector or the improvement of the X-ray laser to obtain shorter wavelength is needed to realize the system. In addition, tighter focusing of the X-ray pulse on the 2D detector is preferred to detect lower-q scattered signal. The scattered signal could be distinguished from the non-scattered direct beam by tighting the X-ray pulse. In this case, we have to set the 2D detector very close to the sample to obtain tightly focused soft X-ray pulse not only on the detector but also on the sample surface, which is essential to overlap the X-ray pulse with the pump pulse for the time-resolved measurements in future.

Acknowledgements We thank Prof. M. Takenaka and Dr. K. Saijo of Kyoto University for their advices for experimental settings. This study was supported in part by the CPhoST program funded by the Special Coordination Funds for Promoting Science and Technology commissioned by MEXT of Japan, and a Grant-in-Aid for Scientific Research C from JSPS (Grant No. 15K04706).

References

1. Anisimov, S.I., Inogamov, N.A., Oparin, A.M., Rethfeld, B., Yabe, T., Ogawa, M., Fortov, V.E.: The physics of ultra-short laser interaction with solids at non-relativistic intensities. Appl. Phys. A **69**, 617–620 (1999)
2. Wu, C., Zhigilei, L.V.: Microscopic mechanisms of laser spallation and ablation of metal targets from large-scale molecular dynamics simulations. Appl. Phys. A **114**, 11–32 (2014)
3. Sokolowski-Tinten, K., Bialkowski, J., Cavalleri, A., von der Linde, D., Oparin, A., Meyer-ter-Vehn, J., Anisimov, S.I.: Transient states of matter during short pulse laser ablation. Phys. Rev. Lett. **81**, 224–227 (1998)
4. Nishikino, M., Kawachi, T., Hasegawa, N., Ishino, M., Minami, Y., Suemoto, T., Ohnishi, N., Ito, A.M., Sato, K., Faenov, A.Y., Inogamov, N.A., Yamagiwa, M.: Observation of dynamics and modification of solid surface using a picosecond soft X-ray laser. Proc. SPIE **9589**, 958902 (2015)
5. Hasegawa, N., Nishikino, M., Tomita, T., Ohnishi, N., Ito, A.M., Eyama, T., Kakimoto, N., Idutsu, R., Minami, Y., Baba, M., Faenov, A.Y., Inogamov, N.A., Kawachi, T., Yamagiwa, M., Suemoto, T.: The observation of a transient surface morphology in the femtosecond laser ablation process by using the soft X-ray laser probe. Proc. SPIE **9589**, 95890A (2015)
6. Kumada, T., Akagi, H., Itakura, R., Otobe, T., Yokoyama, A.: Femtosecond laser ablation dynamics of fused silica extracted from oscillation of time-resolved reflectivity. J. Appl. Phys. **115**, 103504 (2014)
7. Kumada, T., Akagi, H., Itakura, R., Otobe, T., Nishikino, M., Yokoyama, A.: Non-thermal effects on femtosecond laser ablation of polymers extracted from the oscillation of time-resolved reflectivity. Appl. Phys. Lett. **106**, 221605 (2015)
8. Kumada, T., Otobe, T., Nishikino, M., Hasegawa, N., Hayashi, T.: Dynamics of spallation during femtosecond laser ablation studied by time-resolved reflectivity with double pump pulse. Appl. Phys. Lett. **108**, 011102 (2015)
9. Lindenberg, A.M., et al.: X-ray diffuse scattering measurements of nucleation dynamics at femtosecond resolution. Phys. Rev. Lett. **100**, 135502 (2008)

Chapter 61
Development of a High Repetition Rate and High Pulse Energy Nd:YAG MOPA Laser System

K. Mikami, N. Hasegawa, H. Okada, S. Kondo, M. Nishikino and T. Kawachi

Abstract High repetition rate and high pulse energy laser system as a pump source for Ti:sapphire laser used in laser-driven plasma soft X-ray laser system are required. We developed a new Nd:YAG laser system using master oscillator power amplifier platform. The maximum output energy as over 4 J was obtained at an operating frequency of 50 Hz with a negligible amplified spontaneous emission.

61.1 Introduction

A highly brilliant and coherent X-ray source is a promising tool for scientific researches. At the Kansai Photon Science Institute (KPSI) of the National Institutes for Quantum and Radiological Science and Technology (QST), we developed a fully spatial coherent soft X-ray laser (SXRL) generated by Twin Optical Amplifiers using Zigzag slab (TOAPZ). The TOPAZ is operated with two beam lines of 10 J picosecond pulses each, with a 0.1 Hz operating frequency [1, 2]. The second harmonic of high pulse energy Nd:YAG laser is a typical pump source of the Ti: sapphire laser, which can be applied to a laser-driven plasma SXRL system. SXRL operation with high repetition rate is expected as key tools in many scientific and industrial fields. High repetition rate SXRLs have been demonstrated by 5 and 10 Hz operating Ti:sapphire lasers [3, 4]. In this study, we developed a new 4 J and 50 Hz laser system as a pump source in a Ti:sapphire laser system.

K. Mikami (✉) · N. Hasegawa · H. Okada · S. Kondo · M. Nishikino · T. Kawachi
Kansai Photon Science Institute, National Institutes for Quantum and Radiological Science and Technology, Kyoto, Japan
e-mail: mikami.katsuhiro@qst.go.jp

T. Kawachi et al. (eds.), *X-Ray Lasers 2016*, Springer Proceedings in Physics 202, https://doi.org/10.1007/978-3-319-73025-7_61

61.2 Details of the Developed Laser System

The laser system was constructed using a master oscillator power amplifier (MOPA) platform as shown in Fig. 61.1. A commercial Nd:YAG laser was adapted as the master oscillator, and the wavelength, pulse width, and repetition rate were 1064 nm, 14 ns, and 50 Hz, respectively. Laser pulses of the master oscillator were amplified by two flash-lamp-pumped Nd:YAG rod amplifiers.

The concentrations of the doped Nd^{3+} ions in the YAG crystal rods were 1.1 at.%. The rod size was 70 mm in length and 14 mm in diameter. The double-pass scheme at the first rod amplifier was adapted to extract the most stored energy efficiently. Thermal management for a laser medium is important to prevent thermal effects such as the thermal lens effect. Image relay system with vacuum tubes was installed in the laser system and the distances of a lens pair were adjusted to correct the thermal lens. The vacuum is necessary to maintain the wavefront in the laser system. The beam image at the output of the master oscillator was relayed through the center of Nd:YAG rods to the system output. The Nd:YAG rods were installed into flow tubes made from samarium-doped silica glass to inhibit an amplified spontaneous emission (ASE) and cooled with water in a circumferential direction [5].

61.3 Amplification Characteristics

Figure 61.2a shows the input–output characteristics of a single-pass amplification. The diameter of the input laser beam was 13 mm. The Nd:YAG rod was pumped by two flash lamps and the input electrical power was 50 J for each lamp. The small signal gain coefficient and stored energy were estimated to be 0.328 cm^{-1} and 2.36 J, respectively. Figure 61.2b, c shows the near-field and focusing patterns at the single-pass amplification. The focusing length of the thermal lens was evaluated to be 1785 mm for the horizontal direction and 2158 mm for the vertical direction. A divergence angle from the thermal lens effect could be adjusted by optimized vacuum tubes. The initial beam quality at the output of the master oscillator was

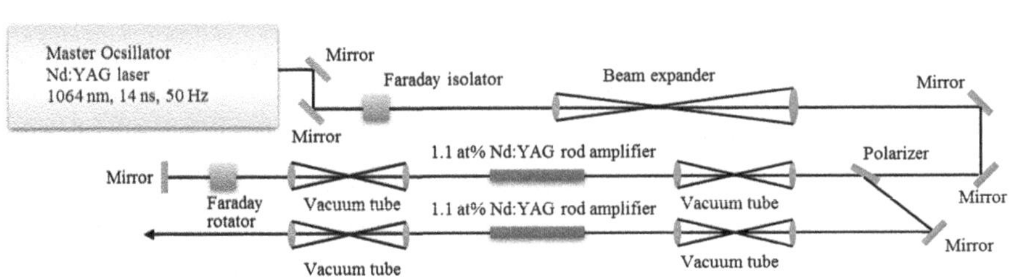

Fig. 61.1 Laser system layout

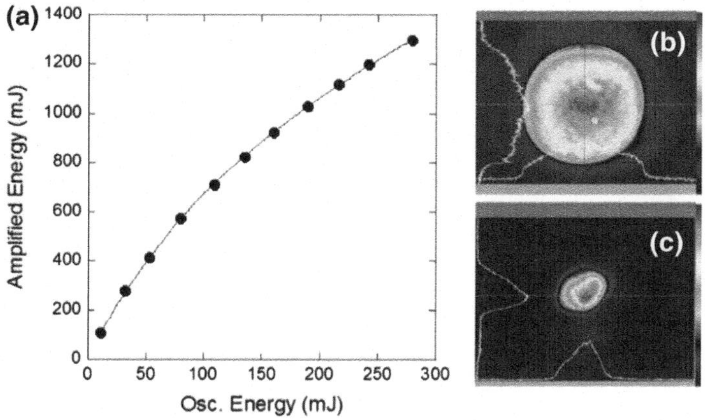

Fig. 61.2 Amplification characteristic at single-pass amplifier **a** input–output characteristics, **b** near-field output pattern, and **c** focusing pattern

maintained after three-pass amplifier. Figure 61.3 shows an input–output characteristics at single-, double-, and three-pass amplifications. The amplifications were in good agreement with the theory of Frantz–Nodvik formula [6]. The maximum value of the output energy was 4.1 J at a 50 Hz operating frequency. The ASE was measured by operating the laser without the master oscillator pulses (free-running) and the value obtained was only 5 mJ, which was considered negligible. The simple layout of the developed laser system and flow tubes made from samarium-doped silica glass was effective to inhibit ASE. By using this laser for pumping the Ti: sapphire laser, the output energy of 1.2 J can be expected under two assumptions;

Fig. 61.3 Input–output characteristics at single-, double-, and three-pass amplifiers

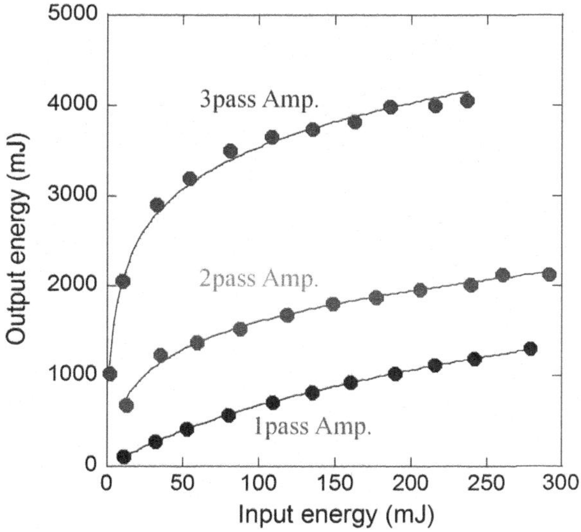

the wavelength conversion efficiency for the second harmonics is 60% and the energy transfer coefficient of a Ti:sapphire amplifier is 50%. However, the laser pulses amplified by the developed laser system still have thermal birefringence as one of their thermal effects. The amplified laser pulses did not provide high conversion efficiency for the second harmonic generation because of inhomogeneous polarization. In future, we will adapt one more Nd:YAG rod amplifier to realize correction of the thermal birefringence with four-pass amplifier. The enhanced output energy is expected to be more than 5 J at an operating frequency of 50 Hz.

61.4 Conclusion

We developed a high pulse energy and high repetition rate Nd:YAG laser system with a MOPA platform to realize a pump source for a Ti:sapphire laser system in the high-repetition-rate laser-plasma driven SXRL. The maximum output energy was 4.1 J at an operating frequency of 50 Hz. The input–output characteristics were in good agreement with the theory of Frantz–Nodvik formula. This laser system will be upgraded to correct the thermal birefringence for the better wavelength conversion to the second harmonic. The enhanced output energy will be expected to be more than 5 J.

Acknowledgements This work was supported by Council for Science, Technology and Innovation, "Cross-ministerial Strategic Innovation Promotion Program (SIP), Infrastructure Maintenance, Renovation, and Management" (funding agency: JST).

References

1. Nishikino, M., et al.: Characterization of a high-brilliance soft X-ray laser at 13.9 nm by use of an oscillator-amplifier configuration. Appl. Opt. **47**, 1129–1134 (2008)
2. Ochi, Y., et al.: Development of a chirp pulse amplification laser with zigzag slub Nd:glass amplifiers dedicated to X-ray laser research. Appl. Opt. **46**, 1500 (2007)
3. Wang, Y., et al.: Demonstration of high-repetition-rate tabletop soft-X-ray lasers with saturated output at wavelengths down to 13.9 nm and gain down to 10.9 nm. Phys. Rev. A **72**, 053807 (2005)
4. Zeitoun, P., et al.: A high-intensity highly coherent soft X-ray femtosecond laser seeded by a high harmonic beam. Nature **431**, 426–429 (2004)
5. Hatae, T., et al.: Development of a prototype YAG laser amplifier for the edge Thomson scattering system in ITER. J. Plasma Fusion Res. **9**, 253–258 (2010)
6. Frantz, L.M., Nodvik, J.S.: Theory of pulse propagation in a laser amplifier. J. Appl. Phys. **34**, 2346–2349 (1963)

Chapter 62
Evaluation of a Flat-Field Grazing Incidence Spectrometer for Highly Charged Ion Plasma Emission in 1–10 nm

Y. Kondo, T.-H. Dinh, T. Tamura, S. Ohta, K. Kitano, T. Ejima, T. Hatano and T. Higashiguchi

Abstract A flat-field grazing incident spectrometer was built to investigate highly charged ion (HCI) plasmas in the spectral region from 1 to 10 nm. The spectrometer consists of a flat-filed grating with 2400 lines/mm as a dispersing element and an X-ray charged-coupled device (CCD) camera as the detector. In order to produce accurate intensity calibrated spectra of the HCI plasmas, the diffraction efficiency of the grating and the sensitivity of the CCD camera were directly measured by using the reflectometer installed at the BL-11D of the Photon Factory (PF). We also studied the calibrated spectra of Gd HCI plasmas which were produced by an Nd: YAG laser.

62.1 Introduction

Spectroscopy of highly charged ion (HCI) has been received a great interest in the last decade due in part to the development of efficient and powerful extreme ultra-violet (EUV) and soft X-ray (SXR) sources for applications [1]. The use of intense unresolved transition array (UTA) emission from laser-produced HCI plasmas is potentially suitable for the compact sources. Ideally, the UTA emission should lie within the reflectance bandwidth of a multilayer mirror. The in-band high-energy emission is attributable, in some cases, to hundreds of thousands of near-degenerate resonance lines lying within a narrow wavelength range. The diagnostics of these

Y. Kondo · T.-H. Dinh · T. Tamura · T. Higashiguchi (✉)
Faculty of Engineering and Center for Optical Research and Education (CORE), Department of Electrical and Electronic Engineering, Utsunomiya University, Utsunomiya, Tochigi, Japan
e-mail: higashi@cc.utsunomiya-u.ac.jp

S. Ohta · K. Kitano
Vacuum and Optical Instruments, Tokyo, Japan

T. Ejima · T. Hatano
Institute of Multidisciplinary Research for Advanced Materials, Tohoku University, Sendai, Japan

T. Kawachi et al. (eds.), *X-Ray Lasers 2016*, Springer Proceedings in Physics 202, https://doi.org/10.1007/978-3-319-73025-7_62

plasmas requires spectral data, with reliable intensity information, of the emission involved over a wide range of wavelengths, from which the plasma parameters such as ion charge state and temperature can be determined [2]. In the following article, we built a flat-field grazing incidence spectrometer (GIS) to observe photon flux emitted from the HCI plasmas in the spectral region from 1 to 10 nm. GIS consists of a flat-field grating with 2400 lines/mm as a dispersing element and an X-ray charged-coupled device (CCD) camera as the detector. The diffraction efficiency and the sensitivity of the CCD are measured at the BL-11D of the Photon Factory (PF) in Japan. To the best of our knowledge, it is the first time the diffraction efficiency of the 2400 lines/mm grating is reported in range 5–10 nm. We also obtain calibrated SXR spectra from Nd:YAG laser-produced Gd plasmas.

62.2 Experimental Result

The schematic setup of the GIS for LPP experiments is presented in Fig. 62.1. The distances between the different components, plasma-slit, slit-grating, and grating-CCD chip, were 610 mm, 237 mm, and 235 mm, respectively. To achieve a high signal-to-noise ratio, the slits were set with a width of 100 μm. The grating was provided by Shimadzu and has a normal incidence angle $\alpha = 88.65°$. The CCD was air-cooled and back-illuminated type provided by Andor. It consists of 2048 × 2048 pixels (each pixel has dimensions 13.5 μm × 13.5 μm) and was directly exposed to the SXR radiation. A resolving power $\lambda/\lambda = 277$ was measured at a wavelength of 3.37 nm.

Figure 62.2(i) depicts the diffraction efficiency of the grating measured at the PF. The incident beam with a divergent angle less than 0.1° and a full width at half maximum (FWHM) diameter of approximately 200 μm was incident on the gating located at the center of a vacuum chamber. The angular profiles of the incident and diffracted beams were scanned by a detector, which consisted of an X-ray diode and a slit with a width of 100 μm. The distance from the grating to the detector was about 255 mm and was fixed due to the configuration of the chamber. A first-order diffraction efficiency exceeding 3% was observed for the wavelength region around 2.75 nm and above 8.0 nm. The efficiency is higher than that from a previous report of about 2% [3] due to improvements in the processing technology of the grating. On the other hand, the presence of a dip apparent in the short wavelength region

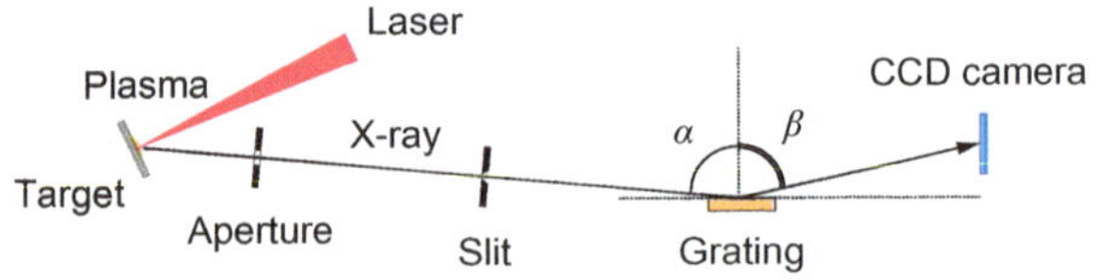

Fig. 62.1 Schematic view of the spectrometer

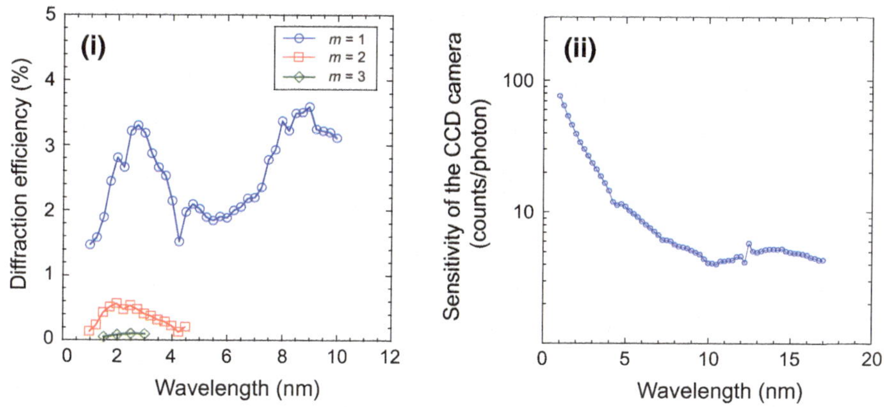

Fig. 62.2 The measured diffraction efficiency (**i**) and the sensitivity of CCD (**ii**)

could be observed in both first- and second-order spectra. These dips are caused by *K*-absorption of oxygen and carbon originating from hydrocarbon contamination on the surface of the grating. Figure 62.2(ii) depicts the sensitivity of the CCD camera. A calibrated AXUV100G X-ray photodiode was used to determine the number of photons incident onto the X-ray CCD camera. The small dip and peak could be due to the *L*-absorption of silicon.

In Fig. 62.3(i), we present time-integrated spectra of the SXR photon flux from LPPs of Gd. An Nd:YAG laser providing a maximum pulse energy of 1 J at a duration of 10 ns (FWHM) was focused on a flat target of Gd in vacuo. The focal spot of 50 μm was kept while changing the pulse energy. The corresponded laser intensities were ranging from 3×10^{11} to 3×10^{12} W/cm^2. Detail equations for

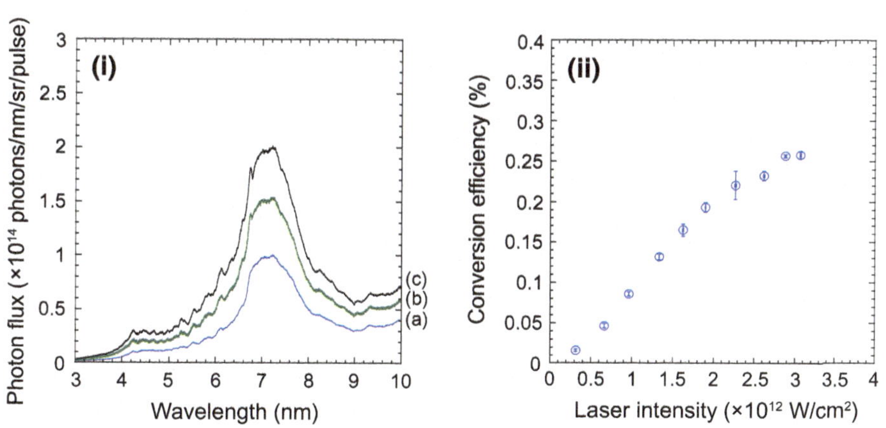

Fig. 62.3 (**i**) Calibrated soft X-ray spectra from Gd HCI plasmas produced by a 10-ns Nd:YAG laser at laser intensities of 1.9×10^{12} (a), 2.6×10^{12} (b), and 3.1×10^{12} W/cm^2 (c). (**ii**) Laser intensity dependence of conversion efficiency as measured by the GIS

the calibration have been reported in [4]. The UTA emission around a wavelength of 7 nm, mainly due to $n = 4 - n = 4$ ($\Delta n = 0$) transitions in ions with an open 4*f* or 4*d* outermost subshell, increases rapidly with increasing laser intensity. However, the peak wavelength at 7.3 nm is almost unchanged. A small but sharp peak at 6.76 nm and some peak structures below 6.59 nm were observed from a laser intensity of 2.6×10^{12} W/cm^2. According to a collisional-radiative (CR) simulation, the Gd plasma will have strong emission at 6.76 nm when the plasma temperature is close to 80 eV, which optimizes the population of Gd^{17+} and Gd^{18+} ions whose resonance emission lies at this wavelength [5]. The energy conversion efficiency at 6.76 nm in a bandwidth of 0.06% calculated from the spectra of the Gd plasma is displayed in Fig. 62.3(ii). Due to both the presence of a small sharp peak at 6.76 nm and the increase of the conversion efficiency, we expected the increase of the present Gd^{17+} and Gd^{18+} ions with time and space in the plasma. The maximum conversion efficiency of 0.26% was observed at a laser intensity of 3×10^{12} W/cm^2.

62.3 Summary

The evaluation of the absolute response of a flat-field grazing incidence spectrometer for HCI plasma spectroscopy has been reported. The diffraction efficiency of the holographic fabricated grating with 2400 lines/mm and the absolute spectral response of a CCD were directly measured by using the BL-11D beamline of the Photon Factory. This allows for highly accurate spectroscopy of HCI plasmas. We also studied the calibrated spectra of Gd HCI plasmas which were produced by a 10-ns Nd:YAG laser. A maximum energy conversion efficiency of 0.26% was observed at a laser intensity of 3×10^{12} W/cm^2.

Acknowledgements This work was performed under the auspices of MEXT, and "Project for Bio-Imaging and Sensing at Utsunomiya University" from MEXT.

References

1. Greg, T.: Optical lithography: lithography at EUV wavelengths. Nat. Photonics **4**, 809 (2010)
2. Dinh, T.H.: Temporal behavior of unresolved transition array emission in water window soft X-ray spectral region from multiply charged ions. Appl. Phys. Lett. **107**, 121101 (2015)
3. Koike, M.: Performance of laminar-type holographic grating for a soft X-ray flat-field spectrograph in the 0.7–6 nm region. Rev. Sci. Instrum. **74**, 1156 (2003)
4. Dinh, T.H.: Evaluation of a flat-field grazing incidence spectrometer for highly charged ion plasma emission in soft X-ray spectral region from 1 to 10 nm. Rev. Sci. Instrum. **87**, 123106 (2016)
5. Li, B.: Investigation of Gd and Tb plasmas for beyond extreme ultraviolet lithography based on multilayer mirror performance. Appl. Phys. Lett. **101**, 013112 (2012)

Author Index

T. Kawachi et al. (eds.), *X-Ray Lasers 2016*, Springer Proceedings in Physics 202, https://doi.org/10.1007/978-3-319-73025-7

Subject Index

T. Kawachi et al. (eds.), *X-Ray Lasers 2016*, Springer Proceedings in Physics 202, https://doi.org/10.1007/978-3-319-73025-7

Printed by Printforce, the Netherlands